From Newton to Hawking

A History of Cambridge University's Lucasian Professors of Mathematics

Cambridge University's Lucasian professorship of mathematics is one of the most celebrated academic positions in the world. Since its foundation in 1663, the chair has been held by seventeen men who represent some of the best and most influential minds in science and technology. Principally a social history of mathematics and physics, the story of these great natural philosophers and mathematical physicists is told here by some of the finest historians of science. The journey begins with the search for a benefactor able to establish a '*mathematicus professor honorarius*', and travels through the life and work of the professors, exploring aspects from the heroic to the absurd. Covering both the great similarities and the extreme differences in mathematical physics over the last four centuries, this informative work offers new perspectives on world-famous scientists including Isaac Newton, Charles Babbage, G. G. Stokes, Paul Dirac and Stephen Hawking.

KEVIN C. KNOX is Historian at the Institute Archives, Caltech. He has held positions as Visiting Professor at the University of California, Los Angeles and Ahmanson Postdoctoral Instructor in the Humanities at Caltech.

RICHARD NOAKES is a British Academy–Royal Society Postdoctoral Research Fellow in the History of Science, in the Department of History and Philosophy of Science, Cambridge University. He previously held a Leverhulme postdoctoral fellowship at the universities of Leeds and Sheffield.

From Newton to Hawking

A History of Cambridge University's Lucasian Professors of Mathematics

Edited by

KEVIN C. KNOX

California Institute of Technology, Pasadena, CA, USA

RICHARD NOAKES

Department of History of Philosophy of Science, Cambridge, UK

CAMBRIDGE
UNIVERSITY PRESS

PUBLISHED BY THE PRESS SYNDICATE OF THE UNIVERSITY OF CAMBRIDGE
The Pitt Building, Trumpington Street, Cambridge, United Kingdom

CAMBRIDGE UNIVERSITY PRESS
The Edinburgh Building, Cambridge CB2 2RU, UK
40 West 20th Street, New York, NY 10011-4211, USA
477 Williamstown Road, Port Melbourne, VIC 3207, Australia
Ruiz de Alarcón 13, 28014 Madrid, Spain
Dock House, The Waterfront, Cape Town 8001, South Africa

http://www.cambridge.org

First published 2003

Printed in the United Kingdom at the University Press, Cambridge

Typeface Trump Mediaeval 9.5/15 pt *System* LaTeX 2_{ε} [TB]

A catalogue record for this book is available from the British Library

Library of Congress Cataloguing in Publication data

From Newton to Hawking : a history of Cambridge University's Lucasian professors
of mathematics / edited by Kevin C. Knox & Richard Noakes.
p. cm.
Includes bibliographical references and index.
ISBN 0 521 66310 5
1. Mathematicians – England – Biography. 2. Physicists – England – Biography.
3. University of Cambridge. Dept. of Applied Mathematics and Theoretical
Physics – Faculty – Biography. I. Knox, Kevin C. II. Noakes, Richard.
QA28.F76 2003
510′.92′241 – dc21 2003043584

ISBN 0 521 66310 5 hardback

Contents

Illustrations

Contributors

Mordechai Feingold, Professor of History at the California Institute of Technology, is editor of the journal *History of Universities*. His publications include *The Mathematicians' Apprenticeship: Science, Universities and Society in England, 1560–1640* (Cambridge, 1984) and *The Oxford Curriculum in Seventeenth Century Oxford* (Oxford, 1997). He is currently working on Newtonianism and on a history of the Royal Society.

John Gascoigne was educated at Sydney, Princeton and Cambridge universities and has taught at the University of New South Wales since 1980. He has published on the relations between the universities and the early scientific movement, science and imperialism in the late eighteenth century and the nature of the English Enlightenment. His most recent work is *The Enlightenment and the Origins of European Australia* (Cambridge, 2002).

Rob Iliffe is currently Reader in History of Science at Imperial College, London. He has published widely on the history of early modern science and is completing a book *Priest of Nature: the Private Heresy of Isaac Newton*, for Yale University Press, for publication in 2003. He is Editorial Director of the Newton Project and Editor of the journal *History of Science*.

Kevin C. Knox completed his doctoral dissertation on culture and scientific change in Georgian Cambridge and London at Cambridge University before returning to North America in 1996 as a visiting professor at the University of California, Los Angeles. In 1997, Knox became the Ahmanson Postdoctoral Instructor in the Humanities at the California Institute of Technology, where he currently works as the historian of the Institute Archives. He has been involved in a number of multimedia projects related to the history of science and has published numerous works on late-Georgian natural philosophy and mathematics.

Helge Kragh is professor of history of science at the University of Aarhus, Denmark, and a member of the Royal Danish Academy of Science. His research focuses on the history of physics, chemistry and astronomy since 1850. His publications include *Dirac: A Scientific Biography* (Cambridge, 1990), *Cosmology and Controversy* (Princeton, 1996) and *Quantum Generations* (Princeton, 1999).

Hélène Mialet did her Ph.D. in philosophy at the Sorbonne and the C.S.I. (Ecole des Mines) in Paris. She has held postdoctoral positions at Cambridge and Oxford Universities and at the Max Planck Institute for the History of Science in Berlin. She is Assistant Professor in the Department of Science and Technology Studies at Cornell University. She is presently working on two book-length manuscripts on questions having to do with subjectivity, singularity and creativity in science: the first is an empirical philosophical study of invention in a large applied research laboratory, the second is on Stephen Hawking.

Richard Noakes received his Ph.D. from the University of Cambridge in 1998 and subsequently held a Leverhulme postdoctoral fellowship at the universities of Leeds and Sheffield. He is currently British Academy–Royal Society Postdoctoral Research Fellow in the History of Science at the Department of History and Philosophy of Science, Cambridge. He has published on science and spiritualism in Victorian Britain, and the representation of science, technology and medicine in nineteenth-century generalist periodicals. He is presently writing a monograph on Victorian physical and psychical sciences.

Simon Schaffer is Reader in History and Philosophy of Science at the University of Cambridge. He has recently published essays on the work of Charles Babbage and of James Clerk Maxwell, and several studies on nineteenth-century astronomy. His most recent publication is the co-edited book, *The Sciences in Enlightened Europe* (Cambridge, 1999).

Stephen D. Snobelen took his B.A. and M.A. at the University of Victoria in Canada and his M.Phil. and Ph.D. in History and Philosophy of Science at the University of Cambridge. He was a Junior Research Fellow at Clare College, Cambridge from 1999 to 2001. The topic of his doctoral dissertation was the natural philosophical and religious careers of William Whiston, and his publications include papers on Whiston, Isaac Newton's theology and the popularization of Newtonian natural philosophy. He is currently Assistant Professor in History of Science and Technology at King's College, Halifax, Nova Scotia.

Ian Stewart is Senior Fellow at the University of King's College, Halifax, Canada. He has published several articles on aspects of Isaac Barrow in his Cambridge University context. He is also presently preparing, with Stephen Pumfrey, a translation and critical edition of William Gilbert's *De Mundo Nostro Sublunari Philosophia Nova*.

Larry Stewart is co-editor of the *Canadian Journal of History* and Head of the Department of History. Educated in Britain, France and Canada, he holds degrees from McMaster University and the University of Toronto. His publications include *The Rise of Public Science. Rhetoric, Technology and Natural Philosophy in Newtonian Britain, 1660–1750* (Cambridge, 1992). He is currently engaged (with Margaret C. Jacob) on *The Impact of Science from Newton to the Industrial Revolution* (Harvard University Press) and is writing a study of the influence of chemistry and medicine in the radical revolutions of the late eighteenth century.

Andrew Warwick is Senior Lecturer in History of Science at Imperial College London, where he is also Head of the Centre for the History of Science, Technology and Medicine. His research interests include the history of mathematics and the physical sciences since 1750, and the philosophy and sociology of the sciences. From 1997 to 2001 he was co-editor of *Studies in the History and Philosophy of Modern Physics*.

David B. Wilson is professor of history, mechanical engineering and philosophy at Iowa State University. He received his Ph.D. in the history of science from the Johns Hopkins University. He is the author of *Kelvin and Stokes: A Comparative Study in Victorian Physics* (Bristol, 1987) and editor of *The Correspondence between Sir George Gabriel Stokes and Sir William Thomson, Baron Kelvin of Largs* (Cambridge, 1990). He co-edited *Did the Devil Make Darwin Do It?* (Ames, 1983) and *Physics and Physic: Essays in Memory of John M. A. Lenihan* (Glasgow, 2001). He is currently researching natural philosophy during the Scottish Enlightenment.

Foreword

In December 1663, the executors of Henry Lucas drew up the statutes for the Lucasian professorship, and the following month King Charles II granted the letters patent. These statutes contained many formal instructions, and until the University abolished them in 1857 they comprised the job description of the Lucasian professor.

An important early distinction was that the chair is a University office, established at a time when almost all teaching positions were in the colleges rather than the University. This meant of course that the professor was not required to be an ordained clergyman, in an age when college fellowships were often a means of advancement within the Church of England, and fellows moved into and out of college posts. The founding statutes made sure that the professor could still enjoy a college fellowship, without being subjected to restrictive college conventions.

The founding statutes required the professor to give a minimum of ten lectures. They specify that these must be neatly written out, and then deposited in the University archives. If the professor failed to do so, the Vice-chancellor could withhold the stipend and pay the money saved to the University Library. The University Library does indeed have original lecture notes from Isaac Newton, as part of its extensive holdings of Newton's manuscripts. The letters patent required all undergraduates past the second year and all Bachelors of Arts up to the third year to attend the lectures.

The professor also had to receive visitors for two hours on two days a week, and be willing to respond willingly to questions from curious scholars. For teaching purposes the professor was required to make available globes and mathematical instruments. The letters patent forbade the professor from taking on distracting duties in a college, such as serving as a bursar or tutor.

Through the inventiveness and achievement of many of the seventeen holders of the professorship, it has become one of the most famous chairs in mathematics. In the centuries following the foundation of this chair, Cambridge became the greatest centre for mathematics, and this remains the case today. The Isaac Newton Institute for Mathematical Sciences, opened in 1992, is a national and international visitor research institute. It attracts leading mathematical scientists from the UK and from overseas to interact in research over an extended period.

Isaac Newton used his tenure of the professorship to think privately about the reform of mathematics education at Cambridge. He wanted a mathematically based natural philosophy, with the students well-grounded in geometry and mechanics. He did not attempt to push through these reforms himself, but his eighteenth-century successors to the chair did make changes, and these positioned Cambridge to become a major centre for mathematical natural philosophy in the nineteenth century. From about 1748 the Mathematics Tripos became more searching, and candidates were ranked in order of merit. By 1772 a written examination was introduced. The college teaching officers used the English language editions of Newton's *Principia mathematica* and *Opticks* as set texts.

Charles Babbage, elected in 1828, sought to bring the benefits of the machine age to Cambridge mathematics and the computation of astronomical tables. He felt that mathematical computation could be reduced to a series of machine operations. His difference engine, which he used to generate polynomial tables, was never completed in his lifetime, although a working version was finally produced in 1991: it works flawlessly. His more ambitious analytical engine, for sophisticated calculations, was never built. The history of modern computing can fairly be said to begin with Babbage, and with his papers read at the Cambridge Philosophical Society. In 1949, at

Cambridge, EDSAC became the first stored programme computer to offer a regular computing service, thus realizing the early ambition of the Lucasian savant, Babbage.

During Paul Dirac's long tenure of thirty-seven years, fundamental physics again became the focus of attention of the Lucasian incumbent. He of course brought together quantum mechanics and special relativity. Like his predecessor, Newton, Dirac produced one of the greatest textbooks in physics, the renowned *Principles of Quantum Mechanics*. Dirac received the Nobel prize for physics in 1933, the year after his election to the Lucasian.

It is through the achievements of Newton, Babbage, Dirac and the other Lucasians that Cambridge mathematics continues to be at the centre of the world stage. Through their research papers, textbooks, lectures and graduate students, the Lucasians have stimulated generation after generation of mathematicians to work for the highest standards. With the Isaac Newton Institute for Mathematical Sciences now firmly established, the solid foundation left by successive Lucasians means the University of Cambridge is positioned to continue as a world-class centre excellence in mathematical science.

Stephen W. Hawking
Cambridge

Preface

From Newton to Hawking is about the most celebrated professorship in the world – Cambridge University's Lucasian professorship of mathematics – and it recounts the stories of many of the greatest natural philosophers and mathematical physicists from the Restoration to the twenty-first century. Principally a social history of mathematics and physics, the book has assembled some of the finest historians of science to explore aspects of the professors' lives from the heroic to the absurd. A journey through the professors' great triumphs (and a few humiliating defeats), *From Newton to Hawking* begins with the hunt for a benefactor to endow a '*mathematicus professor honorarius*', and then travels from Isaac Newton's quest to recover uncorrupted ancient truths of natural philosophy to the remarkable life of 'the prophet of the black hole', Stephen Hawking. On its tour of the 'holy city of mathematics', the book revives the mission of the eighteenth-century professors to broadcast the gospel of Newton, the obsession of Charles Babbage to mechanize reason, the commitment of George Stokes and Joseph Larmor to the luminiferous æther and the 'pure soul' and minimalist aesthetic of Paul Dirac.

This book began in summer 1994 when the editors were struck by the fact that so much first-rate scholarship was being written about individual holders of the Lucasian professorship, but that there was not a single work that tied all this together. Encouraged by colleagues and friends, we mobilized an international team of historians and invited them to think about 'their professor(s)' in a different way – as part of a long and changing institution. Fortunately, we were blessed with contributors who enthusiastically accepted the challenge and produced stories that were astonishingly learned, utterly compelling and frequently witty. Unlike the 'book of the conference', this volume does not hail from a scholarly meeting: everybody contributing to this book had a fairly good idea of where other participants stood with regard to their professor(s), and after countless discussions – some face-to-face but most virtual – we were able to put together this volume.

A book covering 350 years in the history of western science and an astonishing range of characters, institutions, artefacts and ideas incurs many debts. Contributors' acknowledgements will be found at the end of each chapter, but we would like to offer our joint thanks here. We are especially indebted to Simon Schaffer, whose encouragement and inspiration throughout this project, and particularly during its early stages, have been indispensable. We would like to express our deepest gratitude to William Ashworth, Andrew Blain, Geoffrey Cantor, Steven French, Graeme Gooday, Jeff Hughes, Andrea Rusnock, Kip Thorne, Alison Winter and Richard Yeo, who gave us their time and wisdom and assisted us in more ways than they know. We would like to thank the academic institutions that have supported this book over the years: the Division of the Humanities and Social Sciences and the Archives (both Caltech); UCLA; the Department of History and Philosophy of Science, Cambridge; the School of Philosophy, University of Leeds; and the Department of English Literature, University of Sheffield. Knox would like to express his indebtedness to colleagues at Caltech and UCLA, especially Susan Davis, Shelley Erwin, Kevin Gilmartin, Judy Goodstein, Adrian Johns, Daniel Kevles and Ted Porter. Similarly, Noakes would like to express gratitude for the invaluable advice provided by colleagues at the universities of Cambridge, Leeds and Sheffield, particularly Anjan Chakravartty, John Christie, Jim Secord, Sally Shuttleworth and Jonathan Topham.

Many connected with Cambridge University's colleges and libraries have helped us in our quest to flesh out the Lucasians and their world. We are especially indebted to Elisabeth Leedham-Green, for sharing her limitless knowledge of Cambridge University: to

Godfrey Waller, Jayne Ringrose and Adam Perkins, for helping us locate strange boxes of Lucasian ephemera in the University Library; to Jill Whitelock of the Whipple Library and Museum and Jonathan Harrison of the Library, St John's College. At Queens' College we received help and support from the President, Jonathan Holmes, Murrey Milgate, Clare Sargeant and Robin Walker.

A book of this kind could not have been produced without generous financial support. Much of the work for this book was undertaken while Noakes was pursuing doctoral research funded by the British Academy, and postdoctoral fellowships granted by the British Academy, the Royal Society and the Leverhulme Trust. Knox's contribution was made possible by a number of fellowships and awards. During his time at Cambridge, he received financial support from the Cambridge Commonwealth Trust. He also appreciated a number of grants and bursaries from Queens' College, the Cambridge Historical Society, the Lightfoot Fund and the A. J. Pressland fund. A fellowship from the Social Sciences and Humanities Research Council of Canada enabled him to conduct further research in the U.S.A. From UCLA he was awarded a fellowship to the William Andrews Clark Library. There he received tremendous aid from Peter Reill, Bruce Whiteman and the library staff. A significant portion of the fifth chapter was written while Knox was a Mayers' fellow at the Huntington Library, where he was encouraged by Roy Ritchie and just about everybody in the Ahmanson reading room. Further writing and much editing were completed while Knox enjoyed a visiting fellowship at the Humanities Research Centre at the Australian National University. In Canberra Iain McCalman, John Gage, Caroline Turner and the entire staff were extremely helpful.

In April 1998 we invited Sir James Lighthill to contribute an afterword to this work on the professorship of which he was the sixteenth holder. His tragic death months later deprived the world of one of the most distinguished applied mathematicians of the twentieth century and this volume of a valuable reflexive conclusion. We hope this book offers new insights into a mathematical genius who, like, the Lucasian with whom he felt particular affinity, Stokes, certainly helped 'keep up the reputation of the chair'.

We are indebted to many people at Cambridge University Press for realizing our Lucasian ambitions, and we would like to offer special thanks to Simon Mitton, for his astonishing patience and for making this book ask more searching questions than we originally posed; to Alice Houston, whose industry in the early stages of the project kept us on an even keel; and to three anonymous referees whose valuable suggestions on the original book proposal prompted us to rethink the balance, structure and detail of the work.

Finally, for being there for us and putting up with more talk about Lucasian professors than they deserved to hear, we would like to thank our respective partners, Fiona and Louise.

Timeline of the Lucasian professorship

Professor	Chair
Isaac Barrow	1663–1669
Isaac Newton	1669–1702
William Whiston	1702–1710
Nicholas Saunderson	1711–1739
John Colson	1739–1760
Edward Waring	1760–1798
Isaac Milner	1798–1820
Robert Woodhouse	1820–1822
Thomas Turton	1822–1826
George Airy	1826–1828
Charles Babbage	1828–1839
Joshua King	1839–1849
George Gabriel Stokes	1849–1903
Joseph Larmor	1903–1932
P. A. M. Dirac	1932–1969
M. James Lighthill	1969–1980
Stephen Hawking	1980–

Introduction: 'Mind almost divine'

Kevin C. Knox[1] and Richard Noakes[2]

[1]*Institute Archives, California Institute of Technology, Pasadena, CA, USA*

[2]*Department of History of Philosophy of Science, Cambridge, UK*

> Whosoever to the utmost of his finite capacity would see truth as it has actually existed in the mind of God from all eternity, he must study Mathematics more than Metaphysics.
>
> Nicholas Saunderson, *The Elements of Algebra*[1]

As we enter the twenty-first century it might be possible to imagine the world without Cambridge University's Lucasian professors of mathematics. It is, however, impossible to imagine our world without their profound discoveries and inventions. Unquestionably, the work of the Lucasian professors has 'revolutionized' the way we think about and engage with the world: Newton has given us universal gravitation and the calculus, Charles Babbage is touted as the 'father of the computer', Paul Dirac is revered for knitting together quantum mechanics and special relativity and Stephen Hawking has provided us with startling new theories about the origin and fate of the universe. Indeed, Newton, Babbage, Dirac and Hawking have made the Lucasian professorship the most famous academic chair in the world.

While these Lucasian professors have been deified and placed in the pantheon of scientific immortals, eponymity testifies to the eminence of the chair's other occupants. Accompanying 'Newton's laws of motion', 'Babbage's principle' of political economy, the 'Dirac delta function' and 'Hawking radiation', we have, among other things,

From Newton to Hawking: A History of Cambridge University's Lucasian Professors of Mathematics, ed. Kevin C. Knox and Richard Noakes. Published by Cambridge University Press.

Figure 1 Newton deified. G. Bickham's 1732 illustration propels Newton's radiating cameo into the heavens, where cherubs hold a variety of philosophical and mathematical instruments that helped the professor construct a new vision of the universe.

'Barrow's proof' of the fundamental theorem of calculus, the 'Saunderson board' (a calculating instrument for the vision-impaired), 'Waring's theorem' of integers, 'Airy's criterion' of telescopic resolving power, 'Stokes's law' of fluid resistance, 'Larmor frequency' of atomic precession in a magnetic field and 'Lighthill's fourth law of engine noise'. Small wonder, then, that scientific and historical literature, as well as a huge array of statuary, tombs, stamps, money, relics and the like, bears tribute to these colossal giants of science. Many of these homages, like the inscription on Newton's tomb in Westminster Abbey, put us mere mortals in our place:

> Here is buried Isaac Newton, Knight, who by a strength of *mind almost divine*, and mathematical principles peculiarly his own, explored the course and figures of the planets, the paths of comets, the tides of the sea, the dissimilarities in rays of light, and, what no other scholar has previously imagined, the properties of the colours thus produced. Diligent, sagacious and faithful, in his expositions of nature, antiquity and the holy Scriptures, he vindicated by his philosophy the majesty of God mighty and good, and expressed the simplicity of the Gospel in his manners. Mortals rejoice that there has existed such and so great an ornament of the human race![2]

Countless other tributes to Newton and his successors are equally as humbling. Take these lines from the obituary notice of George Gabriel Stokes published in *The Times*:

> We may enumerate his scientific papers, we may expatiate upon his work in optics or hydrodynamics, we may dwell upon his masterly treatment of some of the most abstruse problems of pure mathematics, yet only a select body of experts can readily understand how great he was in these various directions, while possibly not all experts understand how much greater was the man than all his works . . . Sir George Stokes was as remarkable for simplicity and singleness of aim, for freedom from all personal

> ambitions and petty jealousies, as the breadth and depth of his intellectual equipment. He was a model of what every man should be who aspires to be a high priest in the temple of nature.[3]

Today, one can make a pilgrimage to a temple – the Cambridge-based Isaac Newton Institute for Mathematical Studies – to worship these high priests of nature. There, visitors are prompted by a series of artefacts to recapitulate the heroic story of the Lucasian professorship. Outside the building are three symbolic statues, representing intuition, genesis and creation, as well as an arboret descended from the Woolsthorpe apple tree that allegedly inspired Newton to invent his theory of gravitation. Upon entering, visitors are presented with a bust of Dirac and a portrait, bust and death-mask of Newton.

Clearly, we mortals have placed great value on the work of the Lucasian professors, and as a consequence much of their handiwork has entered the common coin of our (corruptible) world. We have put tremendous faith in the professors and their intellectual products. As trustworthy icons corporations even trade on their names. As well as being emblazoned on the old one-pound banknotes, Newton has been used to sell everything from apples to zenith telescopes. A chain of computer software stores is named after Charles Babbage; in the UK, Stephen Hawking, whether aware or not that his predecessor George Biddell Airy had invented a method for correcting astigmatism, has endorsed a spectacles retailer.

Even without spectacles – or Newton's telescope for that matter – the Lucasian professors are understood to see farther and with unparalleled perspicuity. The professors themselves have perceived this legacy. As a young, obstreperous reformer, Charles Babbage deemed Newton's *Principia* the 'mill stone around the necks' of gownsmen; but in later life he reflected how the professorship had been 'the only honour I ever received in my own country'. The fiscally mindful Babbage gushed that 'the names of Barrow and Newton have conferred on the Lucasian chair a value far beyond any which mere pecuniary advantage would bestow'. Almost every other incumbent has shared Babbage's deep affinity with Newton. Upon graduating at the top of his class, Isaac Milner – later to be Cambridge's seventh Lucasian professor – was 'tempted to commit his first act of extravagance. In the pride of his heart, he ordered from a jeweller a rather splendid seal, bearing a finely-executed head of Sir Isaac Newton'. Other Lucasians have worked even harder to memorialize their predecessors. Stokes was asked to arrange and catalogue the unpublished optical papers of Newton bequeathed to Cambridge University Library by the Earl of Portsmouth, while Stokes himself was made part of that monument of late-Victorian hagiography – the *Dictionary of National Biography* – by his successor, Larmor.[4] Most of the professors have been humbled by the gargantuan legacy that their predecessors bequeathed: 'It is nice to feel that one holds the same position as Newton and Dirac', James Lighthill said, 'but the real challenge', he admitted, 'is to do work that is even a small fraction as significant.' Although Stephen Hawking has criticized Newton's 'vitriol and deviousness' he also feels close to the author of the *Principia*. As he has recently quipped, 'Newton occupied the Lucasian chair at Cambridge that I now hold, though it wasn't electrically operated at the time.'[5]

From Newton to Hawking recounts the ways these celebrated scientific thinkers have conceived their place within the history of the prestigious professorship. Of greater import, this book uses the context of the mathematical professorship to examine the extraordinary developments in the physical sciences since 1663. These changes relate not simply to the technical content of mathematical and scientific enterprises but the diverse array of uses to which such work has been put, from contemplating the origins of the universe to the design of quieter jet engines. In addition to this aspect of their work, their astonishing talent, tenacious industry and insatiable curiosity help to explain why the Lucasians have dipped their hands in so many things. *From Newton to Hawking* explores the professors as antiquarians, alchemists, orators (Barrow has been called 'one of the great orators produced by England'), theologians, economists, engineers, politicians and church-music composers, as well as pure researchers.[6]

Accordingly, each chapter of this book provides a social history of mathematics, natural philosophy and physics and in so doing shows how the professors shared an intense preoccupation with the *application* of the sciences, both as reliable accounts of the natural world and as bases for such 'nonscientific' subjects as faith, ethics, politics and aesthetics.

Indeed, what emerges from this book is the significant extent to which these nonscientific topics permeated the enterprises of Lucasian professors at least as much as the research, administrative and pedagogical duties associated with their position. For instance, Isaac Newton and his eighteenth-century successors were as determined to restore the basis of true Christian faith through a scientifically rigorous scriptural exegesis as they were determined to promote the true (i.e. Newtonian) account of the natural world. Conversely, for professors like Charles Babbage and Paul Dirac, their 'pure' mathematical research was a means for expunging the corrupted mathematical techniques which inevitably led to dangerous religious practices and troublesome secular ethics.

What the book is *not* is a reference work detailing the administrative details and tedious minutiae of the careers of the Lucasian professors. Nor is it a hagiographical account of disembodied scientific heroes. Alongside their magnificent triumphs are a number of spectacular failures, while the professors themselves have been the objects of scorn, jest and chastisement. They have had sordid controversies with others and have squabbled amongst themselves. The career of Charles Babbage is illustrative: his calculating engine never functioned during his lifetime while it was said that 'he never functioned as a professor'. With his predecessor, George Biddell Airy, he had vigorous disputes over Britain's railway standards and the financing of his computers. He could also be off-putting, to say the least – the great Victorian historian, Thomas Carlyle, once reflected how 'Babbage continues eminently unpleasant to me, with his frog mouth and viper eyes, with his hide-bound wooden irony, and the acridest egotism looking through it.'[7] These criticisms of Babbage also illustrate how readers of *From Newton to Hawking* will learn how Cambridge's most distinguished professors fit into (or not!) their contemporary cultures. The point that the professors are necessarily products of their time cannot be overestimated. Nevertheless, it is a point that habitually has been overlooked, ignored and suppressed. Through careful management of the history of the professors, previous accounts of the Lucasian chair – by both historians and by the chair's occupants – have made it appear that the professorship transcends time and space. Generally, these accounts have taken it for granted that the current professors inhabit the same mental world as their predecessors and present their work as a unified, cumulative and coherent 'project'.

Like Newton's concept of 'flowing time', this idea of continuity is seductive and it is surprising that no publication hitherto has attempted to provide portraits of these men as part of a continuous history. Not only have the professors inhabited the same town and institution, but many have shared the same laboratories, technicians and research programmes. And each professor, in his own way, has envisaged himself as a cog in the scholarly corporation, a kind of temporary placeholder in the eternal succession of professors. As Hélène Mialet suggests later in the book, the professorial chair is akin to medieval kingship: although monarchs and Lucasian professors alike command great respect in the secular world, their temporal incumbency can seem relatively inconsequential compared to the everlasting corporate body they represent.

If the professors themselves often remark that they are mere markers in a continuum of mathematical splendour, one cannot neglect the elements of discontinuity that problematize this grand narrative. One might try to imagine Stephen Hawking and Isaac Newton engaged in an animated conversation (or as *Star Trek* envisaged, in a poker game), but it is likely that their lives, careers and values would have been utterly alien to each other. While the interests, methodologies, habits and research areas of the different professors have been extraordinarily diverse, the sciences and the university itself have

Figure 2 'MC Hawking'. The current Lucasian professor has become a multimedia rap-master. The 'phat' lyrics of '*E = mc Hawking*' assert ' "E" stands for energy, yo that's me, / I'm a brilliant scientist and a dope MC. / Before you step on me I'd think twice G, / I'm the Lord of Chaos, King of Entropy. / There ain't another motherf**ker hard like me, / I'm a universal constant, I'm a singularity.'

undergone radical transformations that make it difficult to compare professors from different eras. In 1663 the conception of the English university as a site of publicly funded experiment was still over two centuries away. Restoration Cambridge was not a research institution, nor would the varsity become one until the second half of the nineteenth century. Even Newton had trouble demonstrating to the republic of letters the value of mathematics, and the protracted transition of Cambridge from chiefly a religious seminary to a scientific Mecca is an integral part of the professors' history.

A BRIEF HISTORY OF CAMBRIDGE

Henry Lucas, who had studied at St John's College, believed that he recognized a breach in Cantabrigian scholarship. Wanting 'to testifie' to his 'affection' for Cambridge and for learning, he resolved to 'ordaine . . . a yearly stipend and sallerie for a professor . . . of mathematicall sciences in the said Vniversitie'. In his will Lucas said that his endowment would 'honor that greate body', as well as assisting 'that parte of learning which hitherto hath not bin provided for'.[8] Yet, although it is striking that the Lucasian professorship was not endowed until the university was over four centuries old, it is, of course, misguided to suppose that the varsity was somehow incomplete before a professor of mathematics began to grace its schools.

Though it may *seem* so to us, the infiltration of mathematics into the Cambridge curriculum was not inevitable. In many senses Cambridge functioned eminently well without a mathematics professor. On the one hand, Cambridge produced plenty of able mathematicians without an endowed professorship, and there were plenty of tutors capable of guiding undergraduates through the rigours of the curriculum. On the other hand, the university's function had little to do with state-of-the-art mathematics. John Wallis – who studied at Emmanuel College in the 1630s before embarking on a career at Oxford as the Savilian professor of geometry – commented on the general low regard for mathematical studies in relation to the purpose of the English universities: 'Mathematics . . . were scarce looked upon as *Academical* studies, but rather *Mechanical*; as the business of *Traders, Merchants, Seamen, Carpenters, Surveyors of Lands*, or the like.' Wallis realized that this opinion concerning appropriate scholarly learning reflected certain interests which had been formed centuries earlier. For, before the Reformation, Cambridge's central mission revolved around its service to the mighty Roman Church, the university serving to train prospective priests. Following the Reformation, Cambridge became *the* site to seek ecclesiastical preferment within the Anglican Church. Accordingly, all undergraduates – whether preparing to return to their estate, to make their way in London at the Inns of Court or to enter holy orders – embarked on a strict regimen of religious tuition and prayer at their colleges, a tradition that was not short-lived. Charles Babbage reminisced how 'the sound of the morning chapel bell . . . call[ing] us to our religious duties' was the only thing that compelled him and his fellow undergraduates to end their night-long devotions at whist.[9]

Alongside his antipathy towards chapel, Babbage also 'acquired a distaste for the routine of studies'. In his opinion the curriculum of the early nineteenth century was antiquated, despite – or because of – its heavy emphasis on the Newtonian philosophy. For Babbage, the university had suffered from its Elizabethan legacy which from the sixteenth century had greatly influenced the trajectory of learning and examinations. During the reign of Elizabeth and the next two Stuarts college tutors confronted undergraduates with subjects from the *trivium*, the *quadrivium* and the philosophies, their studies likely including logic, rhetoric, Aristotelianism, geometry, astronomy and some natural philosophy. By the middle of the seventeenth century tutors occasionally foisted the new natural philosophies upon their charges, and as an undergraduate it seems that, along with Aristotle and Virgil, Isaac Barrow received a dose of Cartesian philosophy. The mandate produced able scholars, but not professional mathematicians.

The Elizabethan statutes also determined how learning was to be *displayed*. In Barrow's time, oral examinations or 'disputations'– not particularly conducive to testing mathematical skills – dominated the evaluation of hopeful sophomores and seniors. Pomp, ritual and ceremony were the order of the day as students tried to convince examiners, and perhaps the occasional royal observer, that they commanded the emblems of good scholarship: 'To call these disputations merely debates between students', one historian has observed, 'is like describing a Spanish bullfight as the killing of a cow.' Even with the 'Newtonianization' of the curriculum, the rites and the spectacles associated with the Senate House Examination remained vital to the institution. Rather than radically overhauling the examination process, mathematical and scientific subjects came to dominate the exams through a glacial process of accretion. Only gradually did the Lucasian professors come to play a role in the process.[10]

The history of the professorship is also bound together with that of the colleges. Through the benefices of diverse wealthy patrons, the colleges had been founded one by one, sprinkled liberally throughout the commercial town, and each virtually independent from the others. Their wealth determined the extent to which each constructed its chapels, halls, common rooms, libraries, dorms and gardens. Regardless of its assets every college armed itself with a battery of bedmakers, cooks, porters and wine stewards to serve its master, tutors, fellows and students. As every Lucasian professor swiftly ascertained upon his matriculation, collegiate life was decidedly *un*egalitarian; a strict social hierarchy governed even the minutiae of academia life. While aristocratic students wore resplendent garments and dined on high table, lesser-born students tended to be at the bottom of the social heap. Both Isaac Newton and Isaac Milner entered Cambridge as humble 'sizars' at their colleges and, accordingly, they were humiliated with chores ranging from ringing chapel bells to emptying chamber pots. One could rise from such humble beginnings to triumph within the intricate political fabric of the colleges. For instance, colleges ministered a number of parishes, and 'meretricious' fellows could be presented with these 'livings'. Yet, as several Lucasian professors discovered, it took plenty of dexterous politicking to rise within the collegiate ranks. Apparently Barrow was blessed with such dexterity: he managed to procure the mastership of mighty Trinity College. Despite his enormous genius, Barrow's successor was perhaps less savvy in college politics. Newton's attempt to secure the provostship of King's College was a dismal failure. On the other hand, Milner, who like Newton had entered Cambridge as a sizar, ended up the twenty-eighth President of Queens' College. Moreover, as John Gascoigne points out in his chapter, the colleges and the central university have continually grappled for supremacy in Cambridge. In terms of the Lucasians, this incessant ebb and flow between university and colleges could dramatically affect the professors, both in terms of their status and in terms of their role in the instruction of students. Often, in this regard, the colleges were preeminent, and during these periods of collegiate ascendancy the Lucasian professor was somewhat marginalized in the academic community. Historically, most colleges found that it was *not* in their interests to have a Lucasian professor. College tutors were quite happy to set academic standards for young gownsmen and often viewed professors as little else than meddlers.

Nevertheless, the late-Georgian and then the Victorian professors began to leave their mark, first as examiners for the prestigious Smith's prize in mathematics and then as influential proponents of curriculum reform. Even though the professors considered

'Mathematics as the Key to Philosophy, as the Clue to direct us through the secret Labyrinths of Nature', the struggle was always up-hill. Many late-Georgian proponents of liberal education did not see great value of mathematics to those other than 'vulgar artisans'. As a freshman Gilbert Wakefield grumbled that Euclid was nothing more than an 'old carpenter'. Yet, with the prodding of the Lucasian professors, gownsmen came to recognize mathematics as an integral part of the philosophical enterprise. After graduating Second Wrangler in 1776 even Wakefield changed his tune: 'But happy that man! who lays the foundation of his future studies deep in the . . . *mathematical* philosophy: . . . Language sinks beneath contemplations so exalted, and so well calculated to inspire the most awful sentiments of the GREAT ARTIFICER'.[11] Thanks in great part to the professors, the varsity began to see the great value of Newtonianism and the mathematical sciences. Thanks also to them the Mathematical Tripos, which had evolved from the Senate House Examination, was seen as the most 'meritocratic' form of evaluating students and it thus became the most prestigious Cambridge examination. By the 1830s, candidates were judged on their ability to tackle questions written in English on printed papers – the form that pervades most examination systems to this day. Victorian legatees of these exams, such as Stokes and Larmor, sat for approximately twenty hour-long papers over a three-week period. After their exhausting labours they would await very public glory or humiliation as they were ranked as a wrangler (first-class) a senior *optime* (second-class) or a junior *optime* (third-class).[12]

Since high wranglers had better chances of obtaining fellowships and respectable employment, the Mathematical Tripos also fostered the growth of private mathematical coaching, and coaches shaped the minds of future professors like George Gabriel Stokes and Joseph Larmor. Although not formally recognized by the university, 'pupil mongering' became such an important part of mathematical instruction that one distinguished Tripos graduate could reminisce in 1912 that 'had there been no chair in mathematics in the University it is probable that the history of the School [of mathematics] would have been practically unaltered'.[13] Another reason why this may seem so is that in the first half of the nineteenth century, the dominant figure in changing the content of mathematical instruction was *not* a Lucasian professor, but William Whewell. As a young don, Whewell had welcomed the efforts of the Lucasian professors Charles Babbage and Robert Woodhouse to introduce continental methods of mathematical analysis into the curriculum. By the 1830s, however, the administratively omnipotent Whewell had become suspicious of analysis. It may have been suitable for advanced mathematical *research* but it was not suitable for Cambridge mathematical *teaching*, whose principal goal was to furnish the nation's future clergymen, lawyers, civil servants and teachers with a 'liberal education', notably the stable mathematical principles provided by such 'permanent' subjects as Euclid's geometry and Newton's mechanics. Whewell's 're-geometrization' and 're-Newtonianization' of the Tripos in 1848–9 split the Mathematical Tripos into two parts – the first consisting of questions on geometrical and nonanalytical topics, and the second, which could only be taken on succeeding in the first part, embracing the more sophisticated analytical subjects.

With Whewell at the helm, the new and progressive sciences of heat, electricity and magnetism were also excluded from undergraduate teaching. However, by the 1860s the importance of electricity and magnetism in educational curricula had risen sharply owing to the rapid development of the electric telegraph industry, a commercially and imperialistically crucial enterprise. Furthermore, the researches of high-flying wranglers like William Thomson and James Clerk Maxwell gave these new sciences a rigorous mathematical foundation and thus secured them a 'permanent' place in the Cambridge curriculum. This was part of a wider change in the mid-Victorian transformation of science teaching in Cambridge, a change owing much to the efforts of an 1850 Royal Commission to help the ancient British universities provide scientific instruction in line with the 'requirements of modern times'.[14] The university had already responded to the burgeoning need to prepare students for the industrialized modern times by founding the Natural Sciences Tripos (first examined in 1851). It continued to respond from the 1860s by creating new professorships and buildings. Long gone were the days of Isaac Newton and Isaac Milner, both of whom had constructed their laboratories in their private residences. By the end of the nineteenth century the 'New Museums Site' boasted (along with museums of zoology, botany and mineralogy) laboratories, workshops and an optical and astronomical lecture room for the Lucasian professors.[15]

Despite these advancements, the nineteenth-century Lucasians were often exasperated by the sluggish rate of change, as well as their remuneration. In 1857, eight years into his professorship, Stokes frequently wrote to his fiancée, agonizing over how they might achieve that desideratum of bourgeois Victorian society – respectability in married life. Reflecting on the deliberations of a University Council enquiry into the endowment of professorships, Stokes suggested that his position would be improved if fellowships remained open to married dons. 'If I were called into residence and my Fellowship were added to the Professorship', Stokes explained,

> our situation would be far, far pleasanter. I should be in a fixed and highly respectable position instead of being like a 'bookseller's hack' as Airy expressed it to me . . . I should do one thing well (at least I hope so) instead of having so many dissimilar things to attend to that I feel as if I were doing them all badly. I should have (probably) much more leisure for researches, which would then become part of my business, to keep up the reputation of the Chair.[16]

Stokes was initially disappointed since his college, Pembroke, did not abolish its celibacy restrictions on fellowships for another decade. Like many Victorian physicists, he had to provide for his new family with teaching and administrative 'hackwork'. In 1860, however, most of the original Lucasian statutes were officially repealed, bringing the chair in line with professional academic positions elsewhere in the country. While they gave the Vice-chancellor and elected officers the power to 'admonish' or sack the professor if he was 'wilfully neglectful of his duties, or guilty of gross or habitual immorality', the statutes also raised the income of the professorship by dipping into the money from Lady Sadler's benefaction of 1710. In 1886 the university channelled further monies into the Lucasian chair whose income had fallen owing to effects of the agricultural depression on the Bedfordshire estates on which the original endowment depended. By 1914 the professorship was regulated by the same statutes that governed most other university chairs, with the holder's main duty being to 'devote himself to the research and the advancement of knowledge in his department and to give lectures in every year'.[17]

These nineteenth-century transformations, along with Stokes's enthusiasm for both the Mathematical and Natural Sciences Triposes, made Cambridge sciences a popular choice for undergraduates. But, like many Victorians, Stokes was more sceptical of another change that was to affect the work of the Lucasian professorship, let alone that of other Cambridge pedagogues: the admission of women students. As Gillian Sutherland has written, 'Cambridge was initially hostile towards women with academic ambitions, deeply reluctant even to tolerate their presences and for a long time treated them as marginal figures.' Indeed, until the late nineteenth century, most Lucasian professors considered cleaning and cooking to be the only appropriate activities for females within college gates. Before Victoria's reign, most official references to women were in the form of decrees by the Vice-chancellor concerning 'provisions against public-women'. Thus, in an age when 'weaker vessels' were seen as distractions from serious study, Newton once accused John Locke of having 'endeavoured to embroil me wth weomen'.[18]

There were some exceptions to this general anxiety about the participation of females in the philosophical enterprise. In an attempt (albeit patronizing) to include women in the study of mathematics, Nicholas Saunderson and John Colson collaborated on a translation of *The Lady's System of Analyticks* (though not published until

1801), written by their counterpart in Bologna, Professor Donna Maria Gaelana Agnesi. Despite its intended audience, Charles Babbage read the book as a freshman and admitted that from it he 'acquired some knowledge'.[19] Babbage's greater debt was to Byron's daughter, Lady Ada Augusta, who created, promoted and sustained a forum for his analytical engine. Ada Lovelace, however, was never a student at Cambridge, for it was not until Stokes's era that females were first admitted as undergraduates (although they were not granted full university membership until 1948!). And though Stokes himself worried that female students would 'impair the heritage of men', his biographer reported how he 'was much pleased when a Newnham lady who had attended his lectures brought him some original work which he approved'.[20]

The twentieth-century Lucasian professors have responded to changes from other directions, not least the dramatic increases in scale of experimental and theoretical physics, applied mathematics, astronomy, cosmology and computing science. Neither Larmor nor Dirac had the size of international research schools boasted by Ernest Rutherford or Frederick Gowland Hopkins, but their careers exhibited the internationalism that was increasingly pervading the sciences, whether this meant attending international conferences, taking up overseas professorships or managing transatlantic professional relationships. With the notable exception of Lighthill, the work of twentieth-century Lucasian professors could hardly be described as 'big science' as far as the material cultures of their projects are concerned. And yet, their researches have depended on the dramatic development of large-scale research facilities. Larmor's and Dirac's evolving conceptions of the innermost structure of matter were built in conjunction with evidence generated across the globe, from Rutherford's Cavendish to Fermilab. Similarly, Hawking's revolutionary work on general relativity has been bolstered by the Hubble space telescope.

The steady expansion and rising status of twentieth-century Cambridge mathematics – both in terms of numbers of practitioners

Figure 3 The Isaac Newton Institute for Mathematical Studies. Like many buildings devoted to scientific pursuits, the structure is both backward- and forward-looking. The exterior features three symbolic statues, representing intuition, genesis and creation, as well as an inspiring arboret descended from Newton's apple tree in Woolsthorpe.

and subdisciplines – is reflected in the establishment of two separate departments of mathematics, the Department of Applied Mathematics and Theoretical Physics (founded 1959) and the Department of Pure Mathematics and Mathematical Statistics (founded 1964). The increasing independence of Cambridge's departments of mathematics from the colleges is even more strikingly symbolized by one of the newest features on Cambridge's landscape: the Cambridge Centre of Mathematical Studies (completed in October 2001). This lavish new home for Cambridge's mathematics departments is geographically far removed from the colleges and close to those other jewels in Cambridge's crown of physical sciences – the Cavendish Laboratory and the Institute of Astronomy (including the Royal Greenwich Observatory). It is also the place where the current Lucasian professor works and, owing to its physical proximity to its disciplinary brethren, can help the Lucasians develop even closer alliances with experimental physics and astronomy. This shift in geography also signals the diminishing significance of the colleges, the humanities' departments and the School of Divinity to the professors. It begs the question of how their principles have shifted over the course of four centuries.

THE PHILOSOPHICAL PRINCIPLES OF MATHEMATICS PROFESSORS

Shortly after Newton published his *Principia*, Richard Marsh, a divine ensconced in St John's College, delivered a fiery sermon to his Cantabrigian cohorts. While many parts of western Europe were beginning to embrace the Newtonian philosophy, Marsh was distraught that mathematics would usurp revelation. From the pulpit he bristled that in the Mosaic account of creation he met 'with no *Laws of Gravity*'. Rhetorically, he asked the modern philosophers, 'what reason have I to believe the wonders of your *Comet*, more than any other Romance?'[21]

As well as the extent to which Newtonianism would be attacked as a philosophy antithetical to revealed religion, Marsh's brimstone is a telling reminder of the strong relation between religion and the scientific products of the Lucasians. For Newton this relation was, of course, intentional: as he told the Master of Trinity College, Richard Bentley, 'When I wrote my treatise about our Systeme I had an eye upon such Principles as might work wth considering men for ye beleife of a Deity & nothing can rejoyce me more than to find it usefull for that purpose.'[22] Cantabrigian scholars had for some time viewed mathematical philosophy as an efficacious means of contemplating the Grand Artificer. A generation before Newton's arrival in Cambridge, the King's College graduate William Oughtred addressed '*the english Gentrie*' when he was accused of neglecting his calling as an Anglican priest:

> in all ages many of the most eminent in the sublimity of Theologie, have beene also conversent in the study of the Mathematicks; ... And in no other thing, after his sacred word, Almighty God (who creating all things in number, weight, and measure, doth most exactly Geometrize), hath left, more expresse prints of his heavenly & infallible truth, then in these Sciences.[23]

In particular, Newton regarded mathematical philosophy as a powerful instrument for combating the 'pious frauds, false miracles &

juggling tricks in matters of religion'. Assuming that 'gentile astrology and theology were introduced by cunning priests to promote the study of the stars', he presumed that by restoring the pristine natural philosophy of the ancients he could help eradicate the corrupted religious practices that he so despised.

Conceiving his labour to generate the *prisca sapienta* as a process of *re*-discovery, Newton placed himself within a conception of time that, like the sectaries of the English Civil War and many Restoration natural philosophers, located his own lifetime as the critical overture to the millennium. Accordingly, his successor in the Lucasian chair, William Whiston, commented that the *Principia* should be construed as 'an eminent prelude and preparation to those happy times of the restitution of all things'.[24] For Newton and several of the eighteenth-century scientific professors of Cambridge, the time had come for a true reading of the two Books – nature and scripture – and with it, the natural processes that would testify to the 'renovation[,] regeneration or restitution of the ye world and ye second coming of Christ'.[25] He was following a robust Cambridge tradition that the professors of the eighteenth century would perpetuate. While Newton had helped Thomas Burnet compose his 1681 *Sacred Theory of the Earth*, Whiston availed himself of the Master's mature mathematical philosophy to pen his 1696 *New Theory of the Earth*. Notably, Whiston used the properties of comets to exegete scripture and bring Burnet's natural theology up to date. Adding to Newton's suggestion that comets deposited revitalizing æthers to a spiritually depleted earth, Whiston equated comets with hell: combining their apogaeic 'Darkness of Torment' and their perigaeic 'ungodly Smoak of Fire', comets became 'the Place of Punishment for wicked Men after the general Resurrection'.[26]

In comparing Whiston's confident exegesis to the work of his twentieth- and twenty-first-century successors, one might envisage the history of the chair as a reflection of the increasing secularization of scientific knowledge. This interpretation is tempting. Take the materialism of Charles Babbage. In promising to reduce 'Miracles' to

Figure 4 Just six years before Newton sent the *Principia* to press, Thomas Burnet published the *Sacred Theory of the Earth.* Newton commended Burnet's work, but also offered him biblical exegeses based upon his own chemical experiments.

pre-set 'irregularities' in his Analytical Engine, Babbage turned prophetic wisdom into a mechanical exercise: 'the maker of the calculating engine', he gloated, 'would thus be gifted with the power of prophecy'. A century later Wolfgang Pauli wittily said of Dirac's spiritual leaning that he 'has a new religion. There is no God and Dirac is his prophet.' Among the numerous comments that Stephen Hawking has made concerning the relation of theoretical physics to religion, he has recently noted, 'General relativity could not predict what should emerge from the Big Bang. Some saw this as an indication of God's freedom to start the universe off in any way God wanted, but others (including myself) felt that the beginning of the universe should be governed by the same laws that held at other times.' Elsewhere Hawking has elaborated on this commitment:

> We are such insignificant creatures on a minor planet of a very average star in the outer suburbs of one of a hundred thousand million galaxies. So it is difficult to believe in a God that would care about us or even notice our existence.... [But] it is difficult to discuss the beginning of the Universe without mentioning the concept of God. My work on the origin of the Universe is on the borderline between science and religion, but I try to stay on the scientific side of the border. It is quite possible that God acts in ways that cannot be described by scientific laws. But in that case one would just have to go on personal belief.[27]

Hawking's religious beliefs certainly contrast markedly with those of most of his predecessors, for most Lucasian professors through the age of Stokes published theological works. While Newton, because of his heretical views, was unwilling to put his religious beliefs into print, the eighteenth-century professors were eager to demonstrate the accord between religion, the natural sciences and mathematics: among his prodigious outpourings Whiston published *A New Theory of the Earth* (1696), *The Accomplishment of Scripture Prophecy* (1708) and *Astronomical Principles of Religion, Natural and Reveal'd* (1717); in the 1730s John Colson demonstrated the breadth of his learning with his *Historical, Critical, Geographical, Chronological, and Etymological Dictionary of the Holy Bible*; to rebut David Hume's knock at English natural science and revealed religion, Edward Waring penned *An Essay on the Principles of Human Knowledge*; similarly, his successor and Doctor of Divinity, Isaac Milner, produced *An Essay on Human Liberty* and his seven-volume *Ecclesiastical History* in his bid to apply 'knowledge of natural philosophy and mathematics to ... stem the torrent of scepticism and infidelity ... inundating this Empire'.[28]

Nineteenth-century Lucasians were also active in producing religious ruminations. Alongside his hymns, Thomas Turton produced *Natural Theology Considered with Reference to Lord Brougham's Discourse on that Subject* (1836) and was also immersed in the debate concerning dissenters' access to higher learning, providing his *Thoughts on the Admission of Persons, Without Regard to their Religious Opinions, to Certain Degrees in the Universities of England* in 1834. Irritated by the *Bridgewater Treatises*, eight best-sellers of the 1830s that upheld natural evidence for divine design and intervention, Charles Babbage penned his unofficial and fragmentary *Ninth Bridgewater Treatise* (1837), which retorted that miracles were not the result of divine whim, but the product of natural law programmed by God. A generation later, attempts to reconcile the two Books (the *Book of Nature* and the Bible) by George Gabriel Stokes, a lifelong evangelical Anglican, demonstrates that the history of the Lucasian chair is *not* a straightforward story of secularization. Stokes's tenure coincided with potent challenges – by the likes of the authors of *Essays and Reviews* (1860) and apologists of Charles Darwin – to the cultural authority of the established church and the plausibility of the Biblical narrative. He responded with numerous religious tracts and books, including his Gifford lectures of 1891–3, and addresses to church congresses. He was also President of the Victoria Institute, established in 1865 to uphold the belief that the claims of science and scriptural truths were in harmony. This would lead his arch opponent, Darwinian champion and high priest of the 'Church Scientific', Thomas Henry Huxley, to criticize angel 'Gabriel' for abusing scientists by allying them with 'everything Churchy & reactionary'.[29]

Even where physics had been invoked by Darwin's guardians, Stokes did not think it had been done so legitimately. Instead, he evoked an immaterial gravitational force or luminiferous æther acting on ponderable matter in order to make plausible the notion of God guiding nature. Joseph Larmor was less vocal than Stokes, but he shared his predecessor's Protestant Irish upbringing and, as Andrew Warwick has suggested, protestant values underpinned his belief in detecting the motion of the earth relative to the æther. While the Irish Protestant physicist Frederick Trouton accused Larmor of becoming 'much more catholic' in his 'scientific beliefs' in accepting Einstein's relativity, Larmor remained a 'sturdy Protestant of Science' since he shared little with Einstein, remaining convinced that a dynamic æther was indispensable to an intelligible electromagnetic theory.[30]

Critical too for the 'sturdy protestants of science' was a sharp distinction between the body and the intellect. Until Dirac's tenure, the Lucasian professors were 'naturally led to observe a remarkable difference between the operations of matter and of the mind'. Early-modern professors, like Isaac Milner, believed that it was critical to show that 'immaterial substances are essentially different from material ones; and... seem to be possessed of certain active principles'. Paradoxically, while the professors suggested that this ontological principle made the sciences subservient to religion, it also elevated the importance of their experimental work. Because of their unique understanding of the material world, the professors used their knowledge of the 'established principles of Experimental Philosophy' to comment on 'brute' matter's passivity and its dependence upon thinking substances. Newton's self-experimentation with a knitting needle upon his own eye is illustrative: by showing that vision was contingent upon the voluntary actions of the mind (and *not* upon the manual manipulation of one's eyeball!) he felt that he had provided evidence against Thomas Hobbes's atheistic materialism. Similarly, during Victoria's reign, when physicist John Tyndall associated the relatively new principle of energy conservation with a materialistic and a deterministic account of man's evolution, Stokes retaliated with his notion of 'directionism' which buttressed his intense Pauline dualism. He argued that an immaterial mind could direct energy flow in a material body and still be consistent with the new energy physics, making plausible the notion of mind independent of the corporeal body.

Hitched to this commitment to Cartesian duality has been a general wont by commentators to emphasize the difference between the professors' bodies and their minds. Disembodiment has become a vital ingredient in Lucasian lore. Famously, the Marquis de l'Hôpital wondered, 'Does Newton eat, drink and sleep as other men do, or is he a genius deprived of bodily form?' More recently, Michael White has noted of Stephen Hawking's illness that 'the disease has not touched the essence of his being, his mind, and so has not affected his work'. Historians Steven Shapin and Christopher Lawrence have proposed that an underlying source of this familiar trope is the long-standing predilection to disembody knowledge claims since this makes them seem more authentic: 'the worth of knowledge', they note, 'has been linked to its stipulated elevation above the mundane and the corporeal'. Since it has been assumed that physical perceptions by unreliable bodies have been consistent sources of the corruption of knowledge, truth and the body have been 'pervasively set in opposition' and scientific practice has been readily disengaged from embodied investigators.[31]

Arguably, this is a reason why stories about the professors' indifference to the corporeal are plentiful. Paul Dirac's stoicism and monastic habitat are legendary and possibly only rivalled by his Cantabrigian contemporary, Ludwig Wittgenstein: 'living in a simply furnished attic in St. John's College,' Dirac 'had a wooden desk of the kind which is used in schools' at which he apparently wrote his 'great work straight off'. It has been suggested that 'he would have been a very contented martyr'. For Dirac's predecessor, Joseph Larmor, running water and other twentieth-century conveniences seemed superfluous to good scholarship. One obituarist noted that he questioned 'modern trends even in such matters as the installation of baths in the College (1920)', and pooh-poohed demands for plumbing improvements: 'We have done without them for 400 years, why begin now?' he proclaimed in a College committee. (Unsurprisingly, it was often noted that he neglected his appearance in his later age.) Newton was 'so intent' upon his studies 'that he ate sparingly', if he remembered to dine at all. One wit commented that 'his cat grew very fat on the food left standing in his tray'. Apparently, Barrow epitomized the dishevelled professor, being 'scholarlike, negligent of his dress and personal appearance to a fault'. 'Once,' Barrow's biographer continued, 'when he preached for Dr Wilkins at St. Lawrence, Jewry, the congregation were so disgusted with his uncouth exterior that all but a few rushed out of church.'[32]

The professors have not been utterly indifferent to the material world. Along with accounts of their ambivalence to the mundane world are a number of counterexamples. These accounts have served several purposes: first, they have given the professors a human face and have shown that a healthy mind is contingent upon a healthy body; second, they have been used to help explain the mediocrity of particular professors; and third, they recount the heroics that are sometimes needed to pursue truth. So, while he initially saw no reason to have St John's replumbed, Joseph Larmor eventually capitulated to the bliss of hot water on tap: 'once the innovation was made he was a regular user. Morning by morning in a mackintosh and cap, in which he was not seen at other times, he found his way across the bridge to the New Court baths.' In contrast to the Marquis de l'Hôpital's æthereal portrait of Newton, others painted the professor as a robust scholar: 'Sir I[saac] thus exercised at once his body & his mind [a]s the operations of the soul depends upon the condition of the organs of the body . . .'. Paulo Frisi also noted that 'he had lost but a single tooth, he never made use of spectacles, he retained a lively eye, a venerale aspect and an elegant stature'. And while Newton 'gave up tobacco' since 'he would not be dominated by habits', his predecessor, finding that it 'tended to compose and regulate his thoughts', loved smoking.

Barrow christened the New World leaf his 'panpharmacon'. In an age of roast-beef eating and beer swilling, the inaugural professor was also 'inordinately fond of fruit'. Nicholas Saunderson preferred his fruit fermented, and happily succumbed to a number of other worldly pleasures. He was renowned for his 'indulgence of women, wine, and profane swearing to ... a shocking excess'. Half a century later Isaac Milner also proved that he could be a *bon viveur*. It is doubtful if he 'indulged' in women, but he was not one to pass up a good meal or a fine bottle of claret. One astonished visitor to Queens' College reported Milner to be the 'most enormous man I ever encountered in a drawing room'. In contrast to the stoicism of Newton and Dirac, the evangelical Milner was also a whiner: 'my whole life has been one of suffering'.[33]

Milner was also 'fond of describing himself an invalid' and used illness to shirk professorial duties. In this regard he was utterly unlike Stephen Hawking and Nicholas Saunderson, who both amazed the world by overcoming their disabilities. In his own lifetime the blind Saunderson confounded Europeans with his extraordinary memory, his impeccable hearing, his remarkable sense of touch and his ability to teach *optics*. Such were Saunderson's amazing skills that Denis Diderot imagined the professor to be the ultimate test of John Locke's theory of perception. Like Hawking, Saunderson's disability led him to develop novel techniques for manipulating equations in his head. While most scientists use reams of paper in their careers, the fact that neither Hawking nor Saunderson put pen to paper has meant that both found innovative methods to tackle, and produce novel solutions for, intransigent problems. Yet, while Saunderson's disability led Diderot to question God's benevolence, Hawking's amyotrophic lateral sclerosis has led twenty- and twenty-first-century commentators to think of him as a kind of angel. Deprived of a healthy body he has become an example of how a beautiful mind can triumph in a corrupt material world.

With the other professors, Hawking's essence has often been deemed spiritual and therefore not political; but, paradoxically, since the work of the professors has been understood to be unencumbered by social interests it is often deployed to serve very political interests. The four centuries that *From Newton to Hawking* spans were amongst the most dramatic in Britain's technological, economic, social and military history. It encompasses the industrial revolution and the arrival of the information age, the rise and fall of Britain's empire, radical shifts in the social and political status of Britons and countless bloody conflicts. Although they did not always comment on these wider contexts, the meaning of the Lucasian professors' achievements would be distorted if these contexts were not considered.

Beginning with its first incumbent, Lucasian professors have worked in the midst of great political ferment. Isaac Barrow, an exemplary royalist, left England for a number of years while the ravages of England's Civil War and the interregnum played out. Pointing to both his famous publications and his obscure manuscripts, historians now routinely talk of Newton's 'politico-theology'. As Newton himself affirmed, there was a strong 'analogy between the world natural and the world politick'. So, while Newton was composing his *Principia* in the mid-1680s, he was also feverishly penning his *Theologiæ Gentilis Origines Philosophicæ*, which detailed the defilement of ancient natural philosophy for political ends. Moreover, Newton, his followers and even his detractors understood that the 1687 *Principia* could be recognized as a piece of political science. Like Locke's political treatises, Newton's work was used to justify the Glorious Revolution and ensuing Whig hegemony. His mathematical philosophy, like the interpretation of Boyle's pneumatics by his colleague Henry More, at once countered pure mechanism, the pantheism of sectarians and the absolutism of Catholics. Conical sections, cometography, universal gravity and a mostly *empty* universe offered keys to a new Whig order and, though the doggerel will not find its way into the literary canon, J. T. Desaguliers' 1727 *The Newtonian System of the World, the Best Model for Government* does exemplify how Georgian Britons could derive political messages from the Master's work.[34]

Similarly, we cannot properly explain the involvement of the Lucasian professors in the longitude problem (notably Newton, Whiston, Waring and Milner) or in the production of better nautical almanacs (specifically Babbage and Airy) without appreciating the imperial, political and economic importance of a strong Royal Navy. The revolutionary era also weighed in heavily as the professors saw how science could be deployed to attack established rule. The cool reception of continental analysis in Cambridge underlines how early-nineteenth-century Britons associated European mathematics with the bloody French Revolution. The following generation of professors were less apt to cringe at mathematical and chemical works from across the Channel. Babbage's promotion of continental mathematical tools for increasing efficiency in mental labour was inextricably linked to his contributions to fierce debates over the new factory system. By the 1840s the tools of continental analysis were integral parts of the Cambridge Mathematical Tripos. One beneficiary of this system was Stokes, who sought to provide the nation's future masters of industry with the practical and intellectual skills needed to sustain one of Britain's most powerful weapons of long-range imperial control – the electric telegraph. The telegraph helped keep the Empire together, something that both Stokes and Larmor, who, as Irish Protestant Tories, opposing Home Rule, were eager to see.

Other social and cultural shifts that have taken place in Britain in the century since Stokes's death have had a clear impact on the world of the Lucasian professor. The dramatic broadening of the educational opportunities of scholars is traceable in the transformed gender, social and ethnic composition of the people working with the current Lucasian professor. The technological descendants of the Victorian telegraphic network have also contributed to the information revolution that has fed back into Cambridge mathematics. New media technologies have helped make Hawking the centre of a global, cutting-edge communication network in theoretical physics as well as an influential political commentator. Today's culture of mathematics is exemplified by the new Cambridge-based Millennium

Mathematics Project which fully exploits the latest web technologies 'to help people of all ages and abilities share in the excitement of mathematics and understand the enormous range and importance of its applications to science and commerce'.[35]

THE PROFESSORSHIP IN A NUTSHELL

Though none of the seventeen Lucasian professors of mathematics would have described themselves as statisticians, their lives have generated some interesting figures. Thirteen of the Lucasians have been fellows of the prestigious Royal Society. John Colson, Thomas Turton and Joshua King either could not find the backing or muster the energy to gain membership. Notoriously, Newton, as the society's president and being of a 'fearful, cautious and suspicious temper', did not support Whiston's gambit for admittance.[36] Besides Newton, only one other Lucasian professor has been the Society's president – George Biddell Airy – although both George Gabriel Stokes, Joseph Larmor and James Lighthill served as secretary. Edward Waring (1784), Airy (1831), Stokes (1893), Dirac (1952) and Lighthill (1998) all received the Royal Society's prestigious Copley medal, while most others have been prominent within the society.

Other awards, decorations and honours have been showered upon the band. Although only one – Paul Dirac – has received a Nobel prize (the prize was not established until 1901), more time-honoured rewards have been plentiful: five have been knighted (Newton, Airy, Stokes, Larmor and Lighthill) and one can reasonably expect that Hawking might soon be called 'Sir Stephen'. Isaac Barrow (Trinity), Isaac Milner (Queens'), George Gabriel Stokes (Pembroke) and Joshua King (Queens') were all rewarded with the mastership of their respective colleges, while University College London snagged James Lighthill as its provost. As discussed earlier, Newton's inability to secure the provostship at King's College was one of his few failures, though he was, however, elected to Parliament (twice), a triumph that is only slightly overshadowed by the fact that the only record of him speaking within the House of Commons was a request to have shut a draught-causing window. Alongside Newton, George Gabriel Stokes and Joseph Larmor also represented Cambridge as MPs while, famously, Charles Babbage twice stood unsuccessfully for the borough of Finsbury.

Most Lucasians have been concerned with eternal rewards, both for themselves and for their fellow Christians. Although none did so in the twentieth century, seven – Barrow, Whiston, Colson, Waring, Milner, Turton and Stokes – donned the vestments of the Anglican Church. Indeed, Thomas Turton was elevated to the see of Ely after he stepped down from his mathematics chair. Such was his antipathy towards the Church established that Newton sought a special dispensation from the King in order to avoid taking Holy Orders; but only the foolish have suggested that his evasion had anything to do with an inclination towards the secular.

Whiston, like Newton, loathed the doctrine of the holy and undivided Trinity. This deep commitment to Arianism leads also to another statistic: branded a heretic by the University's Vice-chancellor, he is the only professor to have been unwillingly removed from the chair. Besides Whiston, none, to our knowledge, has been suspected of transgressing the professorial statute involving 'treason, heresy, schism, voluntary manslaughter, notable theft, adultery, fornication or perjury'. Nor do any past Lucasian professors seem to have been arrested for any other crimes or misdemeanours. Accordingly, all but one election for the professorship has been precipitated by either wilful resignation or the death of the incumbent, the former being slightly more common. Resignation accounts also for trimming the average length of tenure – almost exactly twenty years. Stokes, weighing in for an astounding 54 years, is almost singularly responsible for driving up the average: meanwhile the tenures of Airy, Woodhouse and Turton combined could not see through the 1820s. These Lucasians and the rest of the professors, however, have shared one obvious characteristic: they have all been white males.

These statistics are illuminating. They fail, however, to uncover the extent to which each professor has been embedded within the cultures in which he deployed his expertise. Along with a journey through the professors' great triumphs (and a few humiliating defeats), *From Newton to Hawking* travels through three-and-a-half centuries to find these diverse scientific cultures. Though readers of this book will discover some fascinating continuities over the duration of the professorship, these cultures will also show how different the professorship of the twenty-first century is from 1663, when the chair was endowed.

In the first chapter Moti Feingold recounts the protracted search for a benefactor with the wherewithal to establish a '*mathematicus professor honorarius* ... with a House of Purpose'. While reminding us that the absence of the mathematical professorship should not be construed as a lack of mathematical activity at the varsity, he also shows why Henry Lucas's endowment and the work of the inaugural professor, Isaac Barrow, were so valuable to the institution. Although some commentators may have felt that Barrow was 'but a child in comparison to his pupil Newton', Feingold convinces us that Barrow's profound and ambitious studies, and particularly his research in optics, cannot be taken lightly. In addition, he addresses an apparent paradox that the first Lucasian professor presents: although Barrow considered mathematics the 'fruitful Mother of all Disciplines, and benign Nurse of all Studies', he was deeply resistant to publishing his mathematical work, even complaining to a fellow divine that he was 'wasting [his] time and intellect' in mathematics. In pointing to the tension between Barrow's love for the mathematical sciences and devotion to theology, Feingold's portrait of Barrow sets the stage for discussion of his tormented successor.

Robert Iliffe's account of Isaac Newton masterly synthesizes the radically diverse activities of the second Lucasian professor. Unearthing the full extent of Newton's intellectual activities and contextualizing these within his Cantabrigian and metropolitan scientific milieux, Iliffe portrays the 'Great Man' as a psychologically troubled mortal, but a mortal who believed that he was on a mission from God. Constantly distinguishing himself from 'the vulgar', Newton

Figure 5 William Blake's 'Newton'. Often construed as the quintessential work depicting the irresolvable conflict between the two cultures – the culture of art and the culture of science, Blake's representation of the Lucasian professor is a renowned example of the countless different meanings, and omnipresence, of 'Newtonianism' in Georgian England.

conceived his divinely sanctioned role to involve the recovery of uncorrupted ancient truths, both scientific and religious. So, though we may remember Newton for his major contributions to mathematics, astronomy and optics, Iliffe tells us we must not forget that the 'Great Man' was just as much a revolutionary in alchemy and theology. Only with an appreciation of these interests can we begin to fathom the truly radical nature of Newton's work, not to mention the extent of his remarkable genius.

Newton was a hard act to follow. In his influential study of 'Enlightenment Cambridge', D. A. Winstanley observed that 'Cambridge in the eighteenth century was sadly lacking in eminent mathematicians'. This pronouncement has been reiterated by the Lucasian professors themselves: 'There is no doubt that there was a stagnation in scholarship in Cambridge throughout the eighteenth century', complained Sir James Lighthill: 'this unreformed Cambridge was really bad. A great pity really.'[37] But, in *From Newton to Hawking*, the chapters devoted to the eighteenth-century professors show that this supposition is unwarranted. Though the age may not have been a heroic one for Cantabrigian natural philosophy, it was none the less one of vibrant activity. Of course much of this activity was directed at interpreting, protecting and disseminating the unparalleled genius of Newton. Newton's work, the eighteenth-century professors believed, had catapulted Britain into a new age. The judicious use of his philosophy would solve scientific, technological, religious and political problems. Along with 'Newtonianizing' other fields of enquiry, from theology to medicine, it was therefore the mission of the eighteenth-century professors to broadcast the existing gospel of Newton. But what, exactly, this gospel *was* was open to debate, even amongst the mathematical professors themselves. Since his corpus was so gargantuan and so enigmatic, Newton's intellectual legacy was fraught with difficulties. Each professor found that he needed to interpret Newton in order to fight the growing number of enemies who found the Newtonian philosophy intellectually and morally bankrupt.

The 'Great Man's' immediate successor, William Whiston, epitomized this ambition to defend Newton's work and to bring his 'Divine Philosophy within Reach' of mortal Britons. Moreover, where Newton had held his theological cards close to his chest, Whiston brazenly – and, perhaps, cavalierly – applied the scientific reasonings of the *Principia* and the *Opticks* to scriptural exegesis. In their treatment of his extraordinary attempt to render both Newton's philosophy and scripture transparent (via Newton's natural philosophy), Stephen Snobelen and Larry Stewart follow Whiston's unconventional path from the private serenity of cloistered Cambridge to the public bustle of Augustan London. Banishment from Cambridge, they argue, was only one of many signs that Whiston was embroiled in the chief religious controversies of the era. By delineating his great success in the metropolis Snobelen and Stewart also show that his expulsion from the university, seemingly paradoxically, enabled a career boost as Whiston found 'fame and fortune' in both metropolitan coffeehouses and in print. Whether 'solving the Longitude' with exploding mortars, linking comets to Noahic catastrophe and 'Divine Vengeance', or galvanizing polite audiences with fantastic electrical phenomena, Snobelen and Stewart show precisely why it was difficult *not* to listen to Whiston.

With the rustication of Newton's successor from Cambridge, Edmund Halley quipped that 'Whiston was dismissed for having too much religion, and Saunderson preferred for having none.' Nevertheless, the story of Nicholas Saunderson and his successor, John Colson, is as much one of continuity as it is of discontinuity. Although neither professor antagonized the Anglican establishment as did 'wicked Whiston', both Saunderson and Colson emulated Whiston's endeavours to make popular the central tenets of Newton's *œuvre* and to vanquish detractors of the 'Great Man'. John Gascoigne pays special attention to the pedagogical enterprises of these two Lucasians. In so doing he also shows how their work chimed in with the other Cambridge Newtonians who, locking horns with the likes of Bishop Berkeley, were anxious to establish that Newton's philosophy led neither to 'absurdity' nor to the 'heresies of infidels'. John Colson, for instance, saw Newton's *Method of Fluxions* to the press, not simply to give Britons better access to a powerful analytical tool but to ensure that the 'visible and sensible form' of the fluxional calculus led directly to godly truths. Although not having access to the 'visible', the blind Saunderson did give his mathematical practice a 'sensible form' and this leads Gascoigne to a discussion of how the professor's physical disadvantage gave the Enlightenment minds of Denis Diderot, Samuel Johnson and Edmund Burke pause to consider relationships between sense experience, ideas and the nature of the deity.

Like Saunderson and Colson, the following two Lucasian professors, the 'awkward' and 'melancholic' Edward Waring and the 'arrogant' and 'incomparable' Isaac Milner, were anxious to preserve the status of Cambridge dons as the authentic representatives of Newton and to use this status to quash the increasingly hostile attacks on the university. Touching upon the intellectual products of the Enlightenment and the major shifts in the sciences, as well as the revolutionary contexts of the *fin de siècle*, Kevin Knox shows how Waring and Milner dealt with the devastating critiques of Cambridge's scientific practice from such luminaries as Joseph Priestley and, later, the irascible Charles Babbage. Although remarkably dissimilar in personality, the two Lucasian professors shared common strategies for preserving the place of spirit in the natural world, the primary articles of the Anglican Church and the 'traditions' of university life. Yet, Knox argues, it would be a mistake to regard these two Lucasians as mere reactionaries, for their participation in the national and international republic of letters signalled some new characteristics of the nineteenth-century don.

With Milner's death in 1820, reform of neither the institution nor of the professorship looked promising. As an undergraduate Babbage had satirized a bitter religious dispute in which Milner was a key player in an attempt to launch a revolutionary mathematical society; but partially due to Milner's resistance, Babbage's 'Analytical Society' fizzled. Nevertheless, in the following decades the Lucasian professors played important roles in making Britain the preeminent scientific state and in changing the university from a 'gentleman's club' to a research institution. Concomitant with these transformations in the 'holy city of mathematics' was the rising eminence of the professorship itself. In his account of the professorship from 1820 to 1839, Simon Schaffer recaptures the complex, and very divergent, interests of four Lucasian professors: Robert Woodhouse, Thomas Turton, Charles Babbage and George Biddell Airy. Along with vivid accounts of vicious electoral campaigns and combination room intrigue, Schaffer places the professors' interests within a precarious university culture that simultaneously insisted upon maintaining its rich if dated scholarly traditions but realized that it needed to come to terms with the new philosophies of manufacture, machinery and political economy. Expertly glossing the careers of Woodhouse and

Figure 6 Woolsthorpe. Along with must-see Trinity College, many admirers of Newton have made the edifying pilgrimage to Newton's family home in Lincolnshire.

Turton – whose tenure was arguably the nadir of the professorship – Schaffer concentrates on the ambitions and anxieties of George Biddell Airy and Charles Babbage. In addition to surveying the instruments and techniques that eventually made Airy a model Astronomer Royal, Schaffer describes Babbage's obsession with improving the efficiency of Britain's imperial economy through rationalizing the emerging mechanisms of the factory system. Literally mechanizing mathematical reasoning with his calculating engine, Babbage forced less progressive Cambridge men like William Whewell to rethink what scholarship meant for both the university and for the empire.

George Gabriel Stokes was one of the new breed of Cambridge scientists that helped Whewell reformulate scientific practice at Cambridge and, indeed, throughout Britain. Contrasting the keen experimentalist with his competent but lacklustre predecessor,

Joshua King, David Wilson portrays Stokes as a key arbiter of science. While producing ground-breaking research in optics and hydrodynamics, Stokes, as professor and secretary of the Royal Society, was in a strategic position to comment upon myriad subjects – both scientific and cultural – that captivated Victorians: the luminiferous æther, spiritualism, the immortality of the soul, X-rays, radioactivity and Darwinian evolution. As for the last, Wilson describes how, for Stokes, Victorian physics not only generated accounts of the cosmos which could be reconciled with Genesis, but also symbolized the high standard of scientific reasoning that Darwinianism, that potent weapon against Creationism, failed to reach.

Stokes may have been the last Lucasian professor of Victorian Cambridge, but it was said of his successor, Joseph Larmor, that his 'heart was in the nineteenth century'. In his account of the twentieth century's first new Lucasian professor, Andrew Warwick examines Larmor's protracted quest to describe what he envisioned as the fundamental essence of the universe – a dynamical, luminiferous and electromagnetic medium. While some have viewed Larmor as a kind of anachronism unwilling to abandon an obsolete and fantastical concept, Warwick suggests that Larmor's work has been gravely misrepresented and unearths the underlying sophistication of the professor's dynamical æther. Larmor's dynamical æther, Warwick shows, was more than a convenient way of unifying electromagnetic and optical phenomena: it represented an ontological reality that made progress in physics possible and revealed the underlying unity of nature. So, in the face of widespread claims that the Michelson–Morley experiment had failed to generate evidence of the æther, Larmor insisted that his æther theory explained why this null result was essential to the construction of theoretical physics. Warwick explains why this in turn enabled Larmor to construct a natural history of physics that placed this æther at the locus of an ineluctable and benevolent process of discovery.

If Joseph Larmor is, somewhat unjustly, remembered for his reluctance to embrace new scientific theories, his successor Paul Adrian Maurice Dirac is often memorialized for revolutionizing physics with audacious claims about the nature of the submicroscopic world. Nevertheless, Dirac shared with Larmor what might be called a nonempirical methodology for favouring physical theories. In his exploration of the life of Paul Dirac, Helge Kragh delves into the unusual mental world of the Nobel laureate, using Dirac's vision of purity, rationality and beauty to excavate both the motives and the processes behind the professor's startling work. As a 'pure soul', Dirac, Kragh explains, was obsessed with dissociating himself from the mire of traditional academia and scientific practice. As such, he usually worked in monastic isolation and was often viewed as an antisocial curmudgeon. Similarly, as a 'fanatic of rationalism', he scorned anything that seemed to him to smack of social interest, be it in reference to an experimental research programme or a political ideology. Kragh explains how this rationality, seemingly paradoxically, was integrated with a deep commitment to mathematical aesthetics. Such was his fixation with this enigmatic aesthetic that Dirac was wont to equate beauty with truth, and even reject experimental evidence if it conflicted with his notion of a beautiful equation. Yet, despite his unorthodox attitudes and working habits, Kragh shows exactly why so much of Dirac's work remains central to modern, orthodox physics.

Central too to the orthodox scientific world – but also to a host of unconventional creeds – is the work of the current Lucasian professor, Stephen Hawking. In the final chapter of this volume Hélène Mialet examines the remarkable and courageous life of 'the prophet of the Big Bang'. Contrasting Hawking's career with that of his predecessor, James Lighthill, and considering both the professor's debilitating illness and the stunning theoretical achievements that helped make him famous, Mialet's ingenious analysis follows the route that turned the seemingly most mortal of men into a celebrated oracle. By virtue of the timeless professorship he represents, the panoply of machines and humans that enable him to work and the fact that his theoretical physics is often deployed in theological speculations, she argues that we can consider Hawking as a kind of angel who is at once seemingly immortal, immaterial and ubiquitous. Once considered a 'stop-gap

Figure 7 The Royal Seal of the Lucasian professorship. With all the regal trimmings, including the iconography of the monarchy and the declaration of Charles II, 'Dieu et mon droit', the Lucasian professorship is given special legitimacy.

professor', Hawking has metamorphosed into a beatified media darling whose opinions are sought from the White House to the Vatican. In so doing he has become the quintessential Lucasian professor.

The recent advances in computing, the technologies that keep Hawking at work and Hawking's own statements concerning 'the end of physics' give Mialet pause to speculate about the future of the Lucasian chair. Is it possible, she wonders, that the mathematical professors will one day become superfluous? Fascinatingly, the professors themselves have from the beginning wondered about this eventuality. In 1675, the first two Lucasian professors – Isaac Barrow and Isaac Newton – were pessimistic about further advances in mathematics and therefore, presumably, what future mathematical professors would do with their time. According to reports, Newton was 'intent upon Chimicall Studies and practises, and both he and Dr Barrow &c [were] beginning to think math[emati]call Speculations to grow at least nice and dry, if not somewhat barren'.

This barrenness was a chimera, for, among other things, Newton's own 'Queries' gave investigators plenty of fertile regions to probe. Nevertheless, questions about the end of mathematical physics have resurfaced at the varsity. In 1874 James Clerk Maxwell reflected on a foreign 'opinion' which 'seems to have got abroad, that in a few years all the great physical constants will have been approximately estimated'. While some foreigners worried 'that the only occupation which will then be left to men of science will be to carry on these measurements to another place of decimals', Maxwell was confident that 'the materials for the subjugation of new regions' were being sown.[38] The discovery of the electron, radioactivity and other dramatic events vindicated his optimism and gave physicists like Stokes new avenues of research. Yet, just a decade later Lord Kelvin, another close colleague of Stokes, speculated that accurate measurement was signalling a very different end for physics. He described two 'clouds' over the dynamical theory of heat and light, a theory which most Victorian physicists – not least the Lucasians Stokes and Larmor – believed provided the most satisfactory unifying account of the physical world.[39] For Kelvin, measurements of the specific heats of gases and the apparent nonmotion of the earth relative to the æther posed serious problems for the equipartition theorem of energy developed for molecular behaviour and the electromagnetic æther.

These problematic cornerstones of classical physics were eventually 'dispersed' by two monuments of postclassical physics – Planck's quantum theory and Einstein's theories of relativity. These monuments gave both Paul Dirac and Stephen Hawking the opportunity to posit startling new conceptions of the universe; but by the end of the twentieth century it seemed that, with the apparent unification of quantum mechanics and relativity theory, postclassical physics was at an end too. In his 1984 best-seller *The Brief History of Time,*

Hawking cautioned that this goal had many 'false dawns', including Max Born's notorious remark – made after Dirac had constructed his relativistic equation for the electron – that 'Physics, as we know it, will be over in six months.' Yet Hawking has also declared that since 'we know so much more about the universe... there are grounds for cautious optimism that we may now be near the end of the search for the ultimate laws of nature'. Part of his optimism may relate to the astonishing advances in computing, advances that have led Hawking himself to quip that he is 'Intel inside'. It is doubtful, however, that Hawking equates the capacities of a silicon chip with artificial intelligence (AI), and it seems that in this millennium the investigations of the Lucasians will continue to be a very human enterprise. As Henry Lucas envisaged, the Lucasian professor will continue to 'be a man [or woman] of good character and reputable life, at least a Master of Arts, soundly learned and especially skilled in the mathematical sciences'.[40]

Notes

1 Nicholas Saunderson, *Elements of Algebra, in Ten Books: To Which is Prefixed, an Account of the Author's Life and Character, Collected from his Oldest and Most Intimate Acquaintance* (2 vols, Cambridge, 1740), vol. II, p. 740.

2 English translation of Newton's Westminster Abbey tomb in G. L. Smyth, *The Monuments and Genii of St Paul's Cathedral, and of Westminster Abbey*, (2 vols, London, 1826), vol. II, pp. 703–4.

3 *The Times*, 3 February 1903, p. 7.

4 Mrs Laurence Humphry, 'Notes and recollections', in Joseph Larmor (ed.), *Memoir and Scientific Correspondence of the Late Sir George Gabriel Stokes* (2 vols, Cambridge, 1907), vol. I, p. 24; J[oseph] L[armor], 'Sir George Gabriel Stokes', in Leslie Stephen and Sidney Lee (eds), *The Dictionary of National Biography* (hereafter cited as *DNB*) (London, Oxford University Press, 1885–1901).

5 Charles Babbage to C. Wordsworth, 14 November 1826, British Library, Babbage Papers, 37183 f. 366; Charles Babbage, *Passages from the Life of a Philosopher* (London, 1864, 1994), p. 24; see also pp. 18–29. Mary Milner, *The Life of Isaac Milner D.D, F.R.S.* (2 vols, London, 1842),

vol. I, p. 8. Lighthill in interview conducted with James Lighthill by Robert Bruen, July 1998: http://exile.ne.mediaone.net/lucas/lighthill-interview.html. Stephen Hawking, *A Brief History of Time* (New York, 1988), p. 182; Stephen Hawking, *The Universe in a Nutshell* (New York, 2001).

6 Barrow's skills in T[homas] F[rederick] T[out], 'Isaac Barrow', *DNB*.

7 For Babbage's malfunction, and reminiscences of him, see *DNB*; for Babbage's controversies, see Babbage, *Passages, passim*.

8 Lucas's will is in the Registry of Wills in Somerset House: cited in John W. Clark (ed.), *Endowments of the University of Cambridge* (Cambridge, 1904), pp. 165–71.

9 John Wallis, 'Autobiography', in *Notes and Records of the Royal Society of London* 1970, pp. 17–46, 27; Babbage, *Passages*, p. 26.

10 For disputations, see William Costello, *The Scholastic Curriculum at Early Seventeenth-century Cambridge* (Cambridge, MA, 1958), p. 15. For the curriculum and examinations of Cambridge see also Elisabeth Leedham-Green, *A Concise History of the University of Cambridge* (Cambridge, 1996), pp. 120–6.

11 Saunderson, *Elements of Algebra*, vol. xv. Gilbert Wakefield, *Memoirs of the Life of Gilbert Wakefield* (London, 1792), pp. 100–1.

12 For the early history of the Mathematics Tripos see Leedham-Green, *Concise History*, pp. 147–86.

13 W. W. Rouse Ball, 'The Cambridge School of Mathematics', *Mathematical Gazette*, 6 (1912), 311–23, p. 319.

14 Cited in Leedham-Green, *Concise History*, p. 152.

15 *Cambridge University Calendar for the Year 1905–1906* (Cambridge, 1905), pp. 597–601, 725–57.

16 Stokes to Mary Stokes [née Robinson], 3 February 1857, in Larmor, *Memoir and Scientific Correspondence*, vol. I, pp. 55–7, 56–7.

17 For the Statutory Commission see D. A. Winstanley, *Early Victorian Cambridge* (Cambridge, 1955), pp. 314–38; see also *Statutes of the University of Cambridge with the Interpretations of the Chancellor and Some Acts of Parliament Relating to the University* (Cambridge, 1914), p. 62; Cambridge University Archives, MSS UA CUR.39.8, f. 17 (1), p. 3; f. 20 and 21.

18 Gillian Sutherland, 'Emily Davies, the Sidgwicks and the education of women in Cambridge', in Richard Mason (ed.), *Cambridge Minds*

(Cambridge, 1994), pp. 34–47, 34. Isaac Newton, *The Correspondence of Isaac Newton* (7 vols, Cambridge, 1959–77), vol. III; p. 280.

19 Babbage, *Passages*, p. 19.

20 Mrs Laurence Humphry, 'Notes and recollections', in Larmor, *Memoir*, p. 28.

21 Richard Marsh, *The Vanity and Danger of Modern Theories: A Sermon Preach'd in Cambridge, 1699* (Cambridge, 1701).

22 Newton to Bentley, 10 December 1692. Newton, *Correspondence*, vol. III, p. 233.

23 William Oughtred, *To the English Gentrie, and all Other Studious of the Mathematicks* (London, 1632[?]), p. 8.

24 Whiston cited in Simon Schaffer, 'Comets & idols: Newton's cosmology and political theology', in Paul Theerman and Adele F. Seeff (eds), *Action and Reaction: Proceedings of a Symposium to Commemorate the Tercentenary of Newton's Principia* (Newark, 1993), pp. 206–31, 226.

25 Newton, Jewish National and University Library (JNUL), Jerusalem, Yahuda MS 6 f. 11r.

26 Cited in Sara Genuth, *Comets, Popular Culture, and the Birth of Modern Cosmology* (Princeton, 1997), p. 8.

27 Babbage, *Passages*, p. 292. Pauli cited on http://physics.hallym.ac.kr/reference/physicist/dirac_paul.html. Hawking, *Nutshell*, p. 24; Michael White and John Gribbin, *Stephen Hawking: A Life in Science* (New York, 1993), pp. 166–7.

28 Milner, *Life*, vol. II, 698.

29 Adrian Desmond, *Huxley: Evolution's High Priest* (London, 1997), pp. 149, 302 n42.

30 For the æther and metaphysical unity, see Geoffrey Cantor, 'The theological significance of æthers', in G. N. Cantor and M. J. S. Hodge (eds), *Conceptions of Ether: Studies in the History of Ether Theories 1740–1900* (Cambridge, 1981), pp. 135–55; David Wilson, 'The thought of late Victorian physicists: Oliver Lodge's ethereal body', *Victorian Studies* 15 (1971), 29–48. For Larmor, see especially Andrew Warwick, 'The sturdy Protestants of science: Larmor, Trouton, and the earth's motion through the æther', in Jed Z. Buchwald (ed.), *Scientific Practice: Theories and Stories of Doing Physics* (Chicago, 1995), pp. 300–43. Trouton cited in Warwick, 'Sturdy Protestants', pp. 326–7.

31 Paolo Frisi, 'Elogio of Newton', in A. R. Hall (ed.), *Isaac Newton: Eighteenth-century Perspectives* (Oxford, 1999), p. 156; White and Gribbin, *Hawking*, p. 69; Christopher Lawrence and Steven Shapin, *Science Incarnate: Historical Embodiments of Natural Knowledge* (Chicago, 1998), p. 4.

32 Larmor's commitment to plumbing in E. C[unningham], 'Sir Joseph Larmor', *DNB*. Dirac's stoicism in Kragh, *Dirac*, vol. 77, p. 251; Newton's intense study in Steven Shapin, 'The philosopher and the chicken: on the dietectics of disembodied knowledge', in Lawrence and Shapin, *Science Incarnate*, pp. 21–50, 41; Barrow's rank odour in the *DNB*.

33 Larmor's capitulation in the *DNB*; Newton's use of tobacco in Lawrence and Shapin, *Science Incarnate*, p. 40; Barrow's and Saunderson's offensive habits in *DNB*; Milner's girth and sloth cited in John Twigg, *A History of Queens' College Cambridge* (Bury St Edmunds, 1987), pp. 179–182, n. 181.

34 For this reading of the *Principia* see, for example, Margaret Jacob, *The Newtonians and the English Revolution* (Hassocks, Sussex, 1976); Simon Schaffer, 'Newtonian cosmology and the steady state', (unpublished Ph.D. dissertation, University of Cambridge, 1980). For cometography see Schaffer, 'Comets & Idols'. For Newton and the wisdom of the ancients see also J. E. McGuire and P. M. Rattansi, 'Newton and the pipes of Pan', *Notes and Records of the Royal Society*, 21 (1966), 108–43. Newton and the 'world politick' cited in Jacob, *Newtonians*, p. 43.

35 http://mmp.maths.org.

36 James Force, *William Whiston: Honest Newtonian* (Cambridge, 1986), pp. 23–4.

37 http://www.cfm.brown.edu/people/marmanis/lighthill.html.

38 James Clerk Maxwell, 'Introductory lecture on experimental physics [1871]', in W. D. Niven (ed.), *Scientific Papers of James Clerk Maxwell*, (2 vols, Cambridge, 1890), vol. II, 241–55, 244.

39 William Thomson, 'Nineteenth-century clouds over the dynamical theory of heat and light', *Philosophical Magazine*, 2 (1901), 1–40.

40 For Planck and the end of physics, see P. M. Harman, *Energy, Force, and Matter: The Conceptual Development of Nineteenth-Century Physics* (Cambridge, 1982), p. 153; for Hawking and the end of physics see Hawking, *Brief History*, 156. For Lucas's desire, see appendix to this volume.

I Isaac Barrow and the foundation of the Lucasian professorship

Mordechai Feingold

Department of Humanities and Social Sciences, California Institute of Technology, Pasadena, CA, USA

The status today accorded the Lucasian professorship tends to obscure Cambridge's more dubious honour as the last of the medieval and Renaissance universities to secure an endowed professorship in the mathematical sciences. Even rival Oxford, another late beneficiary, received its two professorships nearly half a century before the Lucasian was founded in 1663. Nevertheless, the absence of a professorship should not be taken to mean either the absence of instruction in mathematics or a failure to produce distinguished mathematicians. What it signifies is an embarrassingly acute lag in the European-wide push for the institutionalization and professionalization of the mathematical sciences in the wake of the rapid expansion of the discipline that occurred after 1550.

Indeed, throughout the second half of the sixteenth century alumni of both universities bemoaned the lack of a permanent centre for the study of mathematics in Cambridge. John Dee, Thomas Hood and Richard Hakluyt, to name a few, sought each in his own way to persuade both government and private grandees to endow professorships in mathematics at Oxford and Cambridge, but to no avail. Equally futile proved Petrus Ramus's (tongue-in-cheek) exhortation of Queen Elizabeth, whom he urged not

From Newton to Hawking: A History of Cambridge University's Lucasian Professors of Mathematics, ed. Kevin C. Knox and Richard Noakes. Published by Cambridge University Press.

> to be a pupil of France any longer, but summon the French in their turn over to England . . . Inquiring into the two most erudite universities of your realm, I have learnt that professors of Greek, Hebrew, Medicine, civil law, and theology are honoured with royal stipends . . . but no royal reward has been established for the professors of mathematics . . . And so for you, Your Majesty, I desire Regius Professors of mathematics in both Cambridge and Oxford, to adorn your memory with eternal praise for your magnificent generosity.

Three years later, the young Henry Savile both endorsed and added to Ramus's plea with an eloquent turn of phrase of his own:

> Only England does not have any honours for mathematics, because it does not have any mathematicians. How many illustrious Maecenases do you think there would be, if there were Virgils worthy of their patronage? I remember hearing more than once that some of our foremost men, who had been informed about our universities, or rather were inquiring about mathematics, were extremely aggrieved that in our two most illustrious universities there were not two mathematicians whom they could honour with a Regius professorship and regal generosity.[1]

Further to shame the universities, the first English institution to secure professorships in mathematics was Gresham College, London, which opened its doors in 1597 with generously funded professorships in astronomy and geometry.

While the failure to establish professorships at Oxford and Cambridge may reflect poorly on the English institutions of higher learning, it cannot be equated with conservatism or even hostility toward the mathematical sciences. True, activists campaigning for such endowments often portrayed in bleak terms the state of mathematical instruction at Oxbridge. They did so, however, partly in conformity with prevailing rhetorical practices, hoping thereby to render more effective their message, and partly because they sincerely believed that the absence of a solid and permanent institutional basis for such studies augured ill for the discipline. Typical in this respect is Sir Henry Savile who articulated in the preamble to his 1619 statutes of his foundations at Oxford the rationale for his generosity: 'seeing that Mathematical studies are uncultivated by our countrymen', he wrote, 'and being desirous of supplying a remedy in a quarter almost given up in despair, and to redeem so far as in me lies, almost from destruction, sciences of the noblest kind'.[2]

What had prevented the establishment of professorships (scientific or otherwise) at both Oxford and Cambridge was the peculiar character of the English universities. Unlike their continental counterparts, the English institutions of higher learning had transformed by the second half of the sixteenth century into an amalgam of collegiate corporations that reduced the university – governed by the heads of house – to an administrative agency with duties limited almost exclusively to matriculations and the conferring of degrees. Even the lion's share of undergraduate teaching was taken over by the colleges.

By the late sixteenth century the undergraduate curriculum had become the staple offering at Oxford and Cambridge and it was during this period that the students received their grounding in the entire arts and sciences curriculum. Obviously, it was not expected that such a brisk survey of knowledge would ensure mastery of all subjects or that all students were equally able to keep up the pace. Nor were they recommending a superficial grasp of knowledge. All they intended was to present a panorama of all knowledge, rooted in the interconnectedness of its various constituents, and thereby lay a solid foundation upon which the student could proceed to build, independently, for years to come. The distinctively humanistic colouring of the curriculum does not imply disregard of the mathematical and natural sciences. True, compared with the preoccupation of educators, tutors and students with the arts of discourse and with classical languages and literature, the study of mathematics, as well as of natural philosophy, was less intense. Nevertheless, these latter disciplines were still regarded as integral branches of undergraduate education and were taught by both tutors and college lecturers of mathematics. Such collegiate imparting of the grounding in the mathematical sciences accounts for the ability of the Lucasian professors to dispense with teaching the basics and devote their lectures to more specialized topics.

Understandably, then, the students' loyalties – and subsequent largesse – focused primarily on their colleges, to the neglect and perpetual penury of the universities which, like the colleges, depended on benefactions to support any new initiative. Ultimately, Oxford proved the more successful of the two, acquiring in the four decades that preceded the Civil War a magnificent library, five professorships, three lectureships and a botanic garden – thanks to a group of private benefactors who not only worked in tandem, but served as models for others to emulate and follow. In contrast, the Lucasian professorship was the first successful endowment at Cambridge since Henry VIII had established the five Regius professorships more than 120 years earlier.

Small wonder, then, that frustrated Cambridge men, envying Oxford, redoubled their efforts to seek out prospective donors. Indeed, no sooner had Henry Briggs been installed as first Savilian professor of geometry at Oxford than he began soliciting for a similar foundation for his alma mater. In August 1621, he wrote to Samuel Ward, Master of Sidney Sussex College:

> I should be very glad to see Cambridge as well provided for the stipends of these and other professions . . . If you heare of any that are able and willing to bestowe so muche monye so well, if you please to sende me worde, I will very willingly bestowe some parte of this (6 weekes) vacation in survayinge and plottinge the grounde, for the love I owe to my mother Cambridge.[3]

No able and willing benefactor stepped forward, but in 1639 it seemed as if Cambridge had finally found its Savile. That great intelligencer, Samuel Hartlib, recorded in his diary that 'one of Sir W. Boswell's privat friends is erecting a Mathematicus Professor Honorarius as it were in Cambridge setling a stipend for a 100 lib per annum with a House of Purpose. Only hee advances somwhat slowly in it. The founder will have the Professor of that place to take his corporal oath every 3 years to denominate the fittest successor unto him.'[4] That person was quite likely Henry Lucas, one-time student of St John's College, who was decorated by the university in 1636 with an honorary MA degree and who subsequently represented Cambridge in both the Short and Long Parliaments before the catastrophic Civil War that would tear the country and the university asunder. While Lucas does not appear to have been particularly interested in mathematics, or for that matter the sciences more generally, Sir William Boswell was, and it was at his urging that Lucas – if, indeed, it was to him that Hartlib alluded – contemplated his endowment. In any event, the outbreak of the Civil War brought the endeavour to naught.

A more successful effort, or so it seemed, was made in 1648 when Sir John Wollaston, Lord Mayor of London, was moved at the behest of Thomas Hill, Master of Trinity College, to bestow on the university the yearly stipend of £20 for a reader in mathematics. Wollaston further promised to augment the endowment to £60 per annum if the university approved of the gift. John Smith – among the Cambridge Platonists who rejected the strict Calvinist doctrine of predestination, advocated the use of reason in religion and supported the new science – was appointed the first incumbent and appears to have commenced his tenure in November 1648 with a course of lectures on Descartes's *Geometry*.[5] In all likelihood, among his auditors was a third-year undergraduate, Isaac Barrow. Unfortunately, the lectureship lapsed by the time Smith died in 1652 and Cambridge still awaited its benefactor.[6]

Lucas survived the period of the Civil War and Interregnum, and when he drew up his will on 11 June 1663 – six weeks before his death – he remembered Cambridge. His executors, Thomas Buck and Robert Raworth, were instructed to 'purchase lands to the value of one hundred pounds by the yeare to be imployed and setled as a yearely stipend and sallerie for a professor and reader of the mathematical sciences'. Like Savile earlier, Lucas justified his benefaction with the desire to provide for 'the improvement and encouragement of that parte of learning which hitherto hath not bin provided for'. The trustees swiftly purchased the land, the yearly income from which was a common means to provide a professor's salary, and by 19 December 1663 had drawn up the statutes of the professorship (see Appendix). A month later they also secured the King's letters patent and proceeded officially to appoint Isaac Barrow as professor on 20 February 1663/4.[7] While Barrow's early biographer, Abraham Hill, may be correct to state that John Wilkins recommended him for the position, it is certain that Buck, university beadle since 1626 as well as the Cambridge University printer, was well aware of Barrow's reputation at least as far back as the mid-1650s. Be that as it may, Barrow was actively involved with setting up the professorship even before his appointment, for at the very least he played an important role in drawing up the statutes.

The statutes themselves were loosely modelled on the Savilian statutes, though the duties of the Lucasian professor were made less onerous than those of his Oxford counterparts. Essentially, he was instructed to deliver an hour-long lecture once a week during term on a mathematical topic of his choice as well as to make himself available for at least two hours, in his quarters, for private instruction and the resolution of doubts for those seeking his assistance. Spelled out, too, was the expectation that he have in his possession the necessary mathematical instruments – such as dials, globes, quadrants and telescopes – and that he deposit every year ten lectures in the university library. Again, as in the case of the Savilian professors, the Lucasian professors were not required to take holy orders and were explicitly proscribed from accepting the cure of souls. The King's letters patent further permitted the incumbent to hold a fellowship, notwithstanding college statutes that prohibited fellows from deriving income from other sources. The professor was barred, however, from holding most university offices, such as another professorship or headship of a college.

That Isaac Barrow was eminently qualified for the position is beyond doubt. Born October 1630 in London, Barrow received his early

Figure 8 Portrait of Isaac Barrow by David Logan (1676).

schooling at the Charterhouse. His father intended him to become a scholar but, to his chagrin, the youth distinguished himself instead as 'being much given to fighting, and promoting it in others', so that he was transferred to Felsted School in Essex where he came under the supervision of Martin Holbeach, a noted tutor who had been John Wallis's revered schoolmaster a decade earlier. In February 1646

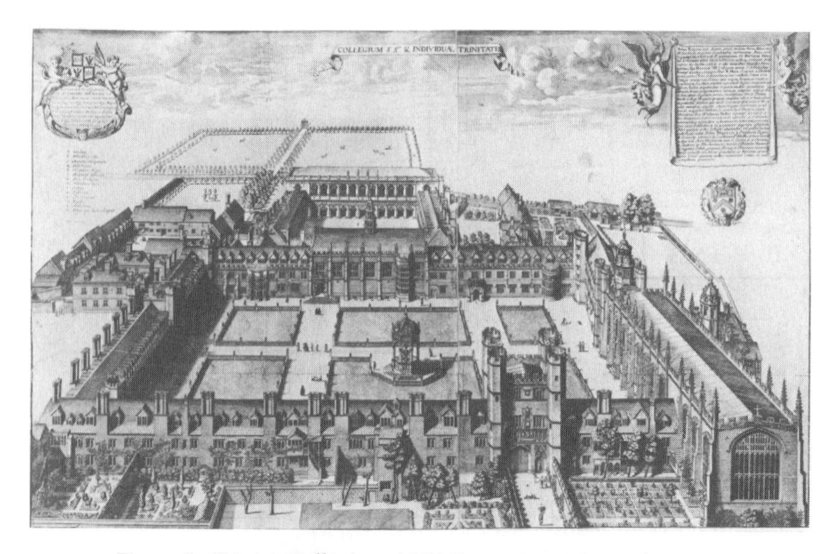

Figure 9 Trinity College, c. 1687, Barrow's residence for many years as student, scholar, professor and Master. The Master's Lodge is located on the far side of Trinity's Great Court, while the set of his protégé, Isaac Newton, was located between the College's main gate and its chapel. From David Logan *Cantabrigia Illustrata* (1690).

Barrow proceeded to Trinity College, Cambridge. His tutor now was James Duport, Regius professor of Greek who, having immediately assessed his charge's exceptional linguistic and literary skills – and sharing in Barrow's royalist sentiments – put him up in his room and taught the virtually penniless scholar gratis. Further testimony to the impression made by the young Barrow can be adduced from the favour shown him by Thomas Hill, the Master, who shielded him from the more godly members of the college even though Barrow openly – and sometimes provocatively – professed royalist and Anglican sentiments. 'Thou art a good lad', Hill supposedly once told Barrow, ''tis pity thou art a Cavalier.' On another occasion he silenced Barrow's critics with the blunt pronouncement that he was 'a better man than any of us'.

Barrow graduated B.A. in March 1649 and was elected fellow of Trinity College shortly thereafter. He proceeded to M.A. in 1652 by which time he was already beginning to make a name for himself through orations, disputations and other public performances, both at Trinity and in the schools. In 1654, his former tutor Duport, now pressurized to step down from the Regius professorship on account of his royalist sympathies, arranged for Barrow to succeed him. However, the well-connected Ralph Widdrington, coveting the office, managed to obtain an injunction from Oliver Cromwell himself which secured for him the professorship. Weary of the charged religious and political atmosphere at Cambridge, Barrow left for the continent. After spending some eight months in Paris, and an equal stretch of time in Florence, he headed east, first to Smyrna and then on to Constantinopole, returning to England in late summer 1659. Within a few months, the Stuart monarchy was restored and Barrow was promptly granted the Greek professorship at Cambridge. Two years later, he was also elected professor of geometry at Gresham College, and for the next year and a half he divided his time between Cambridge and London. In 1662 Barrow was also elected fellow of the Royal Society, but he was never an active member. Abiding by the newly penned Lucasian statutes and comfortable with the remuneration that his new position would bring, Barrow resigned both professorships in spring 1664, shortly after delivering his inaugural lecture as Lucasian professor of mathematics.

Barrow had applied himself earnestly to mathematics by the time he graduated MA and experienced his own *anni mirabiles* – mirroring those of Isaac Newton a decade later – during the two years or so preceding his departure for the continent. He embarked on a profound and ambitious study of the ancient geometricians as well as mastering contemporary developments in mathematics. Indeed, he appears to have travelled virtually the same route that Wallis at Oxford and Christiaan Huygens in Holland traversed at about the same time, immersing himself in issues of rectification and quadrature. Naturally, Barrow, too, devoted an inordinate amount of time to the hotly contested problem of squaring the circle but, as he later told a friend, 'all that he got was a demonstration that it was impossible to be done'. Barrow, however, appears to have gone further still. As Worthington told Hartlib in 1656, Barrow counted his work on Euclid as but 'a meane worke' and that 'hee hath undertaken to doe something upon Archimedes which shall awaken all the World'.[8] Worthington did not elaborate, but among Barrow's discoveries was his method of tangents as well as most of the material that Barrow would discuss when delivering his geometrical lectures during the second half of the 1660s.

As things turned out, Barrow's extended period of travel, followed by his appointment as Regius professor of Greek, interrupted his mathematical studies and he abandoned whatever plans he had to publish the result of his researches. One wonders whether things might have transpired differently had Colonel Drake been persuaded in 1655 to establish a mathematical lectureship earmarked for Barrow – as Samuel Hartlib fantasized – that would have kept Barrow in Cambridge and thus enabled him to continue his pursuit of mathematics.[9] But this was not to be and, as will be shown below, by the time Barrow assumed his duties as Lucasian professor he had lost much of his youthful ardour for the study of mathematics.

Though Cambridge desperately sought an endowed professorship, the university was not devoid of a lectureship in mathematics. By the turn of the sixteenth century the medieval tradition, whereby incepting masters of arts were required to lecture for a full year before receiving grace for their degree, had collapsed and in its place the university had resorted to a system of four salaried readers – in humanities, logic, philosophy and mathematics – chosen from among the new regent masters who usually held their position for a year. Such a concentration of all mathematical instruction in the hands of a single reader somewhat handicapped Cambridge compared with Oxford, for at the sister university each of the mathematical disciplines was staffed with a regent master – also appointed for one year. Nevertheless, many of the Cambridge readers were quite skilled and some, such as Henry Briggs and Thomas Hood, later distinguished themselves as mathematicians. As an educational programme, however, the Cambridge school suffered from the youth of its readers, the one-year limitation on their tenure, and the inevitable fluctuation in their talents and interests – so much so that any continuity or stability was pretty nigh impossible. Curiously, in contrast to Oxford, where the foundation of the Savilian professorships subsumed the other lectureships, at Cambridge the readership persisted at least until the end of the seventeenth century, though we know very little about it save for the names of the incumbents.[10]

None the less, in evaluating mathematical instruction at Cambridge it is important to remember that the Lucasian professor was not the only mathematical act in town. In addition to the university reader, each college had its own mathematical lecturer whose job it was to take charge of all elementary instruction in mathematics, thereby ensuring that the audience for the Lucasian professor had acquired at least a modicum of mathematical knowledge and that the professor could focus on advanced topics of his choice. Hence, regular attendance at the lectures could only be expected from the relatively few who exhibited an aptitude and inclination for such studies. The rest satisfied their requirement by attending lectures at college or those offered by the university reader of mathematics. Furthermore, as I shall discuss below, for those students who sought the Lucasian professor out, the public lectures were not necessarily the most important of his offerings.

Barrow delivered his inaugural oration on 14 March 1665 to a large and enthusiastic audience. Mastering his considerable oratorical skills, Barrow, too, availed himself of the familiar topos which panegyrized the extraordinary deliverance of the mathematical sciences. Considering the lamentable state of learning at the time – exclaimed the new professor – it was veritably a 'wonder that a noble *Maecenas* has at last appeared' whose munificence contributed to 'wiping the Stain of Infamy from off the Age, inspiring Vigour into these languishing Studies, and clearing the Path of doing Benefits which lay covered through long disuse without the least sign of Footsteps'. The dishonour Barrow alluded to, of course, was none other than the lack of an endowed chair, conspicuous considering 'the surprizing Honours' paid to mathematics 'by the present Age'. Further, he continued: 'I know not whether it is more to be accounted a Matter of Wonder, or of Grief, that it had hitherto obtained no Place, no assigned reward, no allowed Privilege in the University, the fruitful Mother of all Disciplines, and benign Nurse of all Studies.'[11]

Barrow also singled out for special praise Thomas Buck, virtually crediting him as the true founder of the professorship:

> For indeed you owe the first Fruits of this huge Benefit, in Part to him: our great Benefactor being excited by his Admonition, persuaded by his Advice, and drawn by his Exhortation, both to institute this Mathematical Profession, and to endow and adorn your Library with a most choice Treasure of Books. For is not he to be accounted the Author of a Benefit who hath gained and as it were given you the Benefactor himself, who hath as well profited for you out of another Person's Wealth, as out of his own Wisdom, his own Good-Will; and who hath derived upon your Grounds the overflowing Streams of another's Munificence? Without whom indeed it had not been, it could not be to Day that I could congratulate you upon the Access of such a Help to your Studies.

Also made explicit is Barrow's gratitude to Buck to whom he owed his election: his 'Kindness towards me has merited a far greater Testimony of Gratitude' than the encomium he had delivered.[12]

As hinted above, by the second half of the seventeenth century the educational import of the public lectures delivered by a professor paled in comparison with his availability for private instruction. Sir Henry Savile, for one, grasped the centrality of such instruction, charging his professors to be 'of easy access to the studious who would consult them on mathematical subjects, and particularly to their legitimate hearers'. And, indeed, in this respect his professors throughout the seventeenth century proved exemplary. Four decades later Thomas Hobbes raised the same issue during his controversy with John Wallis:

> even those men that living in our universities have most advanced the mathematics, attained their knowledge by other means than

> that of public lectures, where few auditors, and those of unequal proficiency, cannot make benefit by one and the same lesson . . . the true use of public professors, especially in the mathematics, being to resolve the doubts, and problems, as far as they can, of such as come unto them with desire to be informed.[13]

Cognizant of this critical, but less formal, aspect of imparting mathematical knowledge, Barrow – ever the conscientious teacher – not only pledged to fulfil the statute regarding private instruction, but offered as testimony his own track record of precisely this sort of assistance:

> And if, while I was a private Person nor otherwise obliged, being enamoured only with the Loveliness of the Thing, I shewed such hearty Desires and Endeavours to have these Sciences in the highest Degree recommended to you; it cannot now be doubted but by Reason of my publick Office, and more solemn Engagement, I will more diligently apply myself to their Promotion according to my slender Ability; since what was then an Inclination becomes now a Duty. But what the laws do strictly require of me, that I have always shewed the greatest Readiness to perform, so that I did not only willingly admit, but earnestly invited you of my own Accord, to familiar Meetings, so that I not only once or twice every Week opened the doors of my private Chamber, but daily published the inward Secrets of my Heart, and unfolded my Breast to all Comers.[14]

That Barrow continued to act in this capacity may be surmised from his repeated apologies to John Collins, whose requests often went unheeded owing to Barrow's professorial duties.

Notwithstanding the auspicious inauguration of the professorship, the early years were somewhat chaotic owing to the shutting down of the university from summer 1665 to April 1666, and again from July 1666 till after Easter 1667, on account of the plague. Barrow appears to have delivered the first fifteen lectures of his *Mathematical*

Figure 10 From Sir Jonas Moore's 1681 *A New System of the Mathematicks*. This shows mathematicians with instruments that Barrow, by statute, had to possess for pedagogical purposes. While Barrow's studies may have been more philosophical than pragmatic, the instruments themselves were deployed for many practical uses, including surveying, ordnance and navigation.

Lectures during the spring and autumn terms of 1664. He then changed course and for the first two terms of 1665 offered what is now the first five of the *Geometrical Lectures* as well as his four lectures on Archimedes. Upon the resumption of classes in April 1666, Barrow took up again his mathematical lectures. The second part of the geometrical lectures were delivered either in the spring of 1667 or the ensuing autumn term. This portion of the geometrical lectures proved quite difficult for his auditors – hardly surprising given that the subject matter was quite an advanced analysis of the properties of various curves – causing Barrow 'to abandon this subject for the moment, and to stray forthwith into pleasanter fields, bright with the flowers of Physics and sown with the harvest of Mechanics, the field of what they called mixed Mathematics'. For the remainder of his tenure, he delivered his optical lectures.

Notwithstanding the statutory requirement, which he himself helped draft, Barrow did not deposit any of his lectures in the Cambridge University library. The reasons for such uncharacteristic – and at first glance puzzling – neglect are not difficult to surmise, however. Barrow, we know, was in the habit of composing his lectures in advance of their delivery and thus could easily have deposited his mathematical lectures and, later, his geometrical lectures without any real exertion on his part. That he nevertheless failed to do so suggests, not wanton delinquency, but an altogether different issue at play. In Barrow's time, depositing a course of lectures in the public library was tantamount to publication. Barrow intimated as much in 1670 when he told Collins – who was eager to obtain sight of Newton's Lucasian lectures – that 'the Mathematick lecturer there is obliged either to print or put 9 Lectures yearly in Manuscript into the publick Library, whence Coppies of them might be transcribed'.[15] It is my contention, then, that Barrow, who as a rule was shy of publication, balked at making public lectures that were designed for university students of varied capacities and not for expert mathematicians.

This is the background for the rather bizarre manner in which Collins connived to overcome the misgivings of his friend. In

November 1668 Collins wrote a letter, purportedly on behalf of the Royal Society, to Edmund Boldero, the Cambridge Vice-chancellor, intimating that Barrow intended to deposit his optical lectures at the university library – the implication being that he had not yet done so. Collins proceeded to urge Boldero to pressure Barrow to publish for 'we fear the author's modesty is such that he will not promote the publication thereof, unless excited thereto'. Such publications would not only benefit the commonwealth of learning, Collins continued, but would fulfil 'the laudable constitution or injunction laid upon' the professor in the statutes as well as set an example for future incumbents. Collins further pressed Boldero to persuade Barrow to publish his other works as well. The ploy succeeded, and four months later Collins received the manuscript of the optical lectures. Nevertheless, Barrow made clear in the 'epistle to the reader' that he had been coerced into publishing and had consented in no small part in order to set a precedent:

> That this little piece of work . . . was not aimed at you, you will at once . . . detect for yourself from many indications; yet I was advised to make it available to you. In the end, though nervous and reluctant, I took this advice, mainly because I thought that to set those who should succeed me in this task . . . an example in promoting its literature.[16]

As it turned out, by the time Barrow's lectures were in press he had been contemplating relinquishing his post for some time. For all his mathematical prowess, Barrow never wavered in his true calling as a divine. Indeed, one of the first things he did upon returning to England in 1659 was to seek ordination from Bishop Ralph Brownrig. But the desired theological vocation eluded him following his appointment as Regius professor of Greek. His subsequent election as first Gresham and then Lucasian professor further diverted Barrow from his true calling, as he saw it, and his frustration was articulated in the note he sent John Tillotson along with a presentation copy of his *Geometrical Lectures*:

> While you, dear man, expound to the people the mysteries of sacred truth, closing the mouths of petulant sophists and, at the same time, waging successful war on behalf of God's law; lo, I am tied miserably to these hooks which you see, wasting my time and intellect. The explanation of my hard lot is manifest, but I will be modest about this unwanted offspring.

Hence, the statutory requirements at Trinity College, which stipulated that a fellow must proceed with theology – the very statute that caused Newton such consternation a few years later – came as a relief to Barrow, who informed Collins on Easter eve 1669 that he had been preoccupied with preparing 'Theological Discourses (as our statutes order) upon the chief points of [the] Catechism, (the Creed, Decalogue, Lord's Prayer, Sacraments, etc.)'.[17]

Barrow's increasing discomfort was undoubtedly alleviated by the growing realization that he had a more than worthy successor waiting in the wings. Already in his inaugural lecture as Regius professor of Greek, Barrow made it clear that he accepted the position only as a caretaker, until such time as a qualified and willing candidate came along. So, too, he accepted the appointment as Lucasian professor in no small part in order to ensure the institutionalization of the mathematical sciences at Cambridge. And by 1669 Barrow was definitely grooming a successor. Despite assertions to the contrary, Barrow and Newton were quite close from at least 1667 – and possibly earlier – and though Barrow may not have been taken entirely into the confidence of his secretive friend, he knew enough about Newton's work, and even more about his potential, to believe that a more suitable candidate could not be found. Before 1669, however, little could be done. Not only did the statutes require that the professor be at least a master of arts, but there existed the expectation that he be of proven erudition. Newton graduated MA in July 1668 and turned twenty-six – the age stipulated in the Savilian statues upon which the Lucasian statutes were modelled – the following Christmas Day.

All that was required now was a public demonstration of Newton's talents. The opportunity presented itself in early 1669 when Barrow encouraged his protégé to compose his *De Analysi* (*On Analysis by Infinite Series*). Not only would the treatise assert Newton's priority in discovering a general method of infinite series (and of the method of fluxions more generally) – following the publication a few months earlier of Nicholas Mercator's *Logarithmotechnia* – but it would also demonstrate his impeccable credentials to a wider audience. Newton complied and this scribal publication was sent first, in July 1669, to John Collins and then to William, Lord Brouncker, president of the Royal Society as well as to John Wallis, Savilian professor of geometry. With the accolades and endorsement of the Royal Society mathematicians in hand, Barrow easily convinced Thomas Buck and Robert Raworth to appoint Newton his successor. This occurred on 29 October 1669 and Barrow resigned the following week. On 25 November Collins informed John Gregory that Barrow 'hath resigned his Lecturers place to one Mr Newton'.[18]

In later years Newton regaled his admirers with stories of how Barrow resigned his chair in recognition of Newton's superior powers. To one, Newton recounted Barrow's amazement with a six-line solution to a problem involving the cycloid he had shown him – and for which Barrow had managed only a tedious one – the upshot being that he declared Newton 'was more learned than he' and promptly resigned his professorship. To another, he related how Barrow 'would frequently say that truly he himself knew something of the mathematics, still he reckoned himself but a child in comparison to his pupil Newton'.[19] Such accounts probably embellish and simplify reality. While Barrow resigned the professorship in favour of Newton, he did so not because he was put to shame by Newton's genius. Nor, as some scholars would have it, did Barrow resign because he was ambitious for lucrative positions at the university or in the Church. All along, Barrow's view of his true calling remained unshaken, so that as soon as he had taken care of the Lucasian professorship (as well as secured Newton's career) he resigned without any position to fall on except his fellowship.

One final task tethered Barrow to the Lucasian professorship: the ongoing business of printing his lectures. The *Optical Lectures* appeared in November 1669. Originally, the volume was envisaged to include Barrow's more sophisticated geometrical lectures as well, but at a certain stage the publisher reconsidered and asked Barrow to publish them separately, adding to them his earlier lectures to make a sizeable volume. Barrow was not particularly pleased with this turn of events as he regarded the earlier lectures as rudimentary. Yet he agreed, unwilling to disappoint Collins. In order to make the book more palatable, however, Barrow and Collins appear to have entertained the idea of asking Newton to append his *De Analysi*, and perhaps some other of his results, to the *Geometrical Lectures*. As late as February 1670 Collins still believed it likely to happen, but soon thereafter Newton made clear his unwillingness to commit himself to print and Barrow's volume appeared without his protégé's work in July 1670. Part of the story Barrow narrated in the preface by way of excusing the publication of a mere 'trifle':

> Of these lectures (which reach you posthumously, as it were), seven (one being excepted) I intended as the final accomplishments and things left over from the *Optical Lectures* recently published. I mention this lest I should be contemned for publishing such sweepings. However, when the stationer thought for reasons of his own that these should be separated from the others, and asked for somewhat [*sic*] else to give the work a distinct quality of its own instead of being no bigger than a pamphlet, I fell in with his wishes (even against my own) and added the first five lectures, cognate in matter and coherent with those that follow. I had prepared them some years previously without any thought of publishing and without the carefulness that such an intention would demand; for they are uncouth in style and unmethodical in arrangement, and contain nothing beyond the comprehension of

> the mere beginners for whom they were intended. Accordingly, I warn experts to ignore those sections or, at least, to view them with generous indulgence. I submit to your view more readily the other seven that I spoke of, hoping that they contain nothing that even the more learned will dislike looking through. The very last lecture a friend (an excellent man but in this sort of transaction a shameless claimant) has extorted from me, or rather has claimed its inclusion as a well-merited right. As for the rest, what these lectures may bring forth or whither they may lead, you can easily judge by just tasting the beginning of each.[20]

As noted above, Collins managed to coerce Barrow into allowing the publication of not only his optical and geometrical lectures, but the mathematical lectures, the lectures Barrow had delivered at Gresham College, and the compact edition of the Greek geometricians that Barrow had prepared in the mid-1650s. Barrow acquiesced primarily from a sense of duty, but he none the less showed himself unwilling to revise his lectures beyond answering certain queries put to him by Collins or pinpointing analogous works of contemporary mathematicians. Consequently, it fell to Collins to carry out the lion's share of the editorial work, assisted with the *Optical Lectures* by Newton. The two sets proved a financial disaster and plans to publish Barrow's other works were interrupted. Eventually, the edition of Apollonius, Archimedes and Theodosius appeared in 1675, while the *Mathematical Lectures* languished until 1683 when the publication of Barrow's much-sought-after theological works gave some hope that there might be a market for his Lucasian lectures as well.

By 1683 Barrow had been dead some six years. In his lifetime he never reneged on his consent to publish his various lectures even though, as a divine, he shivered at the idea of his name associated with mathematical works. Despite his eminent abilities in the discipline, Cambridge's first Lucasian professor would never have considered himself a mathematician, and publishing his works might have implied that he had neglected his real calling. Thus, Collins wrote to

René François Sluse in October 1670 that Barrow had laid aside his mathematical studies so entirely that, though he allowed the publication of his Gresham lectures on perspective, he was 'unwilling to own them as an author'. In April 1670 Barrow went so far as to request Collins to refrain from publishing a laudatory review of his *Optical Lectures* in the *Philosophical Transactions*, 'beyond a short and simple account of their subject. I pray let there be nothing in commendation or discommendation of them; but let them make their fortune or fate, *pro captu lectoris*. Any thing more will cause me displeasure, and will not do them or me any good.'[21]

Barrow's scruples brought about an odd situation whereby his entire output of mathematical publications retained the distinct format of university lectures. And since Barrow refused to revise them, they offer incomplete testimony of his stature as a creative mathematician. But even though the loss of Barrow's papers prevents us from gauging with exactitude the profundity of his knowledge, his determination to ensure the institutionalization of the Lucasian professorship at the very time he himself eagerly waited to quit the field is a moving tribute to a conscientious and renowned divine who, as a mathematician, was widely esteemed as second only to Newton.

Notes

1 Peter Ramus, *Prooemium Mathematicum* (Paris, 1567), pp. 55–9; Bodleian Library, MS Savile 29, 7^{v}, cited in Robert D. Goulding, 'Studies on the mathematical and astronomical papers of Sir Henry Savile', (Ph.D. thesis, Warburg Institute, University of London, 1999), pp. 68, 70–1. I wish to thank Dr Goulding for allowing me to consult his dissertation.

2 *Oxford University Statutes*, tr. G. R. M. Ward (2 vols, London, 1845–51), vol. I, p. 272.

3 Bodleian Library, MS Tanner 73 f. 68.

4 Sheffield University Library, Hartlib papers, 30/4/29A.

5 Thomas Hill, 'An olive branch of peace and accommodation' in his *Six Sermons* (London, 1649), sig. A3–A3^{v}; Sheffield University Library,

Hartlib papers, *Ephemerides* (1648) 15/6/22A; Stephen J. Rigaud (ed.), *Correspondence of Scientific Men of the Seventeenth Century* (2 vols, Oxford, 1841), vol. II, pp. 558–61; John Wallis, *Treatise on Algebra* (Oxford, 1685), pp. 121, 177, 209.

6 In 1652 John Pell had been promised by the Council of State in 1652 the position of mathematical lecturer at Cambridge with a stipend of £200 per annum but his appointment to a diplomatic mission shortly thereafter aborted it. Upon his return to England in 1658, Samuel Hartlib informed Robert Boyle that the Cambridge Vice-chancellor, John Worthington, entreated him (Hartlib) to sound Pell 'whether he would be willing to lay the foundation of a mathematical professorship in that university'. If he were, Worthington told Hartlib, 'the university is as good as resolved to petition' Oliver Cromwell to effect this. Charles Webster, *The Great Instauration: Science, Medicine and Reform 1626–1660* (London, 1975), 1975; Thomas Burch (ed.), *The Works of the Honourable Robert Boyle* (6 vols, London, 1772), vol. VI, p. 112.

7 John W. Clark (ed.), *Endowments of the University of Cambridge* (Cambridge, 1904), pp. 165–71.

8 Hartlib papers, *Ephemerides* (1656) 29/5/64B.

9 Hartlib papers, *Ephemerides* (1655) 29/5/15A.

10 For a list of the late-seventeenth-century lecturers see Cambridge University Archives, Registry 51.

11 Isaac Barrow, *The Usefulness of Mathematical Learning Explained and Demonstrated*, tr. John Kirkby (London, 1735), pp. vii–viii, xiii.

12 *Ibid.* pp. xix–xx.

13 *Oxford University Statutes*, tr. G. R. M. Ward (2 vols, London, 1845–51), vol. I, p. 274; W. Molesworth (ed.), *The English Works of Thomas Hobbes* (11 vols, London, 1839–45, repr. Aalen 1966), vol. VII, p. 346; *Seventeenth-Century Oxford* 'Mathematical sciences and new philosophies', in Mordechai Feingold, Nicholas Tyacke (ed.) (Oxford, 1997), pp. 385–6.

14 Barrow, *The Usefulness of Mathematical Learning*, p. xxv.

15 H. W. Turnbull, J. F. Scott, A. R. Hall and Laura Tilling (eds), *The Correspondence of Isaac Newton* (7 vols, Cambridge, 1959–77), vol. I, p. 54.

16 Rigaud, *Correspondence of Scientific Men*, vol. I, pp. 137–8; Isaac Barrow, *Lectiones Opticae & Geometricae (London, 1674), preface.*

17 Percy H. Osmond, *Isaac Barrow, His Life and Times* (London, 1944), p. 143; Rigaud, *Correspondence of Scientific Men* vol. II, p. 71.

18 H. W. Turnbult *et al.*, *The Correspondence of Isaac Newton*, vol. I, p. 15.

19 Richard S. Westfall, *Never at Rest: a Biography of Isaac Newton* (Cambridge, 1980), p. 206 n. 85; William Stukeley, *Memoirs of Sir Isaac Newton's Life*, A. Hastings White (ed.) (London, 1936), pp. 53–4.

20 Osmond, *Isaac Barrow*, pp. 131–2.

21 Rigaud, *Correspondence of Scientific Men*, vol. I, p. 147; vol. II, p. 74.

2 'Very accomplished mathematician, philosopher, chemist': Newton as Lucasian professor

Rob Iliffe

Centre for the History of Science, Technology and Medicine, Imperial College London

Although the second and most famous incumbent of the Lucasian chair held the position for over thirty years (1669–1701), Isaac Newton's activities as a fellow of Trinity College overshadowed his professorial duties. Undoubtedly the most gifted and accomplished mathematician of the seventeenth century, Newton began his career as professor by lecturing on geometrical optics, but from 1673 to 1687 he lectured yearly on algebra and related topics. However, beyond this his interests were extraordinarily extensive. Philosophy and theology were conventional fare for a university don of the period; other pursuits, such as alchemy, were highly unusual though by no means unknown to apparently orthodox scholars. In each area of study, Newton assailed conventional wisdom and replaced traditional views with his own. In mathematics and optics, he is revered for forming the basis of large portions of modern science. Yet to an overwhelming extent he took theology to be the most significant part of his vocation as a don, and his passionate writing in this field shows him to be as radical a heretic and as revolutionary a thinker as any of his contemporaries. With his startling views about the nature of the Godhead, anonymity within the college of the holy and undivided Trinity was ironically the most suitable form of existence.[1]

From Newton to Hawking: A History of Cambridge University's Lucasian Professors of Mathematics, ed. Kevin C. Knox and Richard Noakes. Published by Cambridge University Press.

Figure 11 Portrait of Isaac Newton by Godfrey Kneller [1702].

Despite the many different forms they took, Newton viewed all his various intellectual and practical activities as being directed towards the goal of truth. His position as Lucasian professor obviously reflected his expertise in mathematics, yet the less than onerous duties associated with it allowed him to pursue whatever intellectual topics took his fancy. Although he lived within the walls of serene Trinity College, he worked during a critical juncture in British history. The ferment of political, cultural and intellectual changes after the Restoration of Charles II in 1660 provided more than a backdrop for his work, for he was deeply influenced by – and in turn responded to – the dramatic events that took place in the period. Indeed, although his *Principia Mathematica* (of 1687) is one of the landmark achievements of human history, it did not appear in a cultural or intellectual vacuum. A child of his time, Newton shared the academic interests of many of his contemporaries but tackled problems and embarked on projects in his own inimitable manner. A relentless perfectionist, he gave short shrift to those who got in his way. A combination of these factors meant that his intellectual interests were characterized by dynamic evolution and by topical diversity. Individual projects were developed and then evolved into other related but different concerns. Treatises were often an amalgam of two previously separate pieces of research. Even when these take on the well-defined form of a treatise, or a published text such as the *Principia*, his dissatisfaction with what he had done thus far forced him constantly to reconsider and rework writings that appear to have been complete.

The award of the Lucasian chair gave Newton the opportunity to develop not merely his mathematical studies but also to pursue the many other activities that are assessed in this chapter. I attempt to link these multifarious interests to his troubled existence, both in Trinity College itself and in his position on the periphery of the metropolitan scientific world of the Royal Society. Despite declaring that he did not 'feign' uncertain 'hypotheses' because they would inevitably lead to barren disputes, he was drawn at various points into revealing his private conjectures about the underlying physical reality of physical phenomena. Moreover, he believed that a unified whole of knowledge and practice, both religious and scientific, had been given to mankind at the beginning of time but had subsequently been corrupted or even lost. Emotionally and intellectually, Newton saw himself as chosen by God to identify corruption and to restore truth wherever it lay.

AN ANGRY YOUNG MAN

Newton's father, also Isaac, died months before his son was born on Christmas Day 1642. When Isaac junior was three, his mother left the family home, Woolsthorpe Manor, to marry Barnabas Smith, the wealthy rector of the local village of North Witham. Newton remained with his maternal grandmother until he was eleven, when his mother – widowed for the second time – returned with three half-siblings for her son. In a period of great political and social unrest, Newton was already capable of shielding himself from the outside world by losing himself in books and other pursuits. As a teenager he stood out as being highly skilful at making mechanical devices and other 'strange inventions', many of the techniques for which were gleaned from reading books on 'natural magic' that were then in vogue. Nevertheless, from about 1655 he followed a more conventional education at Grantham Grammar School and at some point he lodged with a local apothecary where he learned practical skills such as ways of building machines or concocting inks and medicines. From a scholarly point of view, his outstanding talent was sufficiently obvious for both the headmaster of the school and Newton's uncle (a Cambridge MA and, more appropriately, Trinity graduate) to recommend that he continue his studies at university. With such support, he found himself at Trinity College on 5 June 1661.[2]

Although his mother was wealthy by contemporary standards, Newton entered Trinity as a subsizar, a poor student who maintained his existence by acting as a fag or valet for fellows and wealthy students. By all accounts his early life as an undergraduate was not a happy one. Amongst other things, he passed some of his time thrashing fellow students at draughts or lending them money at interest. In 1662 he underwent an emotional crisis and wrote down in code a list of all the sins he had ever committed in thought or deed. Many of these reveal the Puritan world-view that must have been drilled into him as a young boy and they give evidence of an outlook that contrasted vividly with the more carefree and libertine attitude of many of his fellow-students. He berated himself for making pies, swimming, squirting water and talking idly on a Sunday, and for loving Man and money above God. These documents also reveal something of the unhappy life of a virtual orphan. At one point he had threatened to burn down the house of his mother and Barnabas, while after his mother returned he was a constant thorn in the side of acquaintances. His sins included falling out with the servants, 'calling Dorothy Rose a jade', and punching one of his sisters. After two equally depressing years at Trinity, he struck up a friendship with John Wickins, who arrived at Trinity early in 1663. According to his son, Wickins one day found Newton 'solitary and dejected', and after a conversation 'they found their cause of Retiremt ye same, & thereupon agreed to shake off their present disorderly Companions & Chum together'. They kept up this relationship, with Wickins acting as Newton's amanuensis (i.e. secretary) until the former left the college in 1683.[3]

From the outset at Trinity Newton was under the care of his tutor Benjamin Pulleyn and was fed a diet of standard texts expounding the Aristotelian world-view that had been the staple of Cambridge students for centuries. Nevertheless, by the time he attempted to obtain a minor fellowship at Trinity by election in April 1664, he had already begun to immerse himself in the best that seventeenth-century mathematics could offer. In preparation for the examination, Pulleyn sent Newton to see Isaac Barrow, who had begun to deliver his Lucasian lectures in the previous month. As Newton later claimed to John Conduitt, Barrow 'examined him in Euclid wch Sr I. had neglected & knew little or nothing of, & never asked him about Descartes's Geometry wch he was master of'. Being too modest to mention this detail, Newton made a poor impression on Barrow at their first meeting but was nevertheless 'made a scholar of the house'. Humphrey Babington, rector of Boothby (near Woolsthorpe) was a significant member of the college and may well have been instrumental in supporting Newton at this time. Almost certainly, Babington was also crucial when Newton stood for election as fellow of the college in 1667. In September of that year, he faced four days of questioning by senior members of the college and clearly did enough to satisfy his interrogators since he was elected minor fellow on 2 October. By being created Master of Arts on 7 July 1668 he became a major fellow of the college and his outgoings on the trappings of success increased dramatically over the following months. The oath he took for this degree contained the conventional proviso that, like other MAs, he would 'either set Theology as the object of [his] studies and take holy orders when the time prescribed by these statutes arrives [i.e. 7 years], or... resign from the college'.[4]

CROOKED LINES AND PHILOSOPHICAL QUESTIONS

At some point, Newton's unique talent for science and mathematics had prompted more senior fellows to offer him their patronage. In 1664, he turned his attention away from the solid if tedious works of Aristotelian commentators to the books of modern authors such as Robert Boyle and René Descartes. Yet he soon began seriously to question a number of taken-for-granted assumptions in the so-called 'new philosophy', especially the strict mechanism espoused by writers such as Descartes and Thomas Hobbes. In one notebook (from the mid-1660s) in which he recorded his earliest notes and researches, he pondered questions relating to subjects common in texts expounding this 'new philosophy'. In short order Newton moved seamlessly from simple note-taking to making incisive criticisms of contemporary views, and from this to formulating novel ideas of his own.[5]

Topics in the notebook included perpetual motion, the existence or otherwise of atoms, the nature of the soul, gravity, chemistry, 'qualities' (such as 'rarity' and 'fluidity'), magnetism, philosophy and, perhaps most importantly, the nature of light and sensation. As for the last, Newton surmised, *contra* the strict mechanists, that vision must be due in some part to the imagination or fantasy ('fancy') since even with one's eyes shut one could conjure up images by sheer force of will. Indeed, from the start Newton was interested in those very processes that defied crude mechanistic interpretations, such as memory, the existence of the soul, and life. As for the fancy, he adduced evidence that the power of imagination might extend to some form of telekinesis and he discussed how to improve and strengthen that faculty. To test the extent to which vision depended on one's will, he stared at the sun in order to analyse what colours would be produced when he shut his eyes afterwards. Although he remained bed-ridden in a dark room for a couple of weeks as a result of this novel experimental programme, he told people both in the early 1690s and again in the mid-1720s that he could still call the same image of the sun to mind whenever he wanted to. Not long afterwards, he inserted a brass plate and then a bodkin under his eyeball 'as neare to ye backside of my eye as I could', with the double goal of seeing how the experience of colours was affected and to test what after-effects could be conjured up by his imagination.[6]

One extended essay in the notebook was headed 'Of colours', a section that began with a series of notes from Boyle's *Experiments and Considerations Touching Colours* of 1664. Boyle articulated the conventional seventeenth-century view of colours, namely that they were sensations caused by the disposition of bodies to excite motion between the senses and the brain. Colour was held by Boyle – as by virtually all his contemporaries, Aristotelian or otherwise – to be *modifications* of ordinary light as it was reflected or refracted by bodies. Early on in his research, Newton noted that, when examined through a prism, a thread whose top half was coloured blue and whose lower half was coloured red would appear to be separated between its coloured parts. 'Blew-making' rays, he decided, were bent more than red-making rays because they had 'unequall refractions'. At this stage Newton tied his insights to the notion that red-making rays travelled faster than others. He now recognized that white light was compounded of coloured rays, although he still believed that colours could be produced by light being 'modified' in coming into contact with an object.[7]

At some point in early 1666 he brought his different insights together by conducting a series of experiments based on passing light through prisms, perhaps directly as a result of reading either Boyle's book on colours or Descartes's *Dioptrique* of 1637 (whose work with

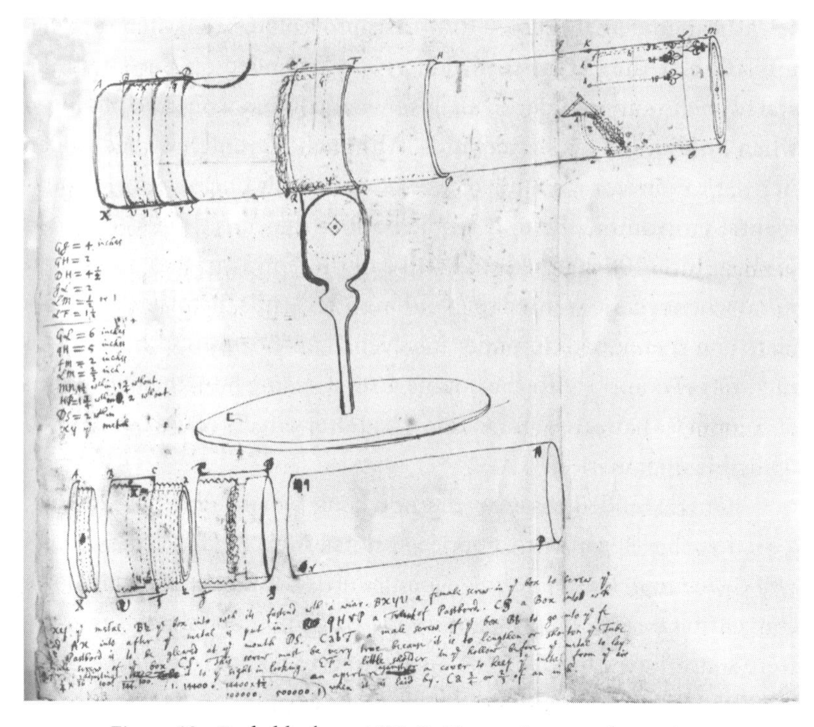

Figure 12 Probably from 1671–2, Newton's own schematic representation of his telescope with measurements and instructions for its construction.

prisms had made use of distances too short for differential refrangibility to become visible). In an essay describing this research, again called 'Of colours', Newton contrived a run of experiments confirming the basic insights that white light is heterogeneously composed of more basic, primary rays, each with its own index of refraction, and – by means of a version of the 'crucial experiment' that would be the centrepiece of his paper in the *Philosophical Transactions* – that each specific colour-making ray is immutable (that is, neither its colour nor its index of refraction can be further 'modified'). Shortly before this he had begun to consider tools for grinding elliptical and hyperbolic lenses in an essay entitled 'Of refractions', again referring back to similar devices mentioned by Descartes. The latter had tried to find the nonspherical shape that would refract incident rays to a point but Newton soon realized that the different refrangibility of specific colour-making rays ('chromatic aberration') meant that no lens could ever produce a clear image, and he gave up his quest. This led him to the construction of a reflecting telescope which used a mirror; but in fact he never fully dismissed the possibility of using a complex lens (made from two media cancelling out each other's tendency to engender chromatic aberration) to improve refracting telescopes. Many of Newton's great advances must have been made in Woolsthorpe, as he left Cambridge because of the threat of plague in the summer of 1665 and returned only in March the following year. After only a few months he returned to Woolsthorpe due to a resurgence of the epidemic and only came back to Trinity for good in April 1667.[8]

At the beginning of 1665 Newton began to investigate the subject of impact. From the beginning he attempted to discover and quantify the change in a body's 'motion' engendered by impact. He based this on the assumption that a body moving or at rest was the passive recipient of some externally impressed 'force'. He came to the view that in collisions bodies act equally on each other, that is, that the amount of force required to generate motion in a body must be equivalent to the amount of force needed to destroy motion. A variant of this notion would recur in the third law of motion in the *Principia*. Of equal significance was his analysis of circular motion. Newton imagined that the 'force' of a body in circulation around a central point could be analysed as if it were subject to reflection in a square whose diagonal was the same as the diameter of the orbit. Since the direction of motions was completely reversed in half a revolution, the force required to achieve this could be understood as equivalent to double the motion required to generate the initial motion. Using similar procedures, he determined that in one revolution the whole force of the orbiting body would be to the force of motion of the body as the length of the perimeter to its radius. That is, if the number of sides and impacts were multiplied so that the reflecting object became a polygon with an infinite number of sides, the force of all the reflections to the force of the body's motion (mv) would be as $2\pi r : r$, or $F = 2\pi mv$. To determine the force that was required to stop a body continuing off into space, the force acting on a body at any moment, Newton divided the total force by the time of a single revolution, that is, $f = 2\pi mv : 2\pi r/v = mv^2/r$, the modern expression for the force acting on a body in circular motion.[9]

This problem could be applied more specifically to measure the force of gravity on earth in relation to the centrifugal force produced by its rotation. Some time after his earlier jottings, he procured Thomas Salusbury's 1665 translation of Galileo's *Dialogue on the Two Chief World Systems* and found from it that, using Galileo's figure for **g** contained therein, the force of the earth from its centre was approximately proportional to the force of gravity as 1 : 144. However, he soon measured **g** indirectly using a conical pendulum and came up with a figure close to the one used today, roughly double that offered by Galileo. A later calculation indicated that the ratio of centrifugal force to gravity at the earth's surface was 1 : 350. Momentously he next compared the 'endeavour of the Moon to recede from the centre of the Earth' with the force of gravity at the surface of the earth, find the ratio was at least 1 : 4000. When he substituted Kepler's third law into his law of centrifugal force, he found that the endeavours of receding were approximately inversely proportional to the square of the distance between the objects (that is, on the assumption that the distance between the earth and moon was approximately 59 or 60 earth radii). Even if the relatively primitive underlying dynamics could not be reconciled with his later notion of universal gravitation, Newton had managed to solve the part of the puzzle that linked the inverse-square law and Kepler's third law.[10]

Whatever the state of Newton's mathematical prowess when he came to Cambridge, neither Descartes's *Geometry* nor the *Arithmetica Infinitorum* of John Wallis was ordinary fare for an undergraduate. His initial engagement with van Schooten's Latin edition of Descartes's *Geometry* probably took place in early 1664 but he was able to pursue original research within a few months. Notably, from Descartes he found out how to investigate tangents to a curve by finding the centre of curvature of a circle that cuts the curve at the tangent. Later, he used techniques derived from Wallis's work on indivisibles to perform integrations of simple curves; in both cases, he developed and generalized the procedures he found in the printed works. Over the next months he created and mastered ways of finding and expressing infinite series, and extended Wallis's techniques for finding areas under certain curves to far more complex situations. Not least, he could now think about how his experience of infinitesimals could be applied to tangents, and in order to do this he used another insight, viz., that lines could be considered as a point traversing a space in time (the so-called generation of figures by motion) and the change in their 'velocities' was equivalent to the rate of change of the curve. By early 1666 he had the ability to show that integration and differentiation, both inverse operations, could be performed on any number of 'curves'. In October 1666 Newton composed a treatise that brought together most of his extraordinary discoveries of the past two years, including the determination of maxima and minima, the expression of the binomial theorem and many other bravura mathematical achievements, all achieved in the form of problems that could be resolved 'by Motion'. Few, if any, historians would dispute the fact that, largely by his own devices but almost certainly with some guidance from the Lucasian professor, Isaac Barrow, Newton had become the leading mathematician in Europe at the age of 24.[11]

Newton later admitted that he had occasionally shown Barrow some of his choicest mathematical labours. In early 1669 the mathematical 'intelligencer' John Collins sent Barrow a copy of Nicolas Mercator's *Logarithmotechnia*, published the previous year. Mercator had derived the series for $\log(1+x)$ by dividing 1 by $1+x$ and squaring each term. When Barrow showed the work to Newton the latter apparently assumed that Mercator knew a great deal more than what he had published, including the binomial theorem, and that he had already extended his technique to other equations. Newton speedily composed a treatise, *De Analysi per Aequationes Numero Terminorum Infinitas* (*On Analysis by Infinite Series*) that summarized his earlier work, in particular the October 1666 tract. *De Analysi* concerned the application of infinite series to quadratures (similar to what we call integration) and showed that the so-called method of tangents (similar to what we call differentiation) was the inverse of the method of quadratures. With the techniques outlined in the essay, Newton claimed that he could reduce any equation to an infinite series of simple terms and, to make good his claim, gave for the first time the series for e^x, sin x and cos x. Once this was communicated to Collins (who, unbeknownst to its author, took a copy), Newton was a bona fide member and indeed leading light of the European mathematical community.[12]

THE PROFESSOR GOES PUBLIC

Before Newton's contacts with the Royal Society, Collins was the most significant conduit for his relations with the London mathematical and scientific community. In July 1669 Barrow mentioned to Collins that he had received *De Analysi* from 'A friend of mine here, who hath a very excellent genius to these things'. As already noted, Newton had been in contact with Barrow for some time before this, although Barrow must still have been taken aback by what *De Analysi* contained. Accordingly, he sent the paper on to Collins with the request that it be sent back as soon as Collins had read it. Barrow revealed the name of the author of *De Analysi* in August, and word of Newton's accomplishments spread fast. In November 1669 Collins told the Scottish mathematician James Gregory – who was also engaging in research into infinite series – that Barrow 'hath resigned his Lecturers place to one Mr Newton of Cambridge, whome he mentioneth in his Optick Præface as a very ingenious person'.[13] As Collins's testimony suggests, Newton's elevation to the Lucasian chair must have been partly orchestrated by Barrow. Newton routinely downplayed his debt to others but towards the end of his life he did reveal that, because of his progess in his studies, Barrow 'procured for me a fellowship in Trinity College in the year 1667 & the Mathematick Professorship two years later'.

Barrow had lectured on optics in 1667 and 1668 (both courses of which Newton must have attended) and though, Newton commented on Barrow's *Eighteen Lectures ... in which the True Causes of Optical Phenomena are Investigated and Explained* of 1669 it is doubtful whether Barrow had a deep grasp of the radical nature of his protégé's work. Yet, given the subject of Barrow's most recent lectures, as well as the imminent publication of a book on the same topic, optics was an appropriate topic upon which to begin Newton's lecturing career. To some extent he dealt with the topics treated in Barrow's text but he brought together the work set down in 'Of refractions' and the second 'Of colours', dramatically extending the insights contained in the earlier essays.[14]

In 1674 Newton submitted a version of his optical lectures (the *Optica*) to the Vice-chancellor for retention in the university library, though another, earlier version of the lectures (the *Lectiones Opticae*) exists. In both versions Newton began with a proof that sunlight consists of unequally refrangible rays but in the earlier *Lectiones* there followed a dissertation on colours, a dissertation on measuring refractions, and finally an analysis of propositions following from his earlier investigations. In the *Optica* he reversed the order of the dissertations, echoing his statement to Henry Oldenburg in the summer of 1672 that the laws of refraction should be investigated *before* the nature of colours was considered. Indeed, the *Optica* appears to be the enlarged and revised version of the original lectures that Newton gave, although there is no evidence that the *Lectiones* represent the state of the lectures as they were actually delivered. A version of the *Optica* was probably finished by early 1672, when he first sent news of his theory to the Royal Society, although the responses to that paper almost certainly forced him to revise whatever version he was considering for publication.[15]

Collins saw a means of publishing Newton's optical and mathematical works together, much as he had done in the case of Newton's predecessor in the Lucasian chair. In late 1669 he met Newton in London, where the two discussed the summing of harmonic progressions and infinite series, although, as for the general method underlying his work, Collins noted 'a wariness in him to impart, or at least an unwillingness to be at the paines of doing'. At about the same time Collins suggested that Newton annotate the *Algebra* of Gerard Kinckhuysen that he himself had just translated from the Dutch. To this end Newton completed a set of notes in the summer of 1670 and told Collins: 'All & every part of what I have written I leave wholly to your choyse whither it shall bee printed together wth your translation or not'. However, he made the apparently bizarre request that if the work were printed Collins should keep Newton's identity anonymous. Newton, as was to be his wont, soon became dissatisfied with what he had done and almost immediately asked for Collins to return the notes so that he could refine them.[16]

Collins had to walk a precarious tightrope, trying both to safeguard the jealously policed property issues surrounding infinite series and also to convince booksellers – who knew the market for mathematical works had collapsed – that there was value in publishing such texts. By the summer of 1671, having heard nothing from Newton for some time, Collins told him that James Gregory was planning to publish a table of logarithms but had dropped plans to publish work on infinite series when he heard how far Newton had progressed in that field. In turn, Newton told him that the previous winter he had revised *De Analysi* and had added some more to his notes on the *Algebra*. The revision of *De Analysi*, now known as the *Tractatus de Methodis Serierum et Fluxionum* (*Treatise on Methods of Series and Fluxions*), expanded his techniques for integrating areas under various curves, while placing the foundations of his calculus on a much more sophisticated basis. In late 1671 or early 1672 Collins still believed that Newton was getting ready to publish twenty of his Lucasian optical lectures, and indeed in December 1671 told a correspondent that Barrow reckoned Newton's optical lectures 'one of the greatest performances of Ingenuity this age hath affoarded'. In April 1672 he wrote to Newton saying that Barrow had told him that Newton was still working on infinite series and integration techniques, and that he had heard that Newton was preparing something 'for the Presse' in Cambridge. Newton replied that he had indeed been thinking of publishing the optical lectures (the *Optica*?) and of adding the discourse on infinite series as an appendix, but that he had since changed his mind, 'finding already by that little use I have made of the Presse, that I shall not enjoy my former serene liberty till I have done with it'. Yet although he had had some hand in producing a new edition of Varenius's *Geography* for the local press, responses to the appearance of his theory of light and colours in the *Philosophical Transactions* had forced him to consider the benefits of publishing at all.[17]

In a more practical vein, Newton had designed and built his reflecting telescope in about 1668, apparently despairing of developing refracting telescopes because of the problems caused by chromatic aberration. It became well known in Cambridge and indeed Newton mentioned it to Collins when they met in London. At the end of 1671 Barrow delivered an example of the instrument to the Royal Society, whereupon it was met with great applause and described and sketched in detail, in order (according to the Secretary of the Society, Henry Oldenburg) to 'prevent the arrogation of such strangers, as may perhaps have seen it here, or even wth you at Cambridge'. Newton replied by giving advice, largely based on his alchemical research, on how to make a suitable mirror for the telescope and then announced that the research that had led him to make the instrument was based on a 'Philosophicall discovery' which in his judgement was 'the oddest if not the most considerable detection wch hath hitherto beene made in the operations of Nature'. He offered to send the Society a report on this discovery and in early February Oldenburg received an extraordinary paper that he duly published in the *Philosophical Transactions*.[18]

Marking Newton's entrance on the European scientific stage, he claimed at the start of the paper that he had bought a glass prism in 1666 and passed sunlight through it, whereupon he was surprised to see that the coloured rays produced thereby made an oblong shape on a wall 22 feet away. He went on to describe an idealized 'crucial experiment', which summarized the doctrines of the Lucasian lectures and claimed to prove that light consisted of rays, each of which had its own index of refraction. However, midway through the paper he discontinued the conventional form of his narrative in which, like other authors in the *Philosophical Transactions*, he related 'historically' how he had come to his discovery and what he had done to corroborate it. Instead, he now stated that since he wished the science of colours to be based on mathematically demonstrated propositions, he would lay down his 'doctrine' in the style of mathematicians and then 'for its examination, give you an instance or two of the *Experiments*, as a specimen of the rest'. As for his thesis, he summarized it as follows: colours were not '*Qualifications of Light*' arising from refractions or reflections but were '*Original* and *connate properties*'. Second, a specific degree of refrangibility was always associated with a specific colour that a given ray was disposed to exhibit; for example, 'The *least Refrangible* Rays are all disposed to exhibit a *Red* colour, and contrarily those Rays, which are disposed to exhibit a *Red* colour, are all the least refrangible'. Third, the species of colour and degree of refraction of primary rays could not be further modified by any further process; if for any reason coloured rays *could* be transformed by refraction or reflection then these were *not* primary rays. Finally, all the primary rays brought together produced white light. For publication, Oldenburg duly removed some of the more overconfident statements in which Newton had claimed that his views were mathematically certain.[19]

Expecting the approbation of 'so judicious & impartial an Assembly' as the Royal Society, instead Newton found himself the butt of what he decried as a 'prejudic't & censorious multitude'. He was seen by some, such as Christiaan Huygens, as intemperately arrogant in claiming a high degree of certainty for his theory, while others such as Robert Hooke agreed with the novel phenomena (typically claiming that he had known about it for ages) but refused to accept the 'hypothesis' of heterogeneity. Other critics proved still more problematic. From Liège, a group of Jesuits made it their business to attack Newton's doctrines as being founded on insufficient and badly designed experiments as well as on faulty logic. In fact, Newton knew right from the beginning that his initial paper had been propounded with excessive brevity, and it proved difficult for many researchers to replicate the main experiments outlined in his text. By the middle of the 1670s, however, Newton had become outraged at what he took to be the obtuseness of the Jesuits and others (mainly Hooke) in misunderstanding or denying his claims. As early as the Lucasian lectures he had condemned the prevalence of a science that rested merely on probabilities, when mathematical foundations could give it absolute certainty. The exchanges with his critics convinced him that absolutely certain doctrines, such as the heterogeneity of white light, had to be detached from mere 'hypotheses' or 'conjectures', such as the view that light was corpuscular. 'Disputing' over things that were uncertain was the enemy of truth while those who accused him of being incompetent or insincere were themselves depraved cavillers who had set out to persecute him for their own devious ends. It is no coincidence that at the same time he was undertaking in private a substantial revision of the history of Christianity in which persecuted true believers were doomed to be harrassed by the ungodly until they could be set free.[20]

CHOSEN BY GOD

Some of the earliest of Newton's theological writings show that he believed himself to be a special individual selected by God to restore true religion. In his theological notebook he noted down from one end relevant passages relating to controversies in the early church, and from the other he listed scriptural passages that pertained to Christ's subordinate role to God. According to Newton, Christ had preexisted his incarnation in a human body in the form of the *logos*, or word, and had then permitted himself to suffer on the cross. For this act he was exalted by his Father: 'the term *logos* before St John wrote, was generally used in ye sense of the Platonists, when applied to an intelligent being, and ye Arrians understood it in ye same sence, &

therefore theirs is ye true sence of St John'. Newton believed his doctrines were close to those of the fourth-century priest Arius, whose 'subordinationist' depiction of the nature of Christ had since appeared to orthodox Protestants and Catholics as one of the worst heresies in existence. Newton denied that Christ was a deity but in so far as he was allowed to share God's will, he could be worshipped as divine. A credal minimalist, he argued that all that was required for a saving faith was the belief that the Christ of the New Testament was the son of God, died on the cross and was resurrected three days later, and would return in due course to rule with the saints. This view, which implied that a Christian polity should permit a wide range of Christian views to flourish (since the true religion was almost certainly not the most popular, or the one directly sanctioned by the state) appears to be in stark contrast to his intolerant attitude to Catholics and trinitarians. Ironically, as Newton was to find out, the English state of the late seventeenth century was no kinder to antitrinitarians than it was to Catholics.[21]

Whatever the initial purpose of engaging in theological research, the nature of his views now made it impossible for Newton to affirm in any conscience his belief in the thirty-nine articles of the Church of England. Indeed, R. S. Westfall has suggested that as early as 1673, Newton had attempted to procure the vacant college law fellowship that did not require its holder to conform to standard Anglican liturgy. However, whether it was because Robert Uvedale enjoyed seniority over him or, more likely, that Barrow (now Master of the College) thought the duties of the law fellowship incompatible with those of the Lucasian professorship, Newton failed to gain the position and retreat from his college loomed. In January 1675 he told Oldenburg that he was about to lose his fellowship and would therefore have to cease his annual fee to the Royal Society, but within weeks he had travelled to London on important business connected with his own position. Whatever happened there, Newton's dispensation from taking holy orders was made official on 27 April. Again, Barrow – probably as ignorant of

Newton's religious beliefs as he had been of his views on optics – almost certainly played a significant role in this unusually right-thinking act by Charles II. In any case, Newton was now completely free to live out whatever role he believed God had chosen for him.[22]

A TRUE HERMETIC PHILOSOPHER

Outside theology, Newton had for some time been engrossed in another topic. Collins told James Gregory in June 1675 that Newton no longer planned to publish anything, but was content to deliver his lectures and do some 'chemistry'. In October he informed Gregory that Newton was 'intent upon Chimicall Studies and practises, and both he and Dr Barrow &c beginning to think math[emati]call Speculations to grow at least nice and dry, if not somewhat barren'. Possibly as early as 1666 or 1667, Newton had begun to take notes on chemistry from Boyle's *Origine of Formes and Qualityes*. He soon turned to practical experimentation and changed from taking notes on more conventional chemistry to extracting information from works that promised more 'noble' insights into the workings of nature. He constructed a furnace in about 1668 and during his trip to London in mid-1669 he bought chemicals and other equipment, as well as a six-volume collation of alchemical writings. For many of his contemporaries, alchemy was an élite art that promised wealth and health to practitioners. For the more élite *adepti*, it was part of a spiritual quest to purify the self and understand the most 'noble' workings of the natural world. Newton certainly believed that alchemy offered a route towards a superior truth, and he took notes from works that described alchemy as a religious quest. Nevertheless, the bulk of Newton's writings are notes or compilations of passages from acknowledged authors. Indeed, apart from two notebooks, which describe chemical experiments, Newton left few obviously 'original' alchemical works.[23]

Although (with the exception of Boyle) Newton's colleagues in the art are largely unknown, the large numbers of transcripts that can only have been taken from other manuscripts testify to a significant circle of alchemists operating in London and possibly Cambridge. In his practical work he sought to liberate an active alchemical form of 'mercury' from its 'fixed' position in metals, extending the approach in the Trinity notebook where he had questioned the intelligibility and plausibility of strictly mechanical accounts of nature. His alchemical researches appear to have been directed towards the discovery of a more 'noble' active principle in nature that perhaps could only be understood by an alchemical adept. In a manuscript entitled *Propositions* he referred to a volatile 'mercurial spirit', the only 'vital agent diffused through all things that exist in the world'.[24]

In a paper that has been tentatively dated to the early/mid-1670s, Newton formulated an explicitly alchemical system of the world. The manuscript begins with what look like a series of chapter headings; one, later deleted, states that the magi agree that metals vegetate according to the same laws as other parts of the natural world. Such vegetation could be promoted by art, but Newton affirmed that it was ultimately 'ye sole effect of a latent sp[iri]t & that this sp[iri]t is the same in all things only discriminated by its different degrees of maturity & the rude matter'. He went on to discuss processes such as 'putrefaction' and 'nourishment', and linked the organic activity by which metals grew or affected each other to the growth of an animal from the egg. In both cases such developments took place by a slow nourishment or 'imbibing', and the connection between the two kingdoms was shown by the fact that nothing had so great an influence on animals as minerals:

> witnes not only the Alkahest to destroy & ye Elixir to conserve but but [sic] their operations in common chymicall physick & in springs &c & therefore since wee live in ye air where their most subtle vapours are ever disperst wee must of necessity have a great dependance on them, witness healthfull & sickly yeares, & barronnes of grownd over mines &c.

Minerals could only unite with our bodies and become part of them if they too had a 'principle of vegetation' within them, and when this happened the two were 'conjoyned like male & female'.[25]

After an analysis of the chemical composition of the world, Newton went on to describe how air gently rising from the bowels of the earth was continuously produced by 'minerall dissolutions & fermentations... boying up the clouds & still (protruded by ye air ascending under it) ris[ing] higher & hiher [*sic*] till it straggle into ye ethereall regions' when they are so lofty as to lose their gravity. Falling rain compresses the rarified æther underneath it so that the latter descends into the earth and 'endeavours to beare along w^t bodys it passeth through, that is makes them heavy & this action is promoted by the tenacious elastick constitution whereby it takes ye greater hold on things in its way; & by its vast swiftness'. Thus there is a 'circulation of all things', the earth resembling 'a great animal or rather inanimate vegetable [that] draws in aethereall breath for its dayly refreshment & vital ferment & transpires again wth gross exhalations'.

The nourishing principle that gave life to the planet, Newton continued, 'is the subtil spirit wch searches ye most hiden recesses of all grosser matter which enters their smallest pores & divides them more subtly then any other material power w^t ever'. Nature's 'universall agent, her secret fire' was the 'materiall soule of all matter'; once this pervaded a terrestrial substance it would 'concrete with it into one form & then if incited by a gentle heat actuates it & makes it vegetate', continuing to ferment until the heat source was removed. The æther was in all probability a vehicle for the working of some more active and subtle spirit and perhaps bodies were 'concreted of both together, they may imbibe æther as well as air in generation & in y^t æther ye s^{pt} is intangled'. Newton suggested that this was perhaps the 'body of light', since both heat and light had a 'prodigious active principle' and both were 'perpetuall workers' – 'heate excites light... & light exites heat, heat excites ye vegetable principle & that excites increaseth heat'.[26]

This text represents a central aspect of the alchemical programme to which Newton was committed at the time, although to some individuals he occasionally professed ambivalence about the entire enterprise. In the spring of 1676 he read a paper by Robert Boyle in the *Philosophical Transactions* concerning the heating of gold by means of a special sort of alchemical mercury and he told Oldenburg that 'ye fingers of many will itch to be at ye knowledge of ye preparation of such a [mercury], and for that end some will not be wanting to move for ye publishing of it, by urging ye good it may do the world'. Newton was happy that Boyle had revealed nothing further about the mercury but, whatever the true nature of its activity, others had tried to conceal it and therefore it 'may possibly be an inlet into something more noble, not to be communicated wthout immense dammage to ye world if there should be any verity in ye Hermetick writers'. Newton told Oldenburg he had no doubt that the 'great wisdom of ye noble Author' would 'sway him to high silence'. This should last until Boyle was properly informed about the wisdom of publication by an expert: this person would be 'a true Hermetic Philosopher', whose views 'would be more to be regarded in this point then that of all ye world beside to ye contrary, there being other things beside ye transmutation of metalls (if those great pretenders bragg not) wch none but they understand'.[27]

Even as he worried about revealing such alchemical truths to a wider public, Newton found himself increasingly entangled in the very public disputes over his experiments on light and colours that he strove to avoid. Gradually, he grew certain that his correspondents had ulterior motives for questioning both his sincerity in reporting experiments and his competence in making them. At some point in late 1675 he decided that he would publish his views on the nature of light and colours, despite his former resolution not to do so, with the proviso that he would not engage in any controversy as a result. He largely but not entirely eschewed alchemical language and concentrated on the nature of a pervasive æther that explained various phenomena such as light and sound.[28]

Newton went on to explain how electricity was produced by an 'ætheriall wind' and argued that gravity was probably caused by 'the continuall condensation of some other such like ætheriall Spirit, not of the main body of flegmatic æther, but of something very thinly &

subtily diffused through it, perhaps of an unctuous or Gummy, tenacious & Springy nature'. All in all, this was a dramatic exposition of his sophisticated cosmology and he again drew from his alchemical work in describing nature as 'a perpetuall circulatory worker, generating fluids out of solids, and solids out of fluids, fixed things out of volatile, & volatile out of fixed, subtile out of gross, & gross out of subtile'. Some things were made to rise and 'make the upper juices, Rivers and the Atmosphere; & by consequence others to descend for a Requitall to the former'. Employing identical concepts to those used in his alchemical manuscript, Newton argued that the sun might also 'imbibe' this spirit to conserve 'his Shining' and prevent the planets from careering off into space, while quite possibly this spirit 'affords or carryes with it thither the solary fewell & materiall Principle of Light'.[29]

Newton's paper was read to the Society in five parts between 9 December 1675 and 10 February 1676. However, Robert Hooke was bound to see close similarities between what was in his own *Micrographia* and what Newton had written at the start of the *Hypothesis* and, having heard only the beginning of Newton's paper, he soon set about telling whoever wanted to hear him that the *Hypothesis* was an exercise in plagiarism. Hooke set up his own rival club to the Society at the beginning of 1676 and the topic of the first meeting (attended by Christopher Wren, amongst others) concerned the nature of Newton's 'borrowings' from himself. Hooke wrote in his diary entry for New Year's day: 'I shewd that Mr. Newton had taken my hypothesis of the puls or wave'.[30]

When he heard of Hooke's claims of intellectual priority, Newton proceeded to show how Hooke had himself taken the central insights of *Micrographia* from Descartes and others. Not only this, but where there *was* something original in Hooke's work, Newton told Oldenburg that the 'experiments I grownd my discours on destroy all he has said about them'. Despite the fact that Hooke had set up the club to further his own interests – doubtless immediately after he had heard Newton's attack on him read to the Society on

30 December – Hooke had good reason to believe that Oldenburg had fomented the trouble between himself and Newton. However, when he heard the central part of Newton's *Hypothesis* read out on 20 January, and probably believing that he had acted far too hastily in condemning Newton earlier, he composed a measured and indeed exceptionally generous letter to Newton and sent it off on the same day. Newton himself responded that he would be pleased to continue a private correspondence and that Hooke had indeed gone beyond Descartes in some matters. If he had in turn gone beyond Hooke, he famously and somewhat arrogantly announced, it was because he was standing on the shoulders of giants. This brutal exchange would be mirrored – with even less palatable results for Hooke – just over a decade later.[31]

CORRESPONDENTS

The extraordinary versatility and breadth of Newton's researches are no better exemplified than by his activities in the period when the disputes surrounding his theory of light and colours were dying down. From one perspective, they represent the wide range of interests that most Cambridge dons, and not just the Lucasian professor, were expected to possess. Yet beyond this, and despite his self-imposed exile from the general European scientific community, Newton was in contact with the most influential British philosophers of the day. In February 1679 he wrote to Robert Boyle regarding a discussion that they had conducted earlier, probably during his visit of spring 1675. Undoubtedly, this letter drew from his alchemical researches, although it was also related to the more conventional philosophical views he had expressed in the *Hypothesis* of 1675. Newton told Boyle that there was 'diffused through all places an æthereal substance capable of contraction & dilatation, strongly elastick, & in a word much like air in all respects, but far more subtile'. This æther was rarer within bodies and denser outside them, although the density gradient of the body's æther merged insensibly into that of the surrounding space and Newton thought this might explain diffraction, only recently brought to his attention. He also believed that his theory could explain why flies were able to walk on water, and why the flat sides of two polished pieces of glass could only be brought into contact with great difficulty. Moreover, he suggested that the 'confused mass of vapors air & exhalations wch we call ye Atmosphere to be nothing els but ye particles of all sorts of bodies of wch ye earth consists, separated from one another & kept at a distance by ye said principle'. He went on to argue that the bulk of the air we breathe was probably metallic, and used his æthereal speculations to account for gravity and other phenomena. Within a few years, however, his views on many of these subjects would have changed dramatically.[32]

Newton's mother died in the spring of 1679 and he only returned to Trinity for short periods of time in the following months. The day after he returned from Woolsthorpe at the end of November, he replied to a letter from Robert Hooke, now a secretary of the Royal Society. Innovative scientific entertainment had virtually ceased at the Society's meetings, and Hooke's attempts to reactivate his erstwhile nemesis smacked of desperation. He virtually begged Newton to communicate to the Society anything 'philosophicall' that might occur to him, and pleaded that 'some' (i.e. Oldenburg) had misrepresented him to Newton in the past. Momentously, he asked Newton what he thought of 'compounding the celestiall motions of the planetts of a direct motion by the tangent & an attractive motion towards the central body'. This was a reference to his own statement of the same notion in his *Attempt to Prove the Motion of the Earth* of 1674, recently reprinted in his Cutlerian lectures. Hooke added that the Astronomer Royal John Flamsteed had determined the annual motion of the earth around the sun by astronomical observations. Newton replied that he'd been busy with family concerns over the previous months but in any case had for 'some years past been endeavouring to bend my self from Philosophy to other studies'. In response to Hooke's reference to his *own* work on orbital dynamics, Newton said that he was sorry that he'd never heard of it (!), and he offered a small 'fansy' concerning the Earth's diurnal motion. If an object fell to Earth, it would not fall behind its original position ('contrary to ye opinion of ye vulgar'), but its west-to-east motion being greater at the height from which it was dropped than at positions closer to the earth, it would fall in front of its original position (the east side). If an object were dropped from a tall tower, diurnal rotation might thereby be proved and on the assumption that the earth offered no resistance, he drew a diagram detailing the spiral path of the object towards its centre.[33]

Hooke realized that Newton's scheme allowed him to refer back to his remarks about orbital dynamics, and to Newton's chagrin revealed in a further letter that this part of Newton's letter had been read at the last meeting of the Society, where Christopher Wren and others had agreed with Newton about the eastward trajectory of the falling object. However, Hooke argued that instead of a spiral, a body such as Newton described would carve out an elliptoidal figure, forever moving according to *afga* (on the diagram) except where it encountered resistance and fell closer to the centre of the earth. Finally, he added that any object dropped from a position above the plane of the equinox (i.e. roughly in the northern hemisphere) would have a southerly as well as an easterly component. Newton in turn replied that, again assuming no resistance, the figure would not be an ellipse but the object would 'circulate wth an alternate ascent & descent made by its *vis centrifuga* & gravity alternately overballancing one another'. He also hinted at a much more sophisticated way of dealing with the problem according to the 'innumerable & infinitly little motions (for I here consider motion according to ye method of indivisibles) continually generated by gravity'. Moreover, he implied that he could deal with a force of gravity that did not remain constant but varied from the centre outwards.[34]

A final letter in this exchange, again from Hooke to Newton, indicated that Hooke had experimentally verified Newton's prediction of a south-easterly trajectory of a falling body. He finished by stating that what now remained was to show what path was carved out by an object centrally attracted according to a force inversely proportional to the square of the distances between them. Having offered Newton

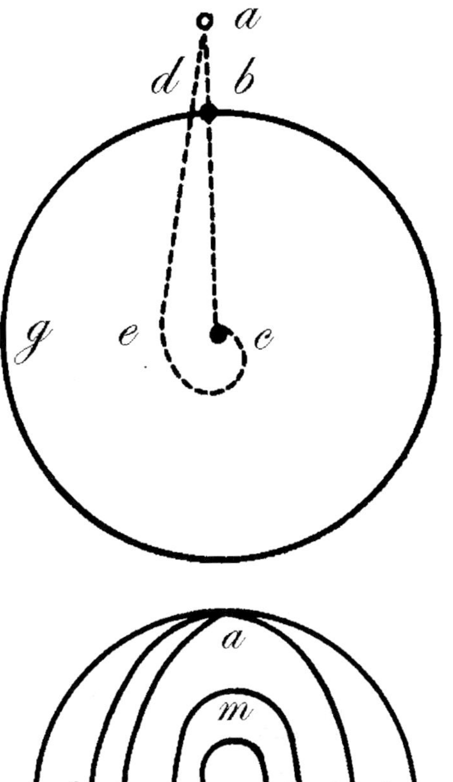

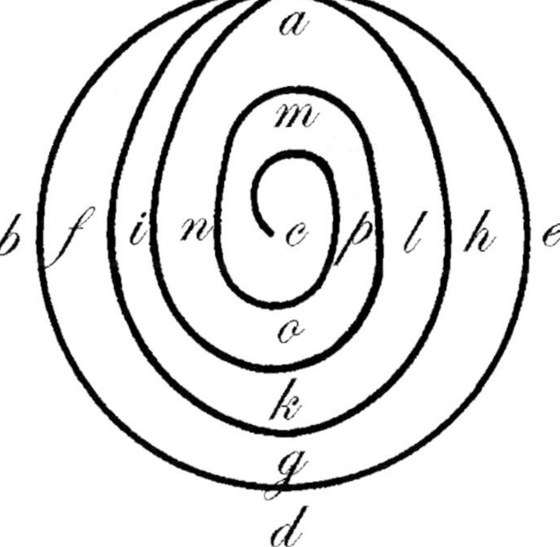

Figure 13 Newton's conception of a body falling towards the rotating earth (assuming no resisting media) with Hooke's response to the Lucasian professor.

the crucial hint regarding a new dynamics of rectilinear inertia and central attraction, Hooke now posed a pertinent question (discussed in London by Wren and Hooke for some years) regarding how to relate the inverse-square law to a resultant orbit – known from Kepler's first law to be an ellipse. 'I doubt not', he told his 'much honoured friend', that 'by your excellent method you will easily find out what that

Curve must be, and its proprietys [*sic*], and suggest a physicall Reason of this proportion'. Despite his later dismissal of Hooke's abilities, and his refusal to continue the correspondence any further, Newton later confessed to Edmond Halley that this correspondence had incited him to think anew about celestial mechanics.[35]

Newton was soon engrossed by another topic. In an initial inquiry that is now lost, Thomas Burnet of Christ's College Cambridge asked Newton some questions about the nature of the initial creation of the cosmos. Burnet's *Telluris Theoria Sacra* (*Sacred Theory of the Earth*) of 1681 would ultimately be the first work in the genre of physico-theology that became popular in the 1690s. Whereas the later works drew heavily from the physics of the *Principia Mathematica* of 1687, Newton's answer to Burnet reveals a great deal about his approach to scriptural exegesis, as well as the state of his pre-*Principia* physics. Newton noted that the creation of mountains and oceans might initially have been caused by the heat of the sun, or by the pressure of either the terrestrial vortex (supposed, following Descartes, to encompass the earth), or of the moon on the primordial waters. The earth would shrivel towards the equator, making the equatorial regions 'hollower' thus allowing water to conglomerate there. Additionally, days in that age would have lasted a lot longer than those of the modern period, giving the process of Creation time to become approximately what it is today. In reply, Burnet pointed out that Moses's description of the world was of the cosmos he was then living in, and not of its original genesis. Otherwise, he continued,

> it would have been a thing altogether inaccommodate to ye people & a useless distracting amusem^{t} and therefore instead of it hee gives a short ideal draught of a Terraqueous Earth riseing from a Chaos, not according to ye order of Nature & natural causes, but in y^{t} order wch was most conceivable to ye people.[36]

Newton replied that Burnet's account of the original production of rocks and mountains from an initially smooth-bottomed sea was the most plausible he had encountered, but belied this flattery by offering a more sophisticated account based on his own chemical experiments. When saltpetre dissolved in water it crystallized in long 'barrs' of salt, even though the initial solution was uniform. If this were like what happened to certain portions of the 'chaos', then one might consider that some 'softer' parts of the air might rise to the surface leaving giant pillars, while leaving channels for water to fall to the ground. When the higher parts of the atmosphere collapsed the water would be expelled with force, causing floods like that of Noah. Along with the water, such exhalations would periodically force out 'subterraneal vapors which then first brake out & have ever since continued to do so, being found be [*sic*] experience noxious to mans health infect the air & cause that shortness of life wch has been ever since the flood'.[37]

As to another question that Burnet had posed regarding the extent of the firmament and the nature of light, Newton remarked that to respond fully 'would require comment upon Moses whom I dare not pretend to understand', but predictably he was unable to resist offering an educated conjecture. All the planets in the solar system were created together, all possessing one common chaos. When the spirit of God moved on it, the planets separated from each other and from the sun, which began to shine even before it became the compact body we now see. Thus the original darkness was followed by the light from the solar chaos and was referred to by Moses as the evening and morning of the first day; in this Newton felt that Moses described the onset of the sun's emanation only in so far as it pertained to the earth. Next, the terrestrial chaos separated into three regions: 'the globe of muddy waters below ye firmament the vapors or waters above the firmament & ye air or firmament it self'. The waters subsided from the higher parts, at which point Moses described the appearance of the sun, moon and stars, *as if humans had then been there to see them*:

> Omit them he could not w^{th}out rendering his description of ye creation imperfect in ye judgment of ye vulgar. To describe them distinctly as they were in them selves would have made y^e

> narration tedious & confused, amused ye vulgar & become a Philosopher more then a Prophet . . . Consider therefore whether any one who understood the process of ye creation & designed to accommodate to ye vulgar not an Ideal or poetical but a true description of it as succinctly & theologically as Moses has done, without omitting any thing material wch ye vulgar have a notion of or describing any being further then the vulgar have a notion of it, could mend that description wch Moses has given us.

Newton finished by mentioning the result of beer poured into milk as another model for the generation of hills on earth, and apologized to Burnet for the 'tedious' letter, while also telling him that he had not 'set down anything I have well considered, or will undertake to defend'.[38]

The Burnet exchange shows that Newton still believed that a vortex surrounded the earth, a view that was to be lambasted in Book 2 of the *Principia*. Another correspondence, this time with John Flamsteed, also indicates how far he was from the doctrines of his masterwork; in this case, what Newton learned was likewise to have far-reaching consequences. Early in November 1680 a brilliant comet became visible to astronomers while another appeared the following month. On 15 December Flamsteed told James Crompton, of Jesus College Cambridge, that he had predicted that the November comet would reappear and that, having looked for it a few days earlier, he had seen it again. Now known as the Great Comet, it was one of the largest ever seen, leaving most Europeans afraid at what it portended, and Flamsteed thought it would last longer than any seen in recent times. At the very moment when Newton was considering how to explain the Creation in physical terms, this extraordinary phenomenon allowed him to think further about the real constitution of the upper firmament. He began to observe it himself on 12 December and recorded his observations in the *Waste Book* he had used for calculations of the mid-1660s. Flamsteed told Crompton (in a passage excerpted by Newton) that he thought the outside of the comet might be 'composed of a liquid wch reflects but little light', and told Edmond Halley (then in Paris) soon afterwards that he thought the sun had attracted the comet within its vortex, as it did the planets.

Flamsteed argued that the comet was turned from its original path by the attraction of the sun, but constantly directed away from this path by the rotation of the solar vortex. The attraction ultimately overpowers the effect of vortex but by the time the comet is at perihelion (i.e. the position when it is closest to the sun) it is travelling directly counter to the flow of the vortex. Now the comet is twisted so that the opposite side is facing the sun and 'having ye contrary End opposite to the Sun hee repells it as ye North pole of ye loadstone attracts ye one end of ye Magnetick needle but repells ye other'. Having turned in front of the sun, the comet would now continue in a straight line away from the sun and contrary to the flow of the vortex, but the vortical sweep actually presses it back towards something like the opposite of its original path. Flamsteed suggested that the comet could have been a stray planet from another 'now ruined' vortex, with a general 'humid' part covering the surface except for odd places where pieces of the solid underlying body poke out above the surface and reflect light. The comet's 'tail' was caused by the sun heating its 'atmosphere'.[39]

One of many Trinity scholars and fellows engrossed by the awesome celestial newcomer, Newton observed the comet until it disappeared in early March, deploying more powerful telescopes as the object faded. Newton differed from Hooke and Flamsteed in holding to the notion that comets travelled in straight lines, and in this he was with the majority of contemporary astronomers. At the end of February Newton proffered some criticisms of Flamsteed's views. He argued that, although he could conceive of the sun continuing to attract the comet on its original path, it would never attract the comet in such a way that it would end up being directly attracted in the direction of the sun. In any case, the vortex would work to push the comet in the opposite direction at perihelion to that indicated by Flamsteed. Newton suggested that the only solution to these problems was to imagine that the comet had turned around *on the other side* of the sun, but then the mechanism for explaining this was unclear. Although he accepted that the sun exerted some centrally attracting force 'whereby the Planets are kept in their courses about him from going away in tangent lines', this could not be magnetic since hot loadstones (natural magnets) lost their power. But even if the attractive power of the sun were like a magnet, and the comet like a piece of iron, Flamsteed had still not offered an explanation for how the sun would suddenly switch from attracting to repelling. As with a mariner's compass, the power of the sun to 'direct' (i.e. influence the direction of) an object was greater than its power to attract or repel, so that 'once so directed the Comet will be always attracted by ye Sun & never repelled'.[40]

Newton later noted that if the comet (considered as one) were magnetically attracted and then repelled by the sun, it must be *continuously* accelerated in both states '& so being continually accelerated would be swifter in its recess then in its access, contrary to what Mr Flamsteed & others beleive [*sic*]'. Second, he argued that if the comet passed around the sun under a similar type of force, at perihelion it would be equally attracted and repelled and so move off at approximately a tangent to its path, orthogonally to the observed trajectory. Moreover, if there did exist a repulsive force it would work on the comet some time *before* perihelion, ejecting it into the heavens close to a line opposite to that seen by astronomers. However, if the comet were subject to a *continuous* directive and attractive (magnetic) force, Newton brilliantly suggested, this attractive or 'centripetal' force would decelerate the comet in its recess and make the comet travel along an orbit close to that observed. The 'centrifugal' repulsive force at perihelion would then have 'overpower'd' the attractive power, allowing the comet to recede from the sun despite the attraction. Magnetism would not for long serve as an adequate mechanism for heavenly phenomena, though the notion of continuous attraction would reappear in the more mature dynamics of the *Principia*. In any case, Newton was satisfied that too few 'pairs' of comets had been seen in the past to make plausible Flamsteed's 'paradoxical' idea that the November and December 1680 comets were one and the same.[41]

THE GREAT APOSTASY AND THE CORRUPTION OF SCIENCE

As Newton communicated with these major natural philosophers, another crisis gripped the nation. By the late 1670s, a number of MPs and other luminaries were attempting to exclude James, brother of the King and the current Duke of York, from assuming the throne on the death of Charles II. This was on the grounds that the Duke had publicly confessed to being a Roman Catholic and thus would be unfit to become the nominal head of the Church of England. Accordingly, the so-called 'Exclusion Crisis' came into being. The same group saw their chance when the 'popish plot' was uncovered. With the support of hundreds of thousands of ordinary Englishmen, the Exclusionists used the supposed threat to Protestantism to indict senior Catholics. Newton needed no prompting to believe that there was a terrible conspiracy to pervert the true religion, although his opinion about what counted as religious truth would have been unpalatable to the vast majority of the exclusionists. Despite being engaged in other theological projects, such as the analysis of the true dimensions of Solomon's Temple and the recovery of the ancient religion, his main energies were now directed towards the exposition of prophecy, in particular the images and descriptions in *Revelation*. In arguing that Roman Catholicism was the religion of the Beast described in the Apocalypse, Newton was not ploughing a new furrow and indeed, the view that the Pope was Antichrist linked him with most Protestants. Nevertheless, the targets of his wrath were not merely Catholics but all trinitarian corrupters of religion and scripture. By extension this included most of his friends and, indeed, every Anglican.

Whatever his lack of objectivity (by our standards), there can be no doubt that Newton devoted himself to studying and mastering the relevant primary sources of early Church history. Following the work of the great exegete Joseph Mede, Newton assumed that the interpretation of prophecy had to pay attention to the various meanings of prophetic terms that had been taken for granted by 'Ancient Sages' in the holy lands before the time of Christ. In one version of an extensive treatise on prophecy he claimed that he had gained prophetic knowledge by the grace of God and was duty-bound to reveal what he had found to others, 'remembering ye judgement of him who hid his talent in a napkin'. This would allow the 'sincere Christian' to progress beyond the standard knowledge of baptism, the laying-on of hands, the resurrection of the dead and eternal judgement on to perfection 'until they become of full age & by reason of use have their senses exercised to discern both good & evil' (from Hebrews 5:12). He drew up a series of rules for interpreting *Revelation* that escaped the twin perils of private 'fansy' (given the status of 'hypotheses') and fraudulent institutional interpretation. Interpretations were to be simple, since truth was always single, while errors and heresies were multifarious. He also cautioned against the overhasty attachment of current events to apocalyptic images, since that had been one of the contributory causes of the civil unrest in the middle of the current century. Scripture could only be interpreted retrospectively, and it endangered the authority of prophecy to make predictions, such as the date of the coming millennium, that had invariably proved false in the past.[42]

In his overarching scheme, Newton had to link a view of how apocalyptic images were interrelated with the historical account of how these images had in fact been fulfilled. Most Protestants assumed that the visions describing the opening of the seals referred to the church in the first centuries after Christ, but that the sixth and seventh seals covered the modern period and indeed the future leading to the Day of Judgement. The sounding of the seven trumpets signalled the battle with the false religion, beginning with the inauguration of Catholicism as the state religion by Constantine the Great and the consequent rise of 'popery' in the fourth century. The trumpets detailed the rise of this fraudulent religion, and the selection by God of a group of elect individuals who might safeguard the truth. In the latter trumpets, God poured vials of wrath on the sinful and idolatrous dwellers on earth, taken by most Protestant exegetes to refer to the Reformation and recent losses by the Catholic Church. Newton agreed with little of this, although he did situate the beginning of the devil's 'game' on earth in the fourth century. Instead, he made the first trumpet and vial refer to the same period (the year 395), the second trumpet and vial to the same period and so on. These images, he contended, had been designed by God to describe the nature of the bestial religion more clearly to those who were to come; the different descriptions in the corresponding vials and trumpets giving distinct perspectives on the major historical events they depicted.

According to Newton, when the seventh seal was opened in 380, it ushered in the worst of times for Christianity. Trinitarians gained control and at the Council of Constantinople the following year antitrinitarian bishops were banished. Accordingly, the Council decreed that the Son had possessed a second soul alongside the *logos*, and that the Son had enjoyed a 'Hypostatical union' with the other soul 'instead of a true incarnation in ye body'. The reason for this was that now it could be said that only the soul and not the Son himself suffered on the cross; this was a way of explaining away the tricky idea that God had in some sense suffered. For Newton, 381 was, 'wthout [*sic*] all controversy', the year when that 'strange religion of ye west wch has reigned ever since first overspread ye world'. Yet for a short period of time, God began to protect those who preached and were prepared to achieve authentic martyrdom for spreading the true word. This did not stop violent persecutions by emperors and Catholic priests against the godly, and the Christian world accordingly became full of hypocrites. On the other hand, the dire persecutions at least had the advantage of allowing those with a true notion of Christ to flee to friendly nations.

Newton's history of the world completely inverted conventional Protestant and Catholic views. The true apostates were trinitarians of all religious persuasions, and he argued passionately that modern historians and indeed all members of the church were engaged in some gigantic conspiracy to rewrite the truth. In short, the real heretics always showed themselves by their attitude to the Godhead and by their persecutory tendencies towards those they falsely condemned as heretics. Newton's contemporaries might say that the doctrine of the Trinity did not deny the essential subservience of the Son, and many might deny that the worshipping of relics and martyrs was idolatry, 'but what', he thundered, 'will they say of whoredom, murder, stealing, lying, perjury, perfidy, drunkenness, gluttony, oppression, pride, voluptuousness, blasphemies, strifes'? In a neat reversal, Newton praised barbarians, vandals and Huns for their attempts to decimate the religion of the Beast. He adeptly reread primary sources to show the state of immorality and unchristian persecution that existed at this time, and took the atrocities committed by his barbarian heroes to be evidence of what moral depths trinitarian Christianity had plummeted. It was perhaps time to restore the true religion, as well as what had been the original, now lost science that had been revealed to Man at the beginning of time.

PRIESTS OF NATURE: TOWARDS THE *PRINCIPIA*

For centuries Christians had believed that religious and scientific knowledge had been given to the first men, but that this had been subsequently lost. In the sixteenth century, Neoplatonists argued that there had been an élite form of learning – the *philosophia perennis* or *prisca sapientia* – that had been revealed to the Ancients before it had been misinterpreted and corrupted. Like many of his contemporaries, such as the Cambridge Platonists Henry More and Ralph Cudworth, Newton subscribed to this view and he created drafts of a substantial treatise (*The Philosophical Origins of the Gentile Theology*) in which the thesis was outlined in detail. In the main, he argued that the Ancients had believed in a sun-centred cosmos but that this had been perverted by misinterpretation. Whereas Pythagoras and others had understood the true meaning of symbolic representations of a heliocentric cosmos, with a central sun encircled by the concentric orbits of the planets, Greeks such as Aristotle assumed that the central object in such a scheme was the earth. The earth was thus wrongly placed at the centre of the cosmos. In that sense, his natural philosophy was an attempt to restore the correct understanding of the natural world, just as his theological work aimed at the restoration of the true religion.[43]

Newton began the second book to a suppressed early version of the *Principia* (written in late 1685) with a reference to the Vestal religion of Numa Pompilius and other Ancients:

> This was the philosophy taught of old by *Philolaus*, *Aristarchus* of *Samos*, *Plato* in his riper years, and the whole sect of the *Pythagoreans*; and this was the judgement of *Anaximander*, more ancient still; and of that wise king of the Romans, *Numa Pompilius*, who, as a symbol of the figure of the world with the sun in the centre, erected a round temple in honor of *Vesta*, and ordained perpetual fire to be kept in the middle of it.

In an extended Latin treatise, almost certainly composed at the same time as he wrote the *Principia*, he argued that religion and science were intimately intertwined, and that priests in the ancient religion had possessed a substantial degree of both religious and scientific knowledge. Via Orpheus and Pythagoras, the Greeks got their correct understanding of the natural world from the Ethiopians and Egyptians, who had concealed these truths from the vulgar. At this time there was a 'sacred' philosophy – communicated only to the *cognoscenti* – and a 'vulgar' version, promulgated openly to the common people. Only the learned knew the meaning of the allegories that the Egyptians used while the Egyptians 'designated the planets in the [correct] order by means of musical tones; and to mock the vulgar, Pythagoras measured their distances from one another and the distance of the Earth from them in the same way by means of harmonic proportions in tones and semitones, and more playfully, by the music of the spheres'. Like the Chaldeans, they also knew that comets were heavenly phenomena and could be treated as if they were a sort of planet. All this learning became perverted when the Greeks began to make incursions into Egypt. The false Aristotelian philosophy was thus the unhappy result of a corrupt reading of the ancient hieroglyphics.[44]

The Egyptians built temples in the form of the solar system and derived the names of their gods from the order of the planets. As such, the ancient religion was modelled on the understanding of the heavens and Newton occasionally referred to the Ancients' 'astronomical theology' in manuscripts from this time – dated by their appearance in the hand of Humphrey Newton (Newton's amanuensis in the mid to late 1680s). If one added the seven known planets (including the moon) to the five elements – air, water, earth, fire and the heavenly quintessence – then one arrived at the twelve basic gods that were common to all the ancient religions. Noah was Saturn and Janus, like Noah, had three sons; this view, known as euhemerism, was common amongst early modern scholars and served to date the history of the world and make all previous accounts consistent. Despite these developments, Newton held that the vestal religion, practised in a monotheistic culture, was the original religion and that all later innovations were corruptions. This project was not some perversely baroque addition to the more rational work of the *Principia*, but a genuine and prolonged commitment by Newton to situate scientific truth in the history of the world. At the same time, however, he was creating a masterpiece of rational mechanics.[45]

THE WORLD OF THE *PRINCIPIA MATHEMATICA*

When Edmund Halley visited Cambridge to see Newton in 1684, it was the result of discussions about celestial dynamics that had been taking place amongst the virtuosi in London for some time. Famously, Halley supposedly asked Newton what curve would result from an inverse-square force law, whereupon Newton immediately replied that he had calculated it to be an ellipse. However, when Newton searched for the demonstration he could not find it, and Halley had to wait until November when he received a short tract entitled *De Motu Corporum in Gyrum* (*On the Motion of Bodies in Orbit*). It showed that an elliptical orbit entailed an inverse-square force law and that, given certain initial conditions, the same force law implied an elliptical orbit out of a number of different possible conic sections. Of equal interest, he also dealt with motion in a resisting medium, indicating that his treatment implied a cosmos that offered no resistance, that is, was empty. Newton, as was his wont, naturally tried to expand the essay and almost certainly the work towards the fully fledged *Principia* that took place over the following months took him away from his beloved theology.[46]

Correspondence with an obliging Flamsteed over the winter of 1684–5 shows that Newton was already trying to link his analysis to a more precise view of the actual motions of the planets and their satellites, as well as of comets, and that he was testing the nature of Kepler's third law. Flamsteed, who had seen it, was aware that Newton's November tract implied that planets could be treated like the sun as centrally attracting objects and indeed Newton mentioned Jupiter's 'action' on Saturn in a letter of December 1684. Flamsteed – still thinking of the attracting force as magnetic – baulked at the idea that planets could influence each other over such long distances. Nevertheless, when Newton composed a revision of *De Motu* in early 1685, there was now a reference to the mutual action of planets, although there was still no mention of universal gravitation.[47]

In any case, Newton was by now working on a number of different fronts. Although he required much more accurate estimations of the paths of Jupiter and Saturn, he also needed to recast completely the dynamics that had served him during his correspondence with Hooke and Flamsteed. Throughout 1685 he worked on the notion of 'force', beginning with the idea of motion being compounded of an 'inherent' force (i.e. a force that kept a body in uniform motion) and an 'impressed' force. There was also a tension between his treatment of continuously acting force and force conceived as a succession of infinitesimally small impulses, although Newton was not to resolve this even in the *Principia*. Throughout 1685 he revised *De Motu* and worked to combine a revised notion of inertia (which did away with the idea that a body had a 'force of motion') with a commitment to absolute space. Only with the latter could one truly say that there was a difference, which had to be visible to God, between absolute and relative motion. Armed with a new definition of inertia, at some point Newton realized that uniform circular motion around a centrally attracting source was not an example of the continuation of the particular state in which the orbiting object found itself, and that this motion and uniformly accelerated motion were the same.[48]

Throughout 1685 he worked on the dynamics that arose from the central insight that bodies continue in their current state unless acted on by the force exerted by another body, and that this force resulted in a change of motion proportional to the force and in the right line of that force. By November 1685 he had completed a work entitled *De Motu Corporum* consisting of two books, viz. the *Lectiones de Motu* (so called because he deposited this text in the University library as his Lucasian lectures) and the *De Mundi Systemate*. Newton was now able to say that the force by which the earth attracted the moon was a kind of 'centripetal force' (a term invented by Newton at this point) that 'very nearly' obeyed the inverse-square law. Similarly, he came to the conclusion that the sun attracted other bodies in the solar system in proportion to their mass.

In many ways the early versions of *De Motu* exemplify what I. B. Cohen has termed the 'Newtonian Style', whereby Newton began with consideration of a simple system in which bodies revolved around a centrally attracting object, and then moved to consider two-body and then many-body situations (which were too complex to give exact solutions). Although the initial assumptions utilize purely mathematical constructs, Newton then goes on – at least in the logical structure of the text – to deal with real physical analogues of these abstractions. Some scholars, such as R. S. Westfall and B. J. T. Dobbs, have argued that the alchemical concept of 'attraction' was a significant influence on Newton's discovery of universal gravitation. Nevertheless, his move from considering the simple idealized situation of one body orbiting around a centrally attracting source to that of a complex (and equally idealized) situation involving many mutually attracting bodies also provides a pathway to his triumphal notion. However historians account for it, towards the end of 1685 he came to the stunning realization that in our cosmos *all bodies attracted every other body* and that this applied to objects near the surface of the earth as much as it did to planets millions of miles apart.[49]

Throughout this period Newton worked on refining the bases of what were to become the three laws of motion. In the *Lectiones* he followed this with a number of propositions that were primarily mathematical in form and that described the forces existing in various possible worlds where different laws operated. He was increasingly careful to refer to 'centripetal force' in an abstract sense rather than to 'gravitation' – the real instantiation of this ubiquitous centripetal force that might be supposed to condition events in this world. At the start of 1686 he decided to split his expanded *Lectiones* into two books, expanding the first by analysing the motions of the moon and adding a proposition (XXXIX) that referred unambiguously to his knowledge of the calculus. Undoubtedly this was in some degree to assert his independence in the development of the calculus from the work of Leibniz, who had published the basic axioms of the calculus for the first time in 1684.

Most of the work of early 1686 dealt with what he termed motions in resisting media, in what was ultimately to be Book Two of the *Principia*. Beyond the treatment of the topic in *De Motu*, Newton added more complex material, both theoretical and experimental, relating to pressure and viscosity. He used the insights from this approach to argue that there were no celestial vortices (the concept was incoherent and there was no evidence from celestial phenomena that they existed) and indeed no such thing as a Cartesian plenum on earth. But if this were the case, then the sort of æthereal explanations he had used in the previous two decades were also compromised; instead, he appeared to be satisfied with appealing to attractive (and, in the short range, repulsive) forces that could explain both micro- and macrophenomena. To some extent, Newton continued to treat the objects of Book Two as though they were merely 'phenomenal' entitites whose real-life microcauses awaited discovery.

The final part of the *Principia*, Book Three, was completed early in 1687 and dealt with the actual system of the world. From basic principles, many of which he devised himself, and from astronomical and physical data, Newton demonstrated that the earth was flattened at the poles (i.e. an oblate spheroid). In great detail and now assuming the reality of gravitation, he dealt with the mutual forces governing the relationship between the earth and the moon and accounted for the minute discrepancies that caused celestial bodies to deviate from 'ideal' elliptical orbits. Finally, he showed how his physics could explain the action of comets and, as a consequence, the comet of 1680–1, which was shown to be one and the same – as Flamsteed had claimed five years earlier.[50]

At some point, probably in late 1683, Newton acquired a new amanuensis to take over the secretarial role that had been played by John Wickins. In the five years he stayed with Newton, Humphrey Newton (no relative) copied work from a number of different areas of Newton's research, including ancient religion and prophecy, while the suppressed book of autumn 1685 and the copy of the great work used by the printer are both in his hand. Whatever Humphrey understood of Newton's real interests, he left the most vivid evidence of Newton's life in college, and that at a time of great political turmoil and during the preparation of the *Principia*. The picture painted by Humphrey was of the classically ill-kempt and obsessive scholar who rarely ventured outside his study for exercise or prayer, rarely touched his food and left his rooms only to give lectures (when sometimes, lacking an audience, he would return after about quarter of an hour). Nevertheless he loved eating apples and at odd times would perambulate around his garden, solicitous of weeds. At such moments he occasionally made 'a sudden stand, turn'd himself about, run up ye Stairs, like another Alchimedes, with an *eureka*, fall to write on his Desk standing, without giving himself the Leasure to draw a Chair to sit down in'. Apart from the Cambridge Platonist Henry More, he apparently received only a select band of scholars to his chambers including John Ellis (of Caius College), John Laughton (librarian at Trinity between

1686 and 1712) and John Francis Vigani (chemistry lecturer and also of Trinity), until the last 'told a loose story about a Nun', doubtless expecting – wrongly, as it turned out – that this was the one area where Newton would acquiesce in politically incorrect ribaldry. Humphrey remembered that Newton constructed his own furnaces, and that despite his other pursuits, he continued to experiment assiduously (especially in spring and autumn): 'Nothing extraordinary, as I can Remember, happen'd in making his experiments, wch if there did, He was of so sedate & even Temper, y^{t} I could not in ye least discern it'.[51]

This inscrutably placid demeanour did not extend to Robert Hooke. In May 1686, just after Book One had been presented to the Royal Society, Halley told Newton that Hooke had 'some pretensions' to the inverse-square law and had claimed that he had brought this to Newton's attention. Although Hooke did not claim any rights to the demonstration that conic sections were generated from such a law, Newton's patience had run out for the last time. He told Halley that Hooke had pestered him in their conversation of 1679–80 and had given him nothing he did not already know. Doubtless thinking the storm had passed, Halley asked for the second part of the work dealing with the application of mathematical principles to the real world. However, Newton had raked over old papers and was not about to let his old adversary off the hook. From the letters that passed between them, he began to rant, Hooke could not judge him ignorant of true celestial dynamics, and in any case there was a publicly available text (a letter to Huygens of 1673) that showed Newton had already compared the forces between the moon and the earth, and the earth and the sun and had at that time known of the inverse-square law. Now, he told Halley, he had decided to suppress the third book (Halley had thought there were only going to be two) that contained propositions, some relating to comets, 'others to other things found out last Winter'. Philosophy was 'such an impertinently litigious Lady that a man had as good be engaged in Law suits as have to do with her'.[52]

Nor did Newton stop here, since upon finishing the main bulk of the letter he heard that at the Royal Society Hooke was making a 'great stir pretending I had all from him & desiring they would see he had justice done him'. A vicious tirade gathered pace and Newton sarcastically noted that, according to Hooke, 'Mathematicians that find out, settle & do all the business must content themselves with being nothing but dry calculators & drudges & another that does nothing but pretend & grasp at all things must carry away all the invention as well of those that were to follow him as of those that went before'. Possibly, Newton implausibly suggested, Hooke may have got the inverse-square law from the copy of his letter to Huygens that was in the Royal Society register. In response Halley put Newton's mind at rest and told him that few others believed that Hooke had either the demonstration relating the elliptical orbit to the inverse-square law or a gigantic system of nature. Newton once more had to admit that Hooke might have been misrepresented to him, and that his letters had at least 'occasioned my finding the method of determining Figures', but there would be no more correspondence between them. Meanwhile, Halley's material reward for saving the *Principia* as we now have it was the award of fifty copies of Francis Willoughby's *History of Fishes*.[53]

THE CAMBRIDGE CASE

Before he had finished the last book of the *Principia*, Newton found himself involved in a new crisis. As feared by more radical Protestants, when he came to the throne in the spring of 1685 James II soon began to relax laws and practices aimed at restricting the ability of Catholics to hold office or attend university. In May 1686 he set up a Court of Ecclesiastical Commission, which claimed the power to alter the statutes of ecclesiastical institutions that had contravened the orders of the King. Some Catholics were inserted into Oxford colleges, especially Magdalen, and Protestant scholars now became fearful that their ivory citadels were being threatened. Things came to a head in early April 1687 when the Magdalen fellows resisted James's choice, ironically a graduate of Trinity College Cambridge, as a replacement for their recently deceased President. Although Anthony Farmer was disqualifed by the 'hanging judge' (and also Trinity graduate) Jeffreys for a number of serious offences, a number of Oxford fellows continued to resist James's choices throughout the following months. Nevertheless, an equally serious threat was now facing the 'other' university.

As early as January 1686–7 James had imposed one Joshua Basset to the Mastership of Sidney Sussex College Cambridge, and in the following month the Vice-chancellor of the university, John Peachell, received an order requiring the university to admit Father Alban Francis to a degree of MA at Sidney Sussex. Debate now raged across the university, and senior members tried to persuade courtiers to change the King's mind. At a university senate meeting of 11 March, Newton was one of a number of 'messengers' deputed by the senate to represent the university in a deputation to the King. This only infuriated James still further, and Newton, Babington, Peachell and others were summoned to appear before the Ecclesiastical Commission on 21 April. Jeffreys harangued them but gave them a six-day extension to prepare their defence further; on the 27th, he was presented with the university's case. When they reconvened on 7 May, Peachell was vigorously cross-examined by Jeffreys and following his abysmal performance the unhappy fate of the institution was sealed. Five days later Newton, Babington and the others were told that, although the commissioners held Peachell largely to blame, the 'sly insinuations' of the others had also invoked their ire. Jeffreys sent them packing with the injunction to sin no more lest a worse fate befall them.[54]

Given what is now known of his private theological researches, it is perhaps unsurprising that Newton became one of the most active defenders of Cambridge University's Protestantism. In the weeks immediately after the danger had been detected he took copious notes on recent cases involving the royal power of dispensation (a topic of which he had some experience) and at a meeting to prepare for the confrontation with Jeffreys on 21 April he pushed strongly for an uncompromising stand on the admission of Father Francis. Humphrey copied out a number of relevant legal and constitutional documents, and Newton did much of his own research to prepare for the defence. In an essay prepared by Newton to deal with the issues involved, he argued that the situation was too important for the university to trust James's promise to safeguard the Protestant religion (as King of England James, despite being a Catholic, was also notionally the most significant defender of the Anglican Church). Indeed, James could not make any such promise, first because it was forbidden by the terms of his own religion and second because he could not in any case legally use his dispensing power in such circumstances. Englishmen would not give up laws governing liberty and property, still less should they give up those guaranteeing religion.[55]

In another essay written before his appearance in front of Jeffreys, Newton went on to examine the limits of the King's dispensing power. Ultimately the issue came down to whether the laws of the land were in some sense the King's property, and whether he could therefore dispense with them as he saw fit. The most significant and, indeed, in this instance relevant case was where there was an apparent 'necessity' to do so, usually in cases where reason of state (*raison d'état*) and the security of the people (*salus populi*) were under threat. Newton argued that the people had at least as much right as the King to dispense in matters of necessity, and the monarch could certainly not claim that events taking place as a result of his own whims came under the relevant heading. This short note, deeply hostile to the practices of the current sovereign, could only have hardened the delegates' attitude to the case. In further documents, this time prepared for the final showdown with Jeffreys on 12 May, Newton concluded that the delegates' stand had been taken to defend their own religion; otherwise, 'we shall be no truer to [our] oaths then a Dutchman would be to his who should one day sweare to do his utmost endeavour to keep out ye sea & ye next day cut ye banks wth his own hands'. Catholics and Protestants could not live 'happily nor long together' in the same university, and if the fountains of priestly education 'be once dryed up ye streams hitherto diffused thence throughout the Nation must soon fall off'.[56]

By 1687 Newton's active life as Lucasian professor almost certainly came to a halt. Having performed in front of what was a studentless (according to Humphrey Newton) audience for many years, in 1684 he deposited a manuscript on algebra in the University library to fulfil his professorial obligations. Published by William Whiston in 1707 under the title *Universal Arithmetick*, in it Newton praised the reliance of ancient mathematicians on geometry while he lambasted the introduction of equations and arithmetical terms into geometry by modern analysts. Yet surprisingly little evidence remains regarding Newton's teaching prowess, or indeed concerning the extent to which he taught at all. One of the two people who recalled attending Newton's lectures was Sir Thomas Parkyns, who attended Trinity in 1680 and later published a book on Cornish wrestling. In this he praised Newton for inviting him to his lectures and for encouraging him to apply mathematics to his analysis of the noble art. As for personal students, evidence remains of only three. In April 1669, probably before he knew of his imminent elevation to the Lucasian chair, Newton had taken on his first pupil, St Leger Scroope. In 1680 he took on his second student, George Markham, though, like Scroope, Markham did not complete his degree and little if any evidence remains of his sojourn at Cambridge. Finally, in 1687 he accepted responsibility for the tuition of Robert Sacheverell, son of the radical Exclusionist William Sacheverell. In all probability Newton's stand against James had brought him to the attention of a group of men whom he would soon join in the House of Commons.[57]

Newton's election to the Commons crowned a dramatic turnaround in his attitude to appearing in public. Beginning with the appearance of the *Principia* and his performance in the Father Francis case, Newton had come to the attention of powerful political grandees, in both Cambridge and London. The arrival of William of Orange and the flight from England by James II in what was to be called the Glorious Revolution gave Newton an opportunity to show his allegiance to the government of the day. Despite the fact that he was described in the most glowing terms on the voting slips, it was still something of a surprise when in January 1689 he was elected as one of the two MPs for Cambridge University. In early February Newton voted with the majority of MPs who determined that James had 'abdicated' from the throne in his retreat. Newton swiftly informed his friend John Covel, Vice-chancellor of Cambridge University, that Englishmen now owed allegiance to William, and he urged Covel to institute a process whereby all new MAs could swear oaths of allegiance to the new monarch.[58]

The issue of imposed oaths naturally disturbed a man with Newton's religious views, and he baulked at forcing those who had already sworn oaths to previous sovereigns to take new ones. However, this distinguished him from many of the more radical parliamentary elements with whom he was now in close contact and, by early April, he had to tell Covel that the House of Lords had decreed that oaths were to be imposed on all office-holders. Throughout the spring and summer Newton had a hand in rewriting a bill relating to the regulation of the statutes of colleges and universities, although ultimately this was not passed into law. More importantly, he served on a committee that drew up the wording for a bill concerning the toleration of various kinds of dissenters. Between March and May radical Whigs and more traditionalist Tories discussed reforming the Test Act of 1673 that for some time had forced office-holders to conform publicly to the doctrines and practices of the Anglican Church. Newton naturally supported the toleration of various shades of Protestantism and believed that the state should allow worthy Protestants of any denomination (such as himself) to hold office. When the bill on this topic was passed into law on 17 May as the Toleration Act, dissenters could now freely engage in public worship. However, the sacramental element of the Test Act had not been repealed and freedom of worship was refused to Catholics and antitrinitarians.[59]

Events like this must have made parliament unpalatable to Newton, and even before it was prorogued on 20 August, he had turned his attention to other things. On a number of occasions in June and

July he met Christiaan Huygens and they discussed orbital dynamics and the nature of light. He also spent a great deal of time with the radical parliamentarian John Hampden and the young Swiss mathematician, Fatio de Duillier. At this point Newton began to think of a senior position for himself, and both Huygens and Hampden recommended him to the King for the provostship of King's College Cambridge. Being Newton, he soon began to collect evidence relating to his fitness and eligibility for the position but finding that the omens were not good, he also contemplated the possibility that William could dispense with the statutes governing the appointment. Accordingly he composed a small tract entitled *The Case of King's College* in which he analysed the past history and constitutional status of the provostship. Newton was duly nominated by William and a meeting to discuss the issue took place at the end of August. Previous exceptions to its general rules were not precedents, the college asserted, and in any case one of the technically unqualified people had at least been a member of the college. The Provost had to be in holy orders since part of his role was to officiate at the altar on certain religious occasions, and in this respect virtually everyone else previously nominated to the post had been properly qualified. Newton did not get the position and the defeat rankled. From now on he turned his attention to London, the King's College affair having almost certainly turned him irrevocably against working in a cloistered environment. Over two years later, when efforts were being made to gain him the position of head of Charterhouse School, he told John Locke that he was 'loath to sing a new song to ye tune of King's College ... a formal way of life is what I am not fond of'.[60]

Even as he suffered setbacks in his quest for public office, Newton's scientific reputation soared. Paradoxically perhaps, the incomprehensibility and difficulty of the *Principia* only served to enhance the standing of the work and its author. Halley referred to his 'divine Treatise', having earlier told him: 'you will do your self the honour of perfecting scientifically what all past ages have but blindly groped after' while in a review of its contents for the *Philosophical*

Transactions, he concluded that 'it may justly be said, that so many and so Valuable Philosophical Truths, as are herein discovered and put past Dispute, were never yet owing to the Capacity and Industry of any one man'. Disciples such as Fatio de Duillier, David Gregory and Richard Bentley vied to be the editor of the next edition of the great work, while others such as Abraham de Moivre devoted themselves to mastering the work's incredibly abstruse contents. In turn, Newton doled out patronage to his followers, such as the Savilian chair of geometry at Oxford that he helped obtain for Gregory.[61]

Newton also revealed to Fatio and Gregory that, as part of his revisions and corrections to the first edition, he was working on a major project that would show that the Ancients had really been Newtonians but had veiled their discoveries in allegories and fables to amuse and bemuse the unlearned vulgar. At one point, he compiled a series of 'classical' additions to the scholia to Propositions 4 to 9 of Book Three, in which he showed that universal gravitation and other doctrines could be divined from a serious reading of the poems of Virgil, Ovid and others. Similarly, he threw himself into a gigantic enterprise that purported to 'restore' the lost geometry of the Ancients, such as Euclid's Porismata. In the 1690s, the *Principia* was hailed by philosophers such as Huygens and Leibniz, although both thought that Newton had neglected the entire purpose of natural philosophy (to uncover the underlying causes and principles of things) by using a concept like 'attraction'. To these men and others, such a notion smacked of the scholastic occult qualities that moderns had been at such pains to reject.[62]

Aside from these efforts, the years after the completion of the *Principia* witnessed possibly the most intense intellectual activity of Newton's life. According to a remark in *Opticks*, Newton's first efforts after the Father Francis affair were directed towards publishing his optical work in an extended form. Most of his work went into recasting the Lucasian lectures, his work on thin films and the *Discourse on Observations* but he also added some more recent work on diffraction. At this point Newton actually planned to produce a work of four books, intending in the concluding book to show how optical effects acted according to small-scale attractive and repulsive forces. In a draft he argued that philosophers should assume that similar kinds of force operated in the micro- as well as in the macroworld, but that this 'principle of nature being very remote from the conceptions of Philosophers I forbore to describe it in [the *Principia* lest it] should be accounted an extravagant freak'. Whatever Newton's original plans, he had reduced the number of books to three by 1694 and indeed *Opticks* ultimately appeared in this form.[63]

Throughout the summer of 1690 Newton worked on another major project dear to his heart, this time the vexed question of how Catholics and trinitarians had corrupted the true text of the New Testament. Due to a relaxation of the licensing laws governing publication, antitrinitarianism had reared its ugly head in the middle of James II's reign and, as we have seen, Newton had unsuccessfully tried to prevent antitrinitarians being classified alongside Catholics as civil undesirables in the Toleration Act of 1689. It remained a dangerous period to air such views for a man seeking high office. In November 1690 he sent a long exposition on this topic to his new friend John Locke, and there can be no doubt that he understood Locke to be sympathetic to his views, despite the thick veil of objective research with which Newton tried to coat his work. This was arguably the most important of a broad range of subjects that passed between them, including natural philosophy, alchemy, toleration, moral philosophy and politics. While Newton's thoughts in many of these areas remained private, Locke was then publishing (or was about to publish) influential works on these themes, such as *A Letter on Toleration*, *An Essay Concerning Human Understanding* and *Two Treatises on Government*.[64]

Locke had been alerted to the issue of textual corruption by the publication of a number of recent works, and he had asked Newton for his view on the disputed text 1 John 5:7 concerning the divine nature of Christ. Newton told Locke that it was yet another Catholic corruption, but that, although they knew this, many humanists and

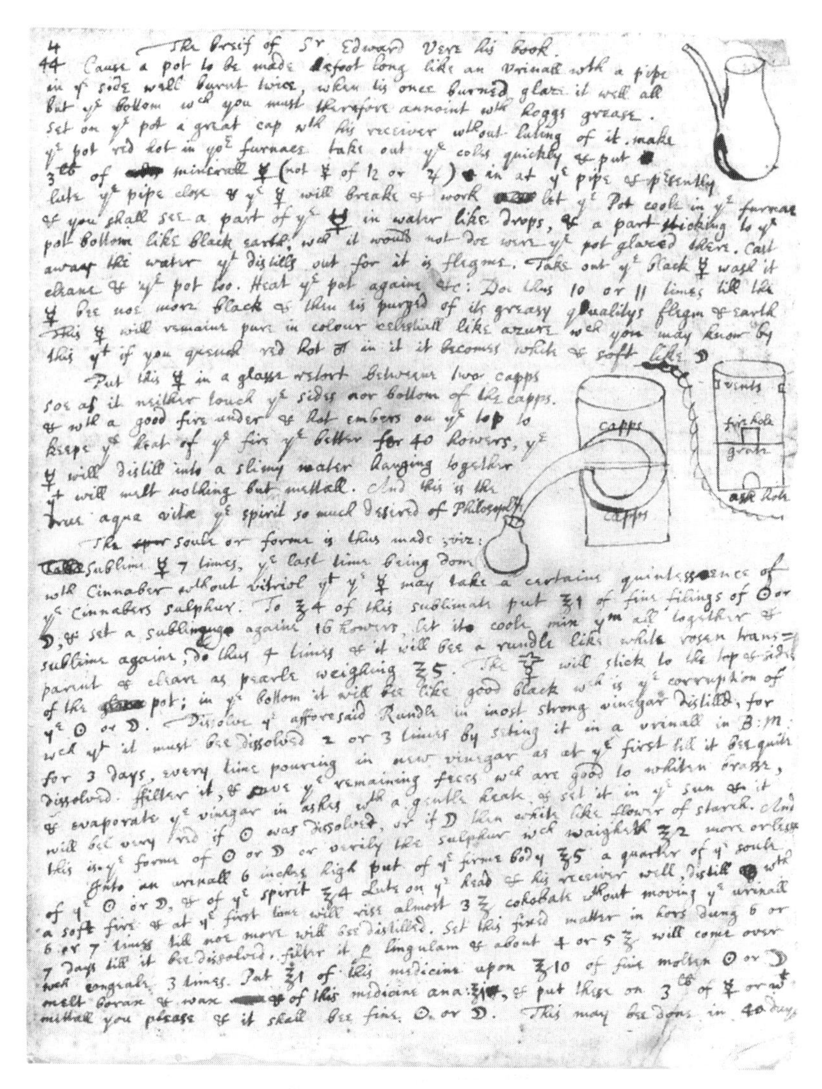
4 The Preif of Sr Edward Vere his book.
44 Cause a pot to be made a foot long like an Urinall wth a pipe
in ye side well burnt twice, when tis once burned glaze it well all
but ye bottom wch you must therefore annoint wth hoggs grease.
Set on ye pot a great cap wth his receiver wthout luting of it. make
ye pot red hot in yor furnace take out ye coles quickly & put
3 lb of minerall ☿ (not ☿ of ♄ or ♃) in at ye pipe & presently
lute ye pipe close & ye ☿ will breake & work let ye Pot cool in ye furnace
& you shall see a part of ye ☿ in water like drops, & a part sticking to ye
pot bottom like black earth, wch it would not doe were ye pot glazed there. Cast
away the water yt distills out for it is flegme. Take out ye black ☿ wash it
cleane & ye pot too. Heat ye pot againe &c: Doe thus 10 or 11 times till the
☿ bee noe more black & then tis purged of its greasy qualitys flegm & earth
This ☿ will remaine pure in colour celestiall like azure wch you may know by
this yt if you grinde red hot ♂ in it it becomes white & soft like ☽
Put this ☿ in a glasse retort betweene two capps
soe as it neither touch ye sides nor bottom of the capps.
& wth a good fire under & hot embers on ye top to
keepe ye heat of ye fire ye better for 40 howers, ye
☿ will distill into a slimy water hanging togather
yt will melt nothing but mettall. And this is the
true aqua vitæ ye spirit so much desired of Philosophers.
The soule or forme is thus made viz:
Sublime ☿ 7 times, ye last time being done
wth Cinnaber wthout vitriol yt ye ☿ may take a certaine quintessence of
ye Cinnabers sulphur. To ℥4 of this sublimate put ℥1 of fine filings of ☉ or
☽, & set a subliming againe 16 howers, let it coole mix ym all togather &
sublime againe, Do thus 4 times & it will bee a rundle like white rosen transparent & cleare as pearle weighing ℥5. The ☿ will stick to the top & sides
of the pot; in ye bottom it will bee like good black wch is ye corruption of
ye ☉ or ☽. Dissolve ye afforesaid Rundle in most strong vinegar distilld, for
wch yt it must bee dissolved 2 or 3 times by siting it in a urinall in B:M:
for 3 days, every time pouring in new vinegar as at ye first till it bee quite
dissolved. filter it, & save ye remaining feces wch are good to whiten brasse,
& evaporate ye vinegar in ashes wth a gentle heat, & set it in ye sun & it
will bee very red if ☉ was dissolved, or if ☽ then white like flowr of starch. And
this is ye forme of ☉ or ☽ or verily the sulphur wch waigheth ℥2 more or lesse
Into an urinall 6 inches high put of ye firme body ℥5 a quarter of ye soule
of ye ☉ or ☽, & of ye spirit ℥4 lute on ye head & his receiver well, distill wth
a soft fire & at ye first time will rise almost 3 ℥ cohobate wthout moving ye urinall
6 or 7 times till noe more will bee distilled. Set this fixed matter in hors dung 6 or
7 days till it bee dissolved. filter it per lingulam & about 4 or 5 ℥ will come over
wch congeale 3 times. Put ℥1 of this medicine upon ℥10 of fine molten ☉ or ☽
melt borax & wax of this medicine ana ℥1, & put these on 3 lb of ☿ or wt
mettall you please & it shall bee fine ☉ or ☽. This may bee done in 40 days.

Figure 14 As an alchemist Newton took copious notes from the works that were circulated among his fellow adepts. Here the Lucasian professor has copied a manuscript entitled *The Work of an old Priest*.

Protestants had preferred to keep the text as it made 'against heresy'. Spurious things had to be purged from the text of scripture, Newton told Locke: '& therefore knowing your prudence & calmnesse of temper, I am confident I shal not offend you by telling you my mind plainly'. What he was about to do, he remarked not a little disingenuously, was 'no article of faith, no point of discipline, nothing but a criticism concerning a text of scripture'. In short, some fourth-century Catholic scholars – the usual suspects – had interpreted the spirit, the water and the blood as the Father, the Son and the Holy Ghost, and this had been inserted by Jerome into his Vulgate. 'Afterwards', Newton added, 'the Latines noted his variations in the margins of their books, & thence it began at length to creep into the text in transcribing, & that chiefly in the twelft & following Centuries when disputing was revived by the Schoolmen'. After printing came into being 'it crept up out of ye Latine into the printed Greek against ye authority of all the greek MSS & ancient Versions'.[65]

Newton's approach to these corruptions was threefold. First he could show how and why the text was inserted into various manuscripts and printed texts. This involved a convoluted but scholarly analysis of texts in which he argued that trustworthy authors would have referred to the text if it existed, but had not done so. There was no evidence that it existed in the oldest Greek texts and indeed some contemporaries had accused Jerome of inserting it according to his own whim. Newton put Jerome on trial and unsurprisingly found him guilty: 'in all that vehement universal & lasting controversy about the Trinity in Jeromes time & both before & long enough after it, this text of the three in heaven was never thought of'.[66] Second, he actually had access to ancient manuscripts and to printed editions that referred to manuscripts where the offending text was missing or flagged as problematic. Prompted by Locke and Jean Le Clerc to look at the recent work of Richard Simon (who had determined that it was impossible to determine the true, original text of the Old and New Testaments), Newton determined that the testimony of the three in one was lacking in all relevant manuscripts. Finally, he claimed, the restored, true text made more sense, and Newton recast the disputed text for Locke's benefit.[67]

At some point after this missive, an emboldened Newton sent Locke an account of many more problematic texts, 'for the attempts to corrupt ye scriptures have been very many & amongst many attempts tis no wonder if some have succeeded'. Whatever Newton's debt to contemporary writings, there is no doubt that he had again immersed himself in primary sources. Yet this reliance on seeing the earliest texts for himself went hand in hand with a passionate commitment to a particular view of the nature of Christ, and with the deeply held conviction that much of the 1611 King James Bible rested on a corrupt manuscript source. According to Newton, all these corruptions had been initially made by Catholics '& then to justify & propagate them [they] exclaimed against the Hereticks & old Interpreters, as if the ancient genuine readings & translations had been corrupted'. The entire age was one farrago of deception: 'such was the liberty of that age that learned men blushed not in translating Authors to correct them at their pleasure & confess openly y^{t} they did so as if it were a crime to translate them faithfully'. To their eternal shame Protestants now collaborated in the crime, and Newton sanctimoniously told Locke that all these deceptions 'I mention out of the great hatred I have to pious frauds, & to shame Christians out of these practises'.[68]

At around the same time as he was undertaking research for the letter to Locke, writing a treatise on the immorality of Athanasius, composing the Classical Scholia, producing a draft of *Opticks* and continuing his alchemical studies, he returned to the *prisca* project and adduced more evidence detailing the incredible early extent of the vestal religion. In a work entitled *The Original of Religions* he asserted that the ancient Chinese, Danes, Indians, Latins, Hebrews, Greeks and Egyptians all worshipped according to the same practices, while Stonehenge in England was clearly another 'prytaneum' or temple. In the Tabernacle and Solomon's Temple the ancient Jews placed the vestal fires in the court of the priests and in the people's court. Like the vestal fires in the prytanea, the Tabernacle was a symbol of the heavens and 'the whole heavens they recconed to be the true & real Temple of God & therefore that a Prytaneum might deserve the name of his Temple they framed it so in the fittest manner to represent the whole system of the heavens'. Nothing could be more 'rational', Newton added, than this aspect of religion. Following his earlier analysis, he argued that this religion was regulated by priests skilled in natural philosophy and indeed he argued that there was no way 'wthout revelation to come to ye knowledge of a Deity but by the frame of nature'. This was the true religion until superstitious elements began to revere the heavenly bodies and the five elements. In time, people honoured their ancestors by applying their names to heavenly bodies, and they came to be worshipped in the prytanea. This corruption of religion (and, by extension, philosophy) was the worst until the perversions of the fourth century took hold.[69]

In the same period Newton had became extremely close to the hypochondriacal Fatio de Duillier, who in late 1692 and early 1693 began to pester the older man with tales of the marvellous cures that could be effected by means of a potion developed by one of his friends. The same friend had apparently alighted on an extraordinary phenomenon involving alchemical gold and mercury that gave rise to a putrefaction and living fermentation of which Fatio thought Newton should be aware. In another letter he asked Newton to invest a substantial amount of money for developing and marketing the alchemical product, and also for training Fatio in its use. Newton left Trinity for a week at the end of May 1693, probably to go to London, and did the same late in June and early July. Yet by now Newton was in the throes of a breakdown, an experience that only became known when he sent a letter to Pepys in the middle of September. In this strange offering, composed while he was still in a great deal of turmoil, Newton was deeply concerned to deny that he had ever tried to use either Pepys or James II as a patron, and he told Pepys that he would have to withdraw from Pepys's acquaintance and indeed never contact any of his friends again. His friend, Locke, received an even more troubling letter, written three days later from a pub in Shoreditch. Like Pepys, this was the first Locke had heard of Newton's concerns. Newton apologized for accusing Locke of trying to 'embroil' him with women, and begged forgiveness for wishing that Locke would die from a sickness from which he had then been suffering. He was sorry for accusing

Locke for being a Hobbist (i.e. a strict materialist) and for saying that Locke undermined the basis of morality in his *Essay Concerning Human Understanding*. Finally, he mentioned his regret 'for saying or thinking that there was a designe to sell me an office, or to embroile me'.[70]

Pepys and Locke reacted with admirable understanding. In late September Pepys wrote to his friend John Millington, who had visited Newton in the midst of his breakdown, saying that he was concerned that Newton had suffered a 'discomposure in head, or mind, or both'. Millington replied that he had met Newton a couple of days earlier and that Newton had claimed he had been ill for a few days when he wrote his letter to Pepys. Millington remarked that the illness did not seem to have affected Newton's reason and, he hoped, never would: 'and so I am sure all ought to wish that love learning or the honour of our nation, *which it is a sign how much it is looked after, when such a person as Mr. Newton lies so neglected by those in power*'. Locke reaffirmed his undiluted friendship for Newton and said that he forgave him unreservedly: 'I do it soe freely & fully y^{t} I truly love & esteem you & y^{t} I have still the same goodwill for you as if noe thing of this had happened'. Like Pepys, he realized that that it was best to leave the next move to Newton, and indeed Newton repeated a line told to Millington that when he wrote his odd letters he had not slept for five entire nights. All this was supposedly caused by sleeping too close to his fire over the previous winter, and exacerbated by a sickness he had suffered throughout the summer.[71]

Whatever the physical cause of his problems, there were other possible contributory factors to his condition and historians have invoked many such to explain this unfortunate incident. One points to the sheer amount of ordinary mercury to which Newton must have exposed himself (he was continuing to experiment avidly in June 1693). Although mercury poisoning exhibits certain effects that are similar to those described by Newton, others, such as the destruction of his teeth, are singularly lacking. In his *A Portrait of Isaac Newton* (1968), Frank Manuel conjectured that Newton might have conceived doubts about the sexual nature of his relationship with Fatio. Newton visited Fatio on a number of occasions in 1693 but after the summer they would only ever enjoy a distant relationship. Newton was undoubtedly emotionally attached to the young man, but further evidence is lacking. Equally plausibly, frustration at his continued lack of success in gaining preferment in London – coupled with his hatred for James II – could have led to the shrill statements in his letters to Locke and Pepys denying he had tried to gain 'anything' by anybody. Finally, there is always the possibility that he drew back from a relationship with a woman, one of the topics in his epistle to Locke. Indeed, in an essay from about this time, Newton launched into a long excursus on the dangers of allowing oneself to fall prey to lust. To struggle with it was to lose the battle, and the best way to deal with the problem was to keep the mind and hands busy on other things. In a diatribe against the deviant sexual practices of monks and nuns he cautioned his anonymous reader that 'he that's always thinking of chastity will always be thinking of weomen, & every contest wth unchast thoughts will leave such impressions upon the mind as shall make those thoughts to return more frequently'.[72]

As he recovered his equanimity and normal life resumed, Newton had one last try at rectifying some of the problems that had attended his treatment of lunar theory in the *Principia*. Arguably, it would be his last major sustained scientific undertaking. From the summer of 1694 he attacked the issue again, and to acquire the latest data he visited John Flamsteed at the start of September. Flamsteed agreed to let Newton see his upgraded observations, although he added the rider that Newton had to promise not to show them to anyone else. In turn, Flamsteed wanted the corrections to lunar theory that Newton could make. Nevertheless, Newton was not about to treat Flamsteed as an equal, virtually demanding that the Astronomer Royal make observations according to his bidding. As it turned out, the three-body problem Newton had to solve in order to make headway with the problem proved too difficult for him, while Flamsteed struggled to provide observations of the type and precision that

Newton demanded and took much of the blame for the latter's failure. In any case, Newton claimed, he had not been merely trying to amend the lunar theory of the *Principia*, but he aimed at something more momentous. He told Flamsteed that he intended to follow a method whereby he gained a 'generall notion of all the equations on w^{ch} [the moon's] motions depend' and afterwards 'by accurate observations to determin them' more exactly. In an atmosphere of increasing mutual suspicion, Flamsteed began to suspect that Newton was showing his own 'corrections' of the data to someone else, while Newton took umbrage at Flamsteed's wish to know the theoretical basis of Newton's emendations. Finally he told Flamsteed that he did not want him to calculate the true position of the moon from a table of refractions that Newton had supplied, but wanted only his raw data. Over the following years the relationship deteriorated still further, never to recover.[73]

CONCLUSION: TO THE CITY

Newton's efforts to find a suitable post in the metropolis finally bore fruit in 1696, and with it he moved into the second part of his career. His erstwhile Trinity colleague and now senior member of the Treasury, Charles Montague, signed a letter confirming Newton's appointment as Warden of the Mint on 19 March 1696. Newton left Cambridge four days later and only returned for a matter of a few days before he finally resigned his professorship in 1701. As Warden, Newton faced a number of challenges. Britain required deep financial reserves to support its military campaign against France, while the practice of 'clipping' coins of the realm had seriously degraded the value of money and the quality of coinage. Furthermore, since they contained a higher proportion of silver than the older 'hammered' coins, the 'new' and heavier milled coins could be melted down at profit and counterfeit money made out of a mixture of clippings and copper. The only remedy to this was to call in all the 'old' money and to increase dramatically the amount of 'new' money produced by the Mint. The 'Great Recoinage' would produce highly standardized coins with a visible edging, all manufactured by state-of-the-art rolling

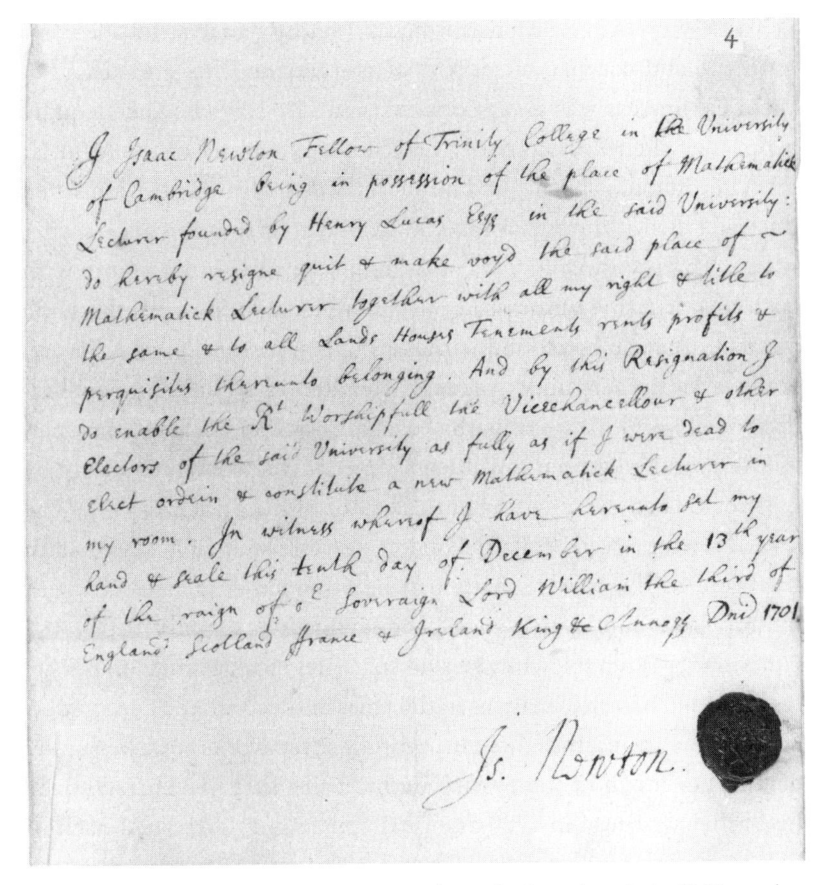

4

I Isaac Newton Fellow of Trinity College in the University
of Cambridge being in possession of the place of Mathematick
Lecturer founded by Henry Lucas Esqr in the said University:
do hereby resigne quit & make voyd the said place of
Mathematick Lecturer together with all my right & title to
the same & to all Lands Houses Tenements rents profits &
perquisites thereunto belonging. And by this Resignation I
do enable the Rt Worshipfull the Vicechancellour & other
Electors of the said University as fully as if I were dead to
elect ordein & constitute a new Mathematick Lecturer in
my room. In witness whereof I have hereunto set my
hand & seale this tenth day of December in the 13th year
of the raign of or Sovereign Lord William the third of
England Scotland France & Ireland King &c Annoq Dni 1701.

Is. Newton.

Figure 15 Newton's resignation from the Lucasian chair, 13 December 1701. In the letter to the Vice-chancellor, Newton declares: 'I do hereby resigne, quit & make voyd the said place of Mathematick Lecturer together with all my right & title to the same & to all Lands Houses Tenements rents profits & perquisites thereby belonging.'

mills. Westfall points out that for some years the wardenship had been treated as a sinecure, and Newton had no need to involve himself to any great degree in the day-to-day affairs of the Mint. Certainly Montague had offered the post to Newton under these terms. However Newton in fact dedicated himself to the recoinage and – to deal with the vast amount of bullion required – the creation of temporary Mints in Norwich, York, Chester, Bristol and Exeter.[74]

Newton apparently performed a Taylorist analysis of the commitment and activity of individual workers and his knowledge of chemical process was exceptionally useful. By 1698, he had virtually taken over the roles ordinarily played by the Master of the Mint, at this time Thomas Neale, and he succeeded to Neale's position when the latter died at the end of 1699. As Warden Newton was also responsible for prosecuting clippers and coiners, pursuing miscreants with gusto and locating witnesses to pecuniary corruption with the same intensity that he had found witnesses to the corruption of scripture. Some jailed coiners understandably threatened to shoot him, and indeed he showed little sympathy to criminals like William Chaloner, whose pitiful pleas to the Warden for mercy were made redundant by his previous double-crossing of the author of the *Principia*.[75]

Newton lived well in London, but showed little interest in literature or the theatre. Although he attended dinner parties with metropolitan luminaries such as John Evelyn, Samuel Pepys and Christopher Wren, Newton was no socialite. He continued to pursue research and to write treatises in the areas of theology and chronology, although by the early eighteenth century these works were becoming denuded of much of the passionate intensity that had characterized his earlier writings. In 1700 one final opportunity presented itself to him to remain in Cambridge. On the eve of his resignation from the Lucasian chair, he was approached by Thomas Tenison, the Archbishop of Canterbury, to take holy orders and become the Master of Trinity. Towards the end of his life Newton told Conduitt that Tenison had pressed him to take the position as Newton knew more (presumably, concerning religion) than the rest of them put together. Newton declined, knowing that he could never publicly affirm religious orthodoxy, but told Tenison that he would do everyone 'the more service by not being in orders'.[76]

In the 1690s, as he later recalled, he had become something of a tourist attraction and he remembered anecdotes from the period that attested to his supposedly bizarre life. Yet by the end of the decade, Newton was fast shedding the traditional image of withdrawn scholar.

At the point where he resigned his chair and severed his major connection with Cambridge, he was about to become one of the most powerful figures in early Augustan Britain, soon acquiring the Presidency of the Royal Society and a knighthood to go along with his position at the Mint. The most private of hermits had become the most public of men, perhaps the most respected of the great and the good in the British establishment. Alongside editions of *Opticks* that were published in 1704, 1706 and 1717, new editions of the *Principia* appeared in 1713 and 1726, consolidating his position as the most influential natural philosopher since Aristotle. Having outlived and triumphed over his enemies, he could now cultivate awestruck and often sycophantic disciples who would in turn work with him on nourishing his legend.

Notes

1 In general, see F. Manuel, *A Portrait of Isaac Newton* (Cambridge, MA, 1968) and R. S. Westfall, *Never at Rest: A Biography of Isaac Newton* (Cambridge, 1984). For comments on previous versions of this text I would like to thank Moti Feingold, Rupert Hall, Kevin Knox and Richard Noakes.

2 For Newton's early life see in general Westfall, *Never at Rest*, pp. 45–66. In fact Newton returned to Grantham Grammar School for extra tuition once the headmaster, Henry Stokes, allowed him to lodge in his house for free.

3 *Ibid.* pp. 58–60, 65, 71–3, 76–8. Other sins relating to earlier periods included theft, jealousy, fighting and slander. Wickins's anecdote about his father's life with Newton is now in King's College, Cambridge, Keynes Ms. 137.

4 Westfall, *Never at Rest*, pp. 102–4, 176–81. Pulleyn was professor of Greek between 1674 and 1686.

5 See J. E. McGuire and M. Tamny (eds) *'Certain Philosophical Questions': Newton's Trinity Notebook* (Cambridge, 1983).

6 *Ibid.*

7 *Ibid.* 'Newton's optical lectures', pp. 5–13.

8 *Ibid*, pp. 10–15. In the second 'Of colours', Newton now passed a beam of light through a prism at minimum deviation, throwing the resultant refracted light on to a wall twenty-two feet away. The oblong shape produced as a result was a stark demonstration of the basic phenomenon of differential refrangibility, and was the first experiment mentioned in Newton's initial paper on light and colours in the *Philosophical Transactions*.

9 See J. Herivel, *The Background to Newton's Principia* (Oxford, 1965) and R. S. Westfall, *Force in Newton's Physics* (London, 1971), chs 2–4. This interpretation of these equations use the modern concept of mass, which Newton had not yet developed.

10 See Herivel, pp. 183–91, 195–7, and Hall, 'Adventurer in thought', 58–63. Kepler's third law holds that the square of the mean period of revolution of a planet is proportional to the cube of its mean radius from the sun.

11 Westfall, *Never at Rest*, pp. 105–37, esp. 137.

12 D. T. Whiteside (ed.), *The Mathematical Papers of Isaac Newton* (hereafter MP) (8 vols, Cambridge, 1967–81), vol. 2, pp. 163–247.

13 Barrow to Collins, 20 July and 31 July 1669, Collins to Gregory, 25 November 1669; H. W. Turnbull, J. F. Scott, A. R. Hall and Laura Tilling (eds) *The Correspondence of Isaac Newton* (hereafter NC) (7 vols Cambridge, 1959–77), vol. 1, pp. 13, 14; Collins to Oldenburg and Oldenburg to Sluse, *c.* 12 September and 14 September 1669; A. R. Hall and M. B. Hall (eds), *The Correspondence of Henry Oldenburg* (13 vols, Madison and London, 1965–86), vol. 6, pp. 227, 233. See also Westfall, *Never at Rest*, p. 203 fn. 75 and 209.

14 A. Shapiro (ed.), *The Optical Papers of Isaac Newton* (3 vols, Cambridge, 1984) Vol. 1. *The Optical Lectures, 1670–1672*, pp. 13–25 esp. 14 n. 46 and 17–20; Newton to Oldenburg, 6 July 1672; NC, vol. 1, p. 209. For Newton's subsequent references to the influence of Barrow see Moti Feingold, 'Newton, Leibniz, and Barrow too: an attempt at a reinterpretation', *Isis*, 84 (1993), 310–338.

15 Shapiro, 'Optical lectures', vol. 1, pp. 17–23.

16 Collins to Gregory, 24 December 1670, Newton to Collins, 11 July and 16 July, 27 September, 1670; NC, vol. 1, pp. 52–60, 30–31, 34–5, 43–4. The notes on Kinckhuysen are in MP, vol. 2, pp. 364–44.

17 Collins to Newton, 5 July 1671, Newton to Collins, 20 July 1671, Collins to Newton, 30 April 1672; NC, vol. 1, pp. 65–6, 67–70, 146–50. See also MP, vol. 3, pp. 32–328 and esp. vol. 3, p. 23.

18 Oldenburg to Newton, 2 January 1671/2; Newton to Oldenburg, 18 January 1671/2; NC, vol. 1, pp. 73, 82–3.

19 Newton to Oldenburg, 6 February 1671/2; NC, vol. 1, pp. 92–107.

20 Newton to Oldenburg, 10 February 1671/2; NC, vol. 1, pp. 108–9.

21 Jewish National and University Library (JNUL), Jerusalem, Yahuda Ms. 14.

22 Newton to Sir Alexander ?, *c.* 1674, Newton to Oldenburg, January 1674/5; NC, vol. 3, pp. 146–7; vol 7, p. 387; Westfall, *Never at Rest*, pp. 330–4. See also Moti Feingold, *Before Newton: The Life and Times of Isaac Barrow* (Cambridge, 1990), pp. 80–8.

23 Collins to Gregory, 29 June and 19 October 1675; NC, vol. 1, pp. 345, 356. See B. J. T. Dobbs, *The Janus Faces of Genius: The Role of Alchemy in Newton's Thought* (Cambridge, 1991); Westfall, *Never at Rest*, pp. 281–99 and K. Figala, 'Newton as alchemist', *History of Science*, 15 (1977), 102–37.

24 Westfall, *Never at Rest*, pp. 293–7 and Dobbs' *Janus Faces*, 25–6; see King's College Cambridge, Keynes Ms. 12A.

25 'Of nature's obvious laws & processes in vegetation', Dibner Ms 1031B, Dibner Library of the History of Science and Technology, Special Collections Branch, Smithsonian Institution; reproduced in Dobbs, *Janus Face*, 256–70, 258–9.

26 *Ibid.* 262.

27 Newton to Oldenburg, 26 April 1676; NC, vol. 2, pp. 1–2. The article in the *Philosophical Transactions* had been signed 'B.R.'

28 Evidence suggests that in late 1677 Newton was once again planning to publish the treatises on fluxions and colours but that fire intervened when he took a rare trip to the bowling green; see Keynes Ms. 130.4, pp. 14–15 and Westfall, *Never at Rest*, pp. 276–7.

29 Newton to Oldenburg, 7 December 1675 and 25 January 1675/6; NC, 1: 360–1, 362–89 ('An hypothesis explaining the properties of light discoursed of in my severall papers') and 413–15. In general see D. C. Kubrin, 'Newton and the cyclical cosmos: providence and the mechanical philosophy', *Journal of the History of Ideas*, 28 (1967), 325–46.

30 See H. W. Robinson and W. Adams (eds), *The Diary of Robert Hooke* (London, 1935), pp. 205–8.

31 Newton to Oldenburg, 21 December 1675, Hooke to Newton, 20 January 1675/6, Newton to Hooke 5 February 1675/6; NC, vol. 1, pp. 404–6, 412–13 and 416–17.

32 Newton to Boyle, 28 February 1678/9; NC, vol. 2, pp. 288–91.

33 Hooke to Newton, 24 November 1679, Newton to Hooke, 28 November 1679; *Ibid.* vol. 2, pp. 297–8, 300–3.

34 Hooke to Newton, 9 December 1679, Newton to Hooke, 13 December 1679; *ibid.* vol. 2, pp. 304–6, 307–8.

35 Hooke to Newton 6 January and 17 January 1679/80, Newton to Halley, 27 July 1686; *ibid.* vol. 2, pp. 309–10, 312–13, 447.

36 Newton to Burnet, 24 December 1680, Burnet to Newton 13 January 1680/1; *ibid.* vol. 2, pp. 319, 321–7.

37 Newton to Burnet, late January, 1680/1; NC, vol. 2, pp. 329–32.

38 *Ibid.* vol. 2, pp. 332–4.

39 Flamsteed to Crompton for Newton, 15 December 1680, Flamsteed to Crompton, 12 February 1680/1, Flamsteed to Halley, 17 February 1680/1; NC, vol. 2, pp. 315–16, 336, 336–9. The observations in the Waste Book are CUL Add. Ms. 4004 fols 97–101v. The comet's path was actually 61 degrees to the ecliptic.

40 Newton to Crompton for Flamsteed, 28 February 1680/1; NC, vol. 2, pp. 340–7.

41 Flamsteed to Crompton for Newton, 7 March 1680/1, Newton to Crompton for Flamsteed,? April 1681; *ibid.* NC, vol. 2, pp. 358–62. Newton to Flamsteed, 16 April 1681, *ibid.* vol. 2, pp. 363–7, esp. 364–5.

42 The following texts come from JNUL, Jerusalem, Yahuda Ms. 1.

43 There are various analyses of the links between ancient science and religion in Yahuda Mss 16, 17 and 41, and New College Oxford Mss 361.1 and 361.3.

44 'The system of the world' in F. Cajori (ed.), *Sir Isaac Newton. Principia* (revision of 1729 translation by Andrew Motte) (2 vols, London, 1962), vol. 2, pp. 549–50.

45 *Ibid.* vol. 2, p. 550.

46 Westfall, *Never at Rest*, pp. 402–68.

47 Newton to Flamsteed, 30 December 1684; Flamsteed to Newton, 5 January 1684/5; Newton to Flamsteed, 12 January 1684/5; NC, vol. 2, pp. 407, 408–9 and 412–3.

48 Westfall, *Never at Rest*, pp. 411–20.

49 *Ibid.* pp. 421–30.

50 *Ibid.* pp. 433–67, esp. 458–63.

51 Humphrey wrote two letters to John Conduitt, one on 17 January and the other on 14 February 1727/8; they are now King's College Cambridge,

Keynes Ms 135. Vigani's faux pas was related to Conduitt by Newton's step-niece and Conduitt's wife, Catherine; see Keynes Ms 130.6, Book 2.

52 Halley to Newton, 22 May, Newton to Halley, 27 May, Halley to Newton, 7 June, Newton to Halley 20 June 1686; NC, vol. 2, pp. 431, 433–4, 434, 435–6.

53 Newton to Halley, June 20, Halley to Newton, 29 June, Newton to Halley, 14 July 1686; *ibid.* vol. 2, pp. 437–9, 441–3 and 444.

54 An account of the 'Cambridge case' can be found in C. H. Cooper, *Annals of the University of Cambridge* (Cambridge, 1853), pp. 614–43.

55 King's College Cambridge, Keynes Ms 149.

56 King's College Cambridge, Keynes Mss 121, 118 and 116.

57 Westfall, *Never at Rest*, pp. 377–8, 398–9, 210 n. 94, 338, 480. In 1687 he deposited a draft of the *Principia* as his lectures over the previous three years.

58 Newton to Covel, 19 February 1688/9; NC, vol. 3, pp. 12–13.

59 Newton to Covel, 6 April [sic], 7 May, 1689; *ibid.* vol. 3, pp. 17, 21–2.

60 King's College, Cambridge, Keynes Ms 117A and B; Newton to Locke, *c.* December 1691; NC, vol. 3, p. 184.

61 Halley to Newton, 24 February, 1686/7 and 5 April 1687; NC, vol. 2, pp. 470, 473–4.

62 See D. Meli, *Equivalence and Priority: Newton versus Leibniz* (Oxford, 1993), pp. 192; I. B. Cohen, 'Introduction' (Cambridge, 1971), pp. 145–61 and E. A. Fellmann, 'The *Principia* and continental mathematicians', *Notes and Records of the Royal Society of London*, 42 (1988), 13–34. For the so-called 'Classical scholia' see J. E. McGuire and P. Rattansi, 'Newton and the pipes of Pan', *Notes and Records of the Royal Society*, 21 (1966), 108–43 and P. Casini, 'Newton: the classical scholia', *History of Science*, 22 (1984), 1–57.

63 Cambridge University Library, Add. Ms 3970.3 fol. 338.

64 Newton to a 'friend' [i.e. Locke], 14 November 1690; NC, vol. 3, pp. 83–129.

65 *Ibid.* vol. 3, pp. 83–4, 88–90. 1 John 5:7–8 reads 'For there are three that bear record in heaven, the Father, the Word, and the Holy Ghost, and these three are one. And there are three that bear witness in earth, the spirit, and the water, and the blood: and these three agree in one.'

66 NC, vol. 3, p. 90.

67 *Ibid.* vol. 3, pp. 96–100, 107–9.

68 Newton to a 'friend' [i.e. Locke], late November 1690 ('The third letter'); *ibid.* vol. 3, pp. 138–9.

69 JNUL, Yahuda Ms. 41.

70 Fatio to Newton, 4 and 18 May 1693; Newton to Pepys, 13 September and Newton to Locke, 16 September 1693; NC, vol. 3, pp. 265–7, 267–70, 279, 280. Newton composed this letter to Locke in the Bull Inn in Shoreditch, which was the terminus of the Cambridge to London carriage service. See also Manuel, *Portrait*, 213–35.

71 Millington to Pepys, 30 September 1693; Pepys to Millington, 3 October 1693; Locke to Newton, 5 October 1693 and Newton to Locke, 15 October 1693; NC, vol. 3, pp. 281–2, 282–3, 283–4, 284.

72 See P. Spargo and C. A. Pounds, 'Newton's "derangement of the intellect": new light on an old problem', *Notes and Records of the Royal Society*; Manuel, *Portrait*, pp. 191–205, 218–23; Westfall, *Never at Rest*, pp. 538–40; R. Iliffe, 'Isaac Newton: Lucatello Professor of Mathematics', in C. Lawrence and S. Shapin (eds), *Science Incarnate: Historical Embodiments of Natural Knowledge* (Chicago, 1998), pp. 121–155, esp. 146.

73 Westfall, *Never at Rest*, pp. 541–8. Flamsteed to Newton, 17 November, 1694; NC, vol. 4, p. 47.

74 Westfall, *Never at Rest*, pp. 549–60, esp. 557.

75 *Ibid.* 561–94, esp. 570, 579–81.

76 King's College, Cambridge, Keynes Ms 130.6 Bk 2.

3 Making Newton easy: William Whiston in Cambridge and London

Stephen D. Snobelen[1] and Larry Stewart[2]
[1] *University of King's College, Halifax, Nova Scotia, Canada*
[2] *Department of History, University of Saskatchewan, Saskatoon, Canada*

FROM AUGUSTUS TO TIBERIUS?

On 31 March 1702, Astronomer Royal John Flamsteed wrote to tell Abraham Sharp that Isaac Newton had 'left his professorship at Cambridge and put Mr Whiston into it'. To this Flamsteed added with evident delight: 'tis sayd by some Malitious people that Augustus left a Tiberius to succeed him purposely to render his own fame the more illustrious'.[1] This analogy of a greater emperor setting up a lesser to succeed him is at once truthful and misleading. As Lucasian professor, Newton produced his first publications on optics and gave the world his *Principia* – works that profoundly revolutionized natural philosophy, astronomy and mathematics. During his decade in the same post, Newton's successor offered no such works of original innovation and discovery. For Whiston, there was to be no uncovering of the laws of physics and no lasting insights into lunar motion. Nor is Whiston's name attached to a school of mechanics, or even a single mathematical theorem.

When judged by the standards of discovery and influence in natural philosophy and mathematics, Whiston is clearly Tiberius to Newton's Augustus. Yet, when the criteria of judgement are shifted to creative pedagogy and a commitment to increasing access to natural

From Newton to Hawking: A History of Cambridge University's Lucasian Professors of Mathematics, ed. Kevin C. Knox and Richard Noakes. Published by Cambridge University Press.

philosophy beyond the mathematical élite, it is Newton who plays the lesser role. This is not to say that innovation and popularization are orthogonal; in the early eighteenth century Edmond Halley demonstrated that it was possible to excel at both, as has Stephen Hawking in our own day. Newton himself was not completely unconcerned with the dissemination of his natural philosophy to a wider audience (his *Opticks*, after all, was first published in English). At the same time, Whiston made a minor, but important, contribution in the field of cosmology. Nevertheless, Newton remains important primarily for his innovation, while Whiston's most noteworthy function was popularization. Whiston was to become one of the most successful among the first generation of Newton's apostles who made careers in London, in the provinces, and even on the continent disseminating experimental philosophy to paying audiences. Whiston's tireless efforts at rendering Newton easy for the nonspecialist and lay public not only helped shape his career path, but also secured him patronage and not insignificant rewards of a pecuniary nature. Such a remarkable dedication also soon revealed itself in his attempts to rescue primitive Christianity from corrupt doctrine, an objective close to Newton's own passion for scriptural purity. Ultimately, in the eyes of many of his contemporaries, this led Whiston close upon the shoals of heresy.

THE ASCENT OF WHISTON'S STAR

Born in 1667 to the rector of Norton-juxta-Twycross, Leicestershire, and the daughter of his father's predecessor, Whiston was intended by his father to be a priest. To this end the young Whiston received his early education at the hands of his father, before attending Tamworth grammar school for a year and a half beginning in 1684. He enrolled as a sizar at Clare Hall, Cambridge, in June 1686 and took up residence the following September. Whiston flourished at Cambridge and was particularly inspired by mathematics, which so animated him that he studied the subject eight hours a day. Although mathematics was only a small part of his examinations, he none the less qualified BA in 1689, ranking twenty-fourth (out of 137) in the *Ordo Senioritatis*

Figure 16 Portrait of William Whiston [c. 1690].

for 1689–90. Thereafter he was elected to the Clare Exeter fellowship in July 1691 and promoted to probationary senior fellow of his college in February 1693, the same year that he received his MA. Ordination followed later that year, with the latitudinarian bishop William Lloyd officiating. In 1694 Whiston became chaplain to John Moore, then Bishop of Norwich. He held this position until 1698, when Moore secured him the living of Lowestoft-cum-Kessingland, a parish on the

Suffolk coast. Whiston initiated something of a reformation in this fishing community, preaching twice daily at Lowestoft, ensuring that his curate preached daily at Kessingland and establishing catechetical lectures in Lowestoft for the spiritual education of the laity. During this time, he delivered his sermons and lectures in a free and easy manner in order to make his teachings accessible to all.

Shortly after his ordination, Whiston returned to Clare to continue his studies, 'particularly the mathematicks, and the *Cartesian* philosophy', which 'was alone in vogue with us at that time'. It was not long after this that Whiston would experience the first of what would be two Newtonian revolutions in his life. After reading a paper by David Gregory, in which the Scottish mathematician gave Newton's *Principia* 'the most prodigious commendations . . . as not only right in all things, but in a manner the effect of a plainly *divine genius*', Whiston, now stirred, to set out to master the inspired text. Whiston had previously encountered the infamously abstruse *Principia* while an undergraduate in 'one or two' of Newton's lectures upon it in the Public Schools – lectures that Whiston later confessed he understood 'not at all at the time'.[2] And so Whiston began, without assistance, the arduous task of mastering a work that even the leading mathematicians of Britain and Europe had difficulty comprehending.

Whiston's swift conversion to Newton's natural philosophy soon bore fruit in his first publication, *A New Theory of the Earth* (1696), a full-length application of Newtonian principles to cosmogony. The *New Theory* employed the gravitational physics of Newton and the geology of John Woodward to demonstrate that the scriptural accounts of creation, the Genesis flood and the final conflagration were 'perfectly agreeable to Reason and Phylosophy'. This was the positive goal. The negative goal was to correct Thomas Burnet's Cartesian and deistic *Telluris Theoria Sacra* (*Sacred Theory of the Earth*) of 1681. Cartesianism was now discredited in Whiston's view as a false romance, and he was passionate about constructing an earth theory on the secure foundation of Newton's *Principia*. But the result was more than simply a theory of the earth, for central to its

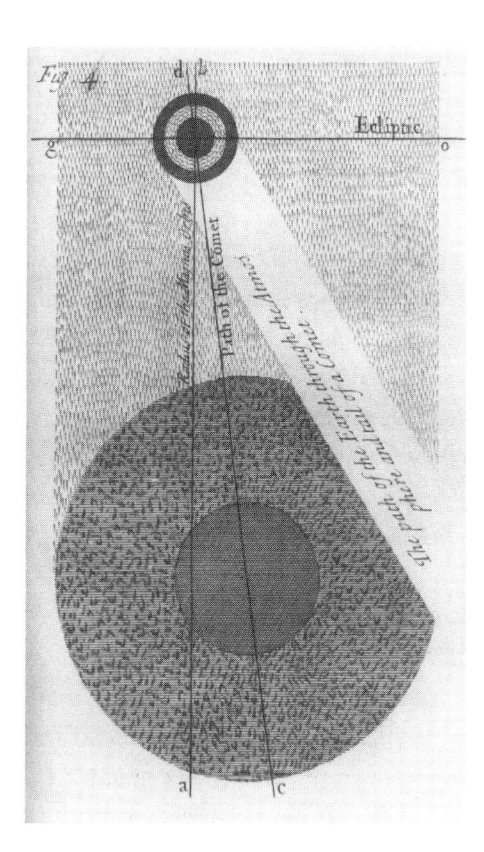

Figure 17 A comet as the source of the Flood. In his *New Theory of the Earth*, a response to Burnet's *Sacred Theory of the Earth*, Whiston argued that his application of Newtonianism to biblical exegesis produced greater awe for the Creator: 'To assign Physical and Mechanical causes for the Deluge, or such mighty judgements of God upon the Wicked, is so far from taking away the Divine Providence therein, that it supposes and demonstrates its Interest in a more Noble, Wise and Divine manner than the bringing in always a miraculous Power wou'd do.'

aims were scriptural exegesis, the design argument and millenarian prophecy.

The chief agency of Whiston's cosmogony was cometary, and he used the close approach of a comet to explain the commencement of the diurnal rotation, the elliptical orbit and the inclined axis of the earth, which he argued had itself originally been a comet. Crucially, cometography also explained the Noahic flood, which, Whiston proposed, began when a comet passed close to the earth, and then distorted and cracked open the crust through gravitational pull, which in turn released the floodgates of subterranean waters beneath the crust. It is in Whiston's use of the then only recently explicated universal law of gravitation that the Newtonianism of the *New Theory* is made explicit. The return passage of the same comet after its perihelion added to the diluvial precipitation water from its vapour trail. Even the final apocalyptic conflagration of the world, Whiston contended, would be brought about by a comet's near passage, and a direct collision at the end of time might also knock the earth out of its orbit to return the terrestrial ball to its original cometary state. The *New Theory* was the first full-length popularization of Newtonian physics and it is significant that the author of the *Principia* had a hand in editing the work and, as Whiston records, 'well approved of it'.[3] The *New Theory* was also the most successful of Whiston's original publications, going through five English editions, summarized in encyclopaedias, translated into German, and reviewed, epitomized and summarized in French. The impact of this work was the primary cause of Whiston's rise to prominence.

WHISTON'S SUCCESSION TO THE LUCASIAN PROFESSORSHIP

Although the precise details have not survived, by no later than January 1701 Whiston was, while at Lowestoft, recalled to Cambridge to begin lecturing as Newton's deputy. Whiston's deputization by Newton had come at the recommendation of Richard Bentley, who, elected Master of Trinity in February 1700 and determined to improve the teaching of mathematics and natural philosophy at the college, was undoubtedly concerned that the Lucasian chair had remained effectively empty for so long. Newton, who may have ceased lecturing as early as 1687, did not even reside in Cambridge after April 1696, when he had been appointed Warden of the Mint in London. Instead, as an absentee office-holder he came to treat the professorship as a sinecure while continuing to receive the annual income from the post. Thus, Whiston's arrival in early 1701 as Newton's deputy brought about a dramatic improvement in the state of natural philosophical and mathematical education at Cambridge. Later that year, Newton formally resigned his post, relinquished his title to the Lucasian lands and requested that a new professor be elected. On the appointed day in January 1702, Whiston was duly elected by a unanimous vote and without so much as a single competitor. Although there is a lack of corroborating evidence, there is no reason to question Whiston's testimony that Newton not only made him first his deputy with 'the full profits of the place', but also made him a candidate and specifically recommended him to the college heads as his successor. Newton's approval of the *New Theory* likely convinced him of Whiston's abilities – or at least assured him that Whiston was a dedicated acolyte only too willing to sing the praises of the great man and to fight battles on his behalf. The professorship, therefore, may have been a reward for Whiston's pro-Newtonian labours. Of course, there may also be some truth in the attribution of motivation in the cynical rumour articulated by Flamsteed.

When Whiston began his Lucasian lectures he was a relatively young man of thirty-three. He and his wife Ruth also had the beginnings of a family in an infant daughter; the family lived in a residence situated across from Peterhouse at the south end of Cambridge. As a married man, Whiston could no longer possess a college fellowship, and it seems he had little, if any, association with his former college during his professorship. Instead, Whiston's affiliations (albeit informal) were with the college of his two predecessors. Intent on making Trinity the centre of natural philosophical and mathematical teaching at Cambridge, Bentley procured rooms for Whiston and his pupils off the south side of Trinity's main gate – a choice perhaps made deliberately since this situated Whiston in almost the same space as that formerly occupied by Newton, whose chambers had been located on the north side of the gate. Clad in a luxurious scarlet gown, the young and confident Whiston would have cut a dashing figure as he proceeded along the same path Newton had trod to the Public Schools to give his lectures.

Whiston began lecturing shortly after his arrival in Cambridge from Suffolk and delivered nine lectures on astronomy during Lent and Easter terms 1701. In all, between Lent 1701 and the end of Michaelmas 1703, he delivered thirty-one lectures on the fundamentals of astronomy. These lectures were not directed at experts or those who had already made extensive preparation in astronomy. Instead, they were intended for the youths of the university, as the title page of the published lectures states (*In Usum Juventutis Academicæ*). As with his Lowestoft parishioners, Whiston employed a relatively 'easy' and 'familiar stile' (albeit in Latin). The focus was also on the practical rather than the theoretical. The lecture syllabus is ambitious and reads like a résumé of Whiston's later career. The range of topics includes geodesy, stellar parallax, the calculation of solstices, the prediction of lunar and solar eclipses, the determination of longitude, the eclipses of the Jovian satellites (as well as the determination of terrestrial longitude by them), Mercurial and Venusian transits and the mathematics of planetary motion. The two final lectures were read on Newton's complex theory of the moon.

Amongst all this, impressionable undergraduates heard appeals to the argument from design. Whiston further touched on the value of past eclipses for establishing the dates of historical events, using a lunar eclipse that appeared in Judaea at the end of the reign of Herod the Great to show that the dating of the Christian era is off by at least three years. Deference is given to Newton throughout, with the text punctuated with laudatory references to the great man. For instance, in lectures on Newton's lunar theory, Whiston employed the complete text of Newton's tract on this subject, but he interspersed it with his own explanatory commentary – a feature demonstrating that Whiston had by this point recognized the need to gloss and explicate Newton's unpalatable writings. As I. Bernard Cohen has observed, Whiston's astronomical lectures provide an astronomical background to Book Three of the *Principia*.

Whiston's first lecture series was followed immediately in Lent 1704 by a second on Newtonian mathematical physics and astronomy. A total of forty-one lectures on these themes were delivered by the end of 1708. Like the *Praelectiones Astronomicae*, these lectures were '*in usum juventutis Academiae*'. The opening words of his first lecture of this series go a long way to revealing his pedagogical methodology:

> AFTER having dispatch'd the Matters of pure Astronomy, we proceed unto the other Part of our Work, the Philosophy of the Famous Sir *Isaac Newton*. For we are purpos'd to trace the Steps of that Great Man, and to set forth his principal and most noble Philosophical Inventions in a more easy Method; that so we may bring that (as I may say) Divine Philosophy within the Reach and Comprehension of those, who are but indifferently perhaps exercis'd in the Mathematicks, and communicate the Knowledge thereof as far as may be.[4]

The slightly more heavy-going material in this series served at once as paraphrase, commentary and expansion of Newton's *Principia Mathematica*. The lectures follow the structure of Newton's *magnum opus*, with Whiston bypassing the more involved mathematics and recasting Newton's calculations as paragraphs of connected prose. These lectures were thus in large part a simplified paraphrase of Newton's *Principia*, although he also drew on material gleaned from Newton's manuscript corrections and additions to the first edition of the *Principia*, as well as the *Opticks* (1704), *Optice* (1706) and Newton's manuscript *Lectiones Opticæ* – then on deposit in the university archives. So close is the paraphrase to Newton's *Principia* that Cohen considers the 1716 English version of these lectures 'the first published version in English of any major portion of the *Principia*'.[5]

The lectures also serve as the first full-length commentary of the *Principia*, and thereby provide modern scholars with a helpful guide as to how the first generation of Newtonians interpreted Newton's text. As he had done previously with his lectures on Newton's lunar theory, Whiston wove into Newton's text his own interpretative gloss. In the case of the published physico-mathematical lectures, however, Whiston's own words are not set off from Newton's source text; nor are references to Newton's *Principia* given regularly in the text of the lectures. The overall effect for the inexperienced student, therefore, was for Whiston's words to blend imperceptibly with Newton's. But Whiston was not beholden to Newton, and his lectures include material from other mathematicians and natural philosophers such as Euclid, Copernicus, Tycho, Kepler, Galileo, Hevelius, Hooke, David Gregory, Flamsteed and Halley. Descartes also appears often, but because one of Whiston's main agendas is to secure Newton's physics as the 'true philosophy', most of the references to the Frenchman are none too complimentary. In fact, Whiston's dual agenda of deifying Newton and demonizing Descartes is made plain from the introductory words of his first lecture, where he says in his opening shot:

> as to the first Laws of Motions and Collisions, *Des Cartes* was so miserably mistaken about them, when he went about to establish them, and hath so boldly impos'd upon the World false Rules concerning Collision and Reflection of Bodies, that it is worth the while to endeavour to root out of the Minds of Men the Prejudices which have sprung from thence.[6]

Apologetics in the physico-mathematical lectures also came in the form of natural religion and an attack on the idolatry of astrology. The two concluding lectures in this series are devoted to a commentary on Halley's cometographical work into which Whiston once again weaves his own explanatory narrative, which in this case is distinguished from Halley's words (the latter being placed within quotations marks). As with his glosses on Newton, Whiston is careful to tease out the meaning of Halley's work, which he sees as 'something too succinct and obscure' and not 'represented plain enough for Beginners'.[7] Whiston was determined to make sure that undergraduates did not feel as alienated from the subject matter of the Lucasian lectures as he had two decades before.

It was only when Whiston had completed his lectures on the *Principia* in Michaelmas 1708 that he began to venture beyond the exposition of the works of others to develop more original material on a subject that had become dear to his heart: lunar and solar eclipses. In February 1709, Whiston commenced what was to be a total of eleven lectures on ancient eclipses. He lectured on this topic through Lent and Michaelmas 1709, and again in Easter term 1710. It is not possible to determine whether Whiston intended more than eleven lectures on this topic, however, since his departure from Cambridge early in Michaelmas 1710 brought an end to his ten years as Lucasian professor.

In May 1707 Whiston and the recently appointed Plumian professor of astronomy Roger Cotes began to deliver the first course of Newtonian experimental philosophy at Cambridge, dividing equally between them twenty-four lectures on hydrostatics and pneumatics that served to elucidate the natural philosophy of Newton, Robert Boyle and others. Robert Smith, Cotes's cousin, assistant and eventual successor, testified that these demonstrations were 'performed before large assemblies at the Observatory in Trinity College Cambridge'.[8] If the assemblies really were large, then Whiston and Cotes likely derived a healthy supplementary income from the course. Although the names of auditors of Whiston's Lucasian lectures have not survived, we know that Joseph Wasse, Stephen Hales and William Stukeley (the latter two becoming prominent Newtonians) attended the Cotes–Whiston experimental course.

Whiston's desire to accommodate the needs of undergraduates must have been in some measure a reaction to his own experiences as a young man, when he sat in the depths of incomprehension at the feet of Newton, who lectured from the ethereal heights. His evangelical promotion of Newtonianism, therefore, reflects not only his passionate commitment to the philosophy, but also his determination to make sure his students were not condemned, like him, to wallow in the fictions of Cartesianism. The level of difficulty of his lecturing syllabus was probably also dictated by Whiston's own mathematical abilities, which, although not inconsiderable, were of a different (and lower) order than those of Newton.

Through his published textbooks, the impact of Whiston's teaching as Lucasian professor extended well beyond Cambridge's lecture halls. The first book he published for academic use was an edition of André Tacquet's *Euclid*, printed by the University Press in 1703. Following in the footsteps of Isaac Barrow, whose own edited version of Tacquet's *Euclid* (1656) had been prepared with his undergraduates in mind, Whiston explained that he published his book 'for the Use of young Students in the University'.[9] Although Whiston's *Euclid* was not to reach the fifty reprints that Barrow's attained by the middle of the eighteenth century, it was issued in a respectable twenty-one (or twenty-two) editions, including a final one retranslated into Greek. In 1707 he published, with the former Lucasian professor's grudging acquiescence, some of Newton's astronomical lectures, and in so doing perhaps revealed an early sign of his irritation with Newton for being so reluctant to publish his own works. Later the same year Whiston published his astronomical lectures (*Prælectiones Astronomicæ*), which he subsequently released in English (*Astronomical Lectures*, 1715, 1728). His *Prælectiones Physico-mathematicæ* appeared in 1710 (second edition, with the lectures on eclipses, 1726) and was published in English as *Sir Isaac Newton's Mathematick Philosophy More Easily Demonstrated* (1716). Whiston also found time to publish a revised second edition of his *New Theory* in 1708.

This was an ambitious publishing schedule for such a short period and, partly because of the later editions, Whiston's textbooks enjoyed a long afterlife. Whiston's *Euclid* was not only used by students at Cambridge, but through its many reprints circulated throughout Britain, Ireland, the continent and America. Even at rival Oxford students were encouraged to use Whiston's texts. They were also used at the dissenting Warrington Academy, while at Yale copies of Whiston's astronomical and physico-mathematical lectures were 'almost intirely worn out' through heavy use.[10]

Whiston's books were not only easier to understand than the *Principia*, but were also cheaper and more plentiful. Counting the first two editions of the *New Theory*, a full 4000 copies of Whiston's popular Newtonian texts were on the market by 1710 – possibly as much as ten times the number of copies of Newton's *Principia* then in circulation. A budding natural philosopher in the early eighteenth century was thus much more likely to encounter Newton through Whiston than in the work of the great man himself. The author of

Whiston's obituary recognized the popularizing role he played while professor, when 'he so clearly explained the Newtonian Philosophy in his *Mathematical and Astronomical Lectures*, which he then published, as to introduce into the University a noble System, which till then was not understood but by a very few'.[11] Similarly, W. W. Rouse Ball credited Whiston with opening up Newton's mathematical works 'to all readers'.[12] Whiston, above all others, rendered Newton easy.

CLERICAL, EXEGETICAL AND CHARITABLE ACTIVITIES

Throughout his tenure as Lucasian professor Whiston remained active as a clergyman. He preached regularly at local churches and, in addition, delivered Sunday-afternoon catechetical lectures in Cambridge's St Clement's. At a meeting of the charitable Society for Promoting Christian Knowledge in 1702, Whiston was nominated by William Lloyd as a Society correspondent, and soon afterwards was promoting in Cambridge one of its main projects: charity schools for poor children – a scheme to which Newton donated. Whiston was appointed steward of the schools, and as part of his duties gave religious instruction to about ninety of the children each month.

Whiston also began to produce works on theology proper in the first decade of the eighteenth century. In fact, his very first publication as Lucasian professor was on a theological theme: an Old Testament chronology and harmony of the Gospel accounts. In addition, Whiston devoted much time to prophetic researches, publishing a full-length commentary on Revelation in 1706, which was the the first of several works on biblical prophecy. Finally, to complete an already full schedule, and as evidence of his growing status as a scholar and theologian, he was also called upon to deliver the prestigious Boyle lectures for 1707. These lectures, established upon a bequest in the will of the illustrious Robert Boyle in 1691, were given annually by clergymen for 'the confutation of atheism', but were 'not to descend into controversies amongst Christians themselves'. But Whiston avoiding doctrinal controversy turned out to be increasingly unlikely. Choosing as his theme the evidence of fulfilled prophecy, a matter of no small moment in an age seeking to legitimate a Protestant succession, Whiston delivered them at no less prestigious a venue than St Paul's Cathedral in London.

The main purpose of Whiston's Boyle lectures was the development of an argument from prophecy for the reality of Providence and the existence of God. Whiston, like others of similar ilk, was motivated by a perceived rise in scepticism and atheism. In his *Accomplishment of Scripture Prophecies*, he amassed 300 examples of prophetic fulfilments to prove his case. It was an exercise in probability and induction, like the experimental lectures he was then giving at Cambridge with Roger Cotes. The future operation of prophecy was guaranteed on the basis of the careful record of past accomplishments, and the more fulfilments that could be accumulated, the greater the likelihood of future success. Thus an active and public theological career was to take up a significant proportion of Whiston's energy while mathematics professor.

HERESY AND EXPULSION FROM CAMBRIDGE

It was Whiston's tireless activity and resolute adherence to a policy of publicity in his theology, however, that led to his star falling from heaven. Whiston's infamous expulsion from Cambridge for his espousal of heresy provides the final chapter of his tenure as Lucasian professor. As with his natural philosophy, Newton was the main inspiration. By 1705 Whiston became aware through fellow Newtonian Samuel Clarke that Newton – a firm believer in the Word – rejected the doctrine of the Trinity, which, as the latter pointed out, was an accretion of postapostolic times. Shortly afterwards, having read Ellies Dupin's account of the first three centuries and Richard Brocklesby's antitrinitarian *Gospel Theism* (1706), Whiston converted to a position similar to Arianism. Not one to hide his lamp under a bushel, Whiston soon began to preach his new-found faith, even including antitrinitarian content in his catechetical lectures at St Clement's. Growing bolder, in August 1709 Whiston published his *Sermons and Essays*, a collection of writings that not only included the confident pronouncement that Christianity was originally Arian, but also a notice that further documentary evidence from the early Church would follow. An indication of Whiston's recent departure from orthodoxy is given in the errata, which give the correction '*dele* Three Persons and One God' for the trinitarian doxology in the sermon he had preached (ironically) at Holy Trinity Church in 1705.

The heresy watchdogs were quick to sniff error and one of those who denounced Whiston's book was the high-flying Dr Henry Sacheverell – virulent opponent of the toleration of dissent. Despite cautions from his friends, further provocations yet came from the Lucasian professor and in mid-October 1710, shortly after the Tories had won the general election, Whiston distributed proposals among the college heads for his *Primitive Christianity Reviv'd*. Within days of this daring action, a majority of the sixteen college heads assembled and moved to silence Whiston. Bentley was not among them. On the basis of his antitrinitarian lectures at St Clement's, Whiston was charged with contravening the forty-fifth of the university's Elizabethan statutes, which forbade the dissemination of teachings opposed to Anglican doctrine. If this statute had not been in place, the college masters could just as easily have convicted him under the Lucasian statutes, which specifically prohibited the mathematics professor from disseminating heresy. Despite his protestations that he should be given time to obtain legal advice and prepare a defence, Whiston was deprived of his professorship and banished from Cambridge on 30 October 1710.

But this was not the end of the matter. Whiston extended the controversy by continuing to publish heretical works after his expulsion. Since he was a priest, Whiston was subject to the discipline of the Convocation of Clergy, and, for the better part of the next four years, had to endure heresy proceedings directed against him. Many in the High Church were concerned that such an outspoken, high-profile heretic like Whiston would prove a dangerous influence on the people. For Whiston, the threat of prosecution was very real. In

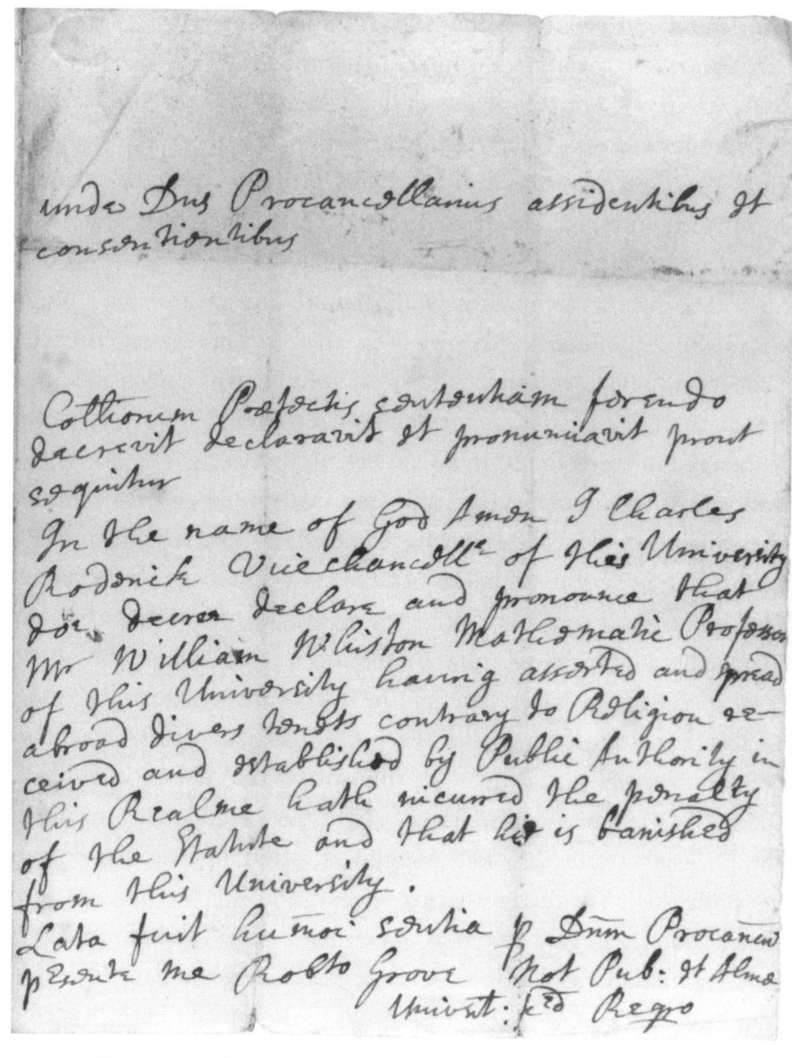
unde Dns Procancellarius assidentibus et
consentientibus

Collegiorum Praefectis sententiam ferendo
decrevit declaravit et pronunciavit prout
sequitur
In the name of God Amen I Charles
Roderick Vicechancellr of this University
doe decree declare and pronounce that
Mr William Whiston Mathematic Professor
of this University having assorted and spread
abroad divers tenets contrary to Religion received and established by Public Authority in
this Realme hath incurred the penalty
of the Statute and that he is banished
from this University.
Lata fuit huiusmodi sententia per Dnm Procancell
praesente me Robto Grove Not Pub: et Almae
Univert: Regro

Figure 18 Whiston's rustication: the Vice-chancellor decrees that Whiston, 'having assorted and spread abroad divers tenets ellipsis contrary to Religion received and established by Public Authority in this Realme hath incurred the penalty of the Statute and that he is banished from this University'.

addition to the desire of many to see him censured, some of Whiston's more reactionary enemies wanted him excommunicated and imprisoned. As it was, he was barred from receiving communion in his local parish church after his arrival in London from Cambridge.

Defenders of orthodoxy also responded with a torrent of printed attacks, warning that his heresy could infest the nation (and thereby guaranteeing that Whiston's name became a household word). Throughout all this, the irrepressible former professor remained steadfast to his convictions. What is more, he also began a tireless publishing campaign and openly advertised his plans to establish not only a 'primitive library' in his home, but also dissenting 'societies for promoting primitive christianity'. Whiston's endeavour was at once daring and inflammatory. His public insinuations that the foundation doctrine of orthodox Christianity had been corrupt since the fourth century cut to the very heart of the institution and hierarchy of England's established church. Nevertheless, an end to his legal trials came finally in 1714–15 with an Act of Grace that followed on the death of Queen Anne, the Hanoverian succession and the return of the Whigs to power.

Throughout his life Whiston viewed his ejection from the Lucasian chair as a grave injustice. In 1711 he laid out the case of his expulsion for all to see in a published account of his prosecution. Legal documents dating from the early 1710s show that he had managed to retain possession or at least some control of the Lucasian lands as late as 1714 – a manoeuvre that may have been intended as a lever to force his readmission. When his successor Nicholas Saunderson died in 1739, the seventy-two-year-old Whiston applied for reinstatement. It was all for naught. Whiston was to remain, as the title pages of many of his publications announce, 'sometime Professor of the Mathematicks at Cambridge'. And so it was that shortly after 30 October 1710, now a disestablished figure cut off from formal clerical and academic careers, Whiston left Cambridge with his family and began to ply his trades in London. His success in securing for himself there both a career and an income is testimony not only to his creativity and tenacity, but also to the very different world of the metropolis, where the culture of the coffeehouse, developing commerce and rising toleration combined to allow astute heretics to bypass the official structures of Church and university to market their religious and secular wares directly to the public.

METROPOLITAN TRIALS

Ejection from a Cambridge professorship in 1710 may have seemed like the nadir of fortunes. Whiston then appeared on the run, denounced from college pulpits to St Paul's and the House of Lords, his writings exhibited in one of the most notorious political trials of the century as proof of the worst vessels of heresy ever to sully the Church Established. The High Church demagogue Henry Sacheverell had been tried by the Lords for sedition – for exclaiming that the church was in danger from 'false brethren', thought to mean in both pulpit and privy council – but his sentence was so light that he was immediately hailed for having challenged the insidious spread of heresy and immorality. Soon Tories were jubilant and a Whig, low-church and Hanoverian succession looked increasingly precarious in the collapse of the ministry of Lord Treasurer Godolphin. William Whiston was to become in these circumstances one of those public figures – more minor than the Duke of Marlborough, also a champion of the Protestant cause, whose victories in battle during Queen Anne's wars contained the continental pretensions of Louis XIV – who also reflected the political alarms of his age. Amidst the rising tensions of church politics and impending crisis involving Hanoverian succession and a Stuart challenge, Whiston's reputation was more commensurate with his High-church nemesis the handsome Sacheverell, around whom Augustan political allegiances would be defined and controversy would swirl.[13] Nevertheless, for Whiston banishment transformed into a new beginning.

Cambridge politics rehearsed the national contest of High and Low. Whiston suspected that Tories lurked behind the charges laid against him in Cambridge, but his utterly uncompromising navigation of politics and religion defined a difficult course. Even before the Sacheverell affair emboldened his enemies, when black robes were on the scent of heresy, the political anxiety of Anne's ministers was palpable. White Kennet, the Whig Dean of Peterborough, believed that Whiston in 1709 was 'sunk at last into what may modestly be called Madness'.[14] Whiston's venture into scriptural exegesis reflected badly in the mirrors of Cambridge colleges and of Anne's court.

Whiston afforded one further opportunity for the battle to be fought. Much to the irritation of Tory sympathizers, his friend Richard Laughton at Clare Hall attempted in 1710 to break up a celebration of Sacheverell's supporters. While Whiston was shown the gate, he immediately contemplated, with Newton's great associate the Reverend Samuel Clarke, a petition to Convocation for a review of the doctrine of the Trinity. The idea was quickly abandoned, as Convocation was then 'grown too hot and too violent'. Politics was absolutely paramount. According to Whiston, Lord Treasurer Godolphin also attempted to prevent Clarke from publishing his *Scripture-doctrine of the Trinity* as 'that would make a great Noise and Disturbance'.[15] Clarke provoked a crisis over heresy by fuelling bonfires already lit. Godolphin had earlier let it be known that the publication of Whiston's *Account of the Primitive Faith, Concerning the Trinity and Incarnation* would spark great heat. Whiston, as ever, treated this notion with contempt: 'If we must never set about a Reformation in Church Affairs, 'till a Lord Treasurer sends us word 'tis a proper Time, I believe it would be long enough before that Time would come'.[16]

The best evidence of the political consequences of Whiston's scriptural speculations is found in the intersections of the Sacheverell affair. On the other hand, Sacheverell had no reservations at all in citing, at his trial before the Lords, Whiston's work as evidence of assaults on church doctrine. Whiston denounced Sacheverell's smokescreen. In a published speech, which he had intended but never gave to Convocation, Whiston complained 'that I and my Designs for the Advancement of Christianity, are immediately and without distinction, ranked with the known Enemies of revealed Religion, with their pernicious Contrivances for its Destruction'.[17] It was an affront to his every design that he should be attested a heretic. But the High Church charged forward, hastening to make the case of a 'church besieged' before the Queen's dying in defence of religion wishes ceased to make any further difference. High Churchmen felt it was increasingly urgent to ensure embarrassment to Whigs who would otherwise inevitably reap the spoils of a Hanoverian succession. Whiston was thus a godsend to the orthodox; he was living proof of apostates let loose, those whose private rhapsodies manufactured the mirages that would be magnified by the hysteria of Sacheverell and in the bonfires of his supporters. In the summer of 1710, just as Sacheverell had ridden in triumph to Magdelen College, Oxford, Whiston refused to bend under High-church hostility, not '*after* most frequent & sincere prayers to God for his direction; *after* an unbiass'd & thorough examination of all ye sacred & authentick writers of ye first times; *after* the hazzard of all my hopes & preferment, of my family, nay of my life it self in ye world'.[18]

Whiston challenged the High Church with its own rhetoric. Enraging his enemies, he cited Sacheverell before the Lords: 'Hear the words of a most noted Person, in a most noted Case, not wholly remote from that before us. "How hard are our Circumstances, if we must be punish'd in this World for doing that, which if we do not, we shall be more heavily punish'd in the next!" What a Condition are we in,' asked Whiston, 'if we are commanded to cry aloud, and spare not, to exhort, rebuke, in season and out of season, on the one hand; and prosecuted, imprison'd, ruin'd on the other!' In a brilliant inversion of the hostility directed against him he took lessons from a papist persecution at the very moment of extreme sensitivity to a Stuart, and Catholic, pretender plotting rebellion in the bosom of Louis XIV:

> Nor, methinks, did the Papists proceed much more absurdly, in not examining, but prosecuting the famous Galileo, for his Assertion about the Motion of the Earth; which was then a monstrous Heresy in Philosophy; and forcing him to recant it, than the Protestants in not examining, but prosecuting Mr. Whiston for his pretended Heretical Assertions now, and under the utmost Penalties the Law can inflict, to oblige him to recant the same.

Where he chose a comparison with Galileo, the contrast was with Sacheverell who had obtained a glorious reputation out of 'a seeming Persecution for Conscience', who was otherwise a man of 'no Character or Consideration at all'. And yet, Whiston asked, 'to what a height of Reputation did he [Sacheverell] rise, and what a prodigious Noise and Bustle has he made in the World?'[19]

As swords might cross, so too would paths. While Whiston removed to London in apparent disgrace, Sacheverell was sentenced before the Bar of the Lords to what was effectively a mild reprimand for such severe charges: his offending sermon was to be burned by the hangman and the doctor barred from the pulpit for three years. Whiston may well have felt, by comparison, that his own honesty had cost him far more. In Whiston's own parish of St Andrew's, Holborn, the High-church champion would surface once again. Whiston set up residence in Cross-Street, Hatton Garden and attended St Andrew's, where at the end of March 1713 Sacheverell ascended the pulpit amid Tory fireworks and rejoicing. In April, Whiston demanded that Sacheverell do him justice on the 'grievous Imputations which, at your Tryal, you laid upon me'. He insisted that Sacheverell, as the legal rector of the parish, ought to judge Whiston's doctrinal opinions 'impartially' in order that he might be readmitted to communion. Two years later, Whiston still had no reply.[20] But he did have an exclusion to reckon with. Sacheverell maintained his fierce hostility and even sought to rid Whiston from the pews at evening service, forcing him in 1719 to stand in the aisles on Sundays.[21]

COMETS AND CATASTROPHE

If some thought Whiston's constant challenges to the High Church an unseasonable madness, others applauded his candour and principle. In the London to which he escaped, Whiston would join his Cambridge experience in natural philosophy and astronomy with his passion for the original foundation of Scripture. In the thicket of metropolitan religion and court politics, words would be his weapons, old ones sought in Scripture, with truth discerned and sifted from ancient Christian texts. This had led him in 1696 to join the furious debates on the origins of the earth as he published his own *New Theory of the Earth* based on Newton's recent *Principia*. The Newtonian philosophy he later described as 'an eminent Prelude and Preparation to those happy

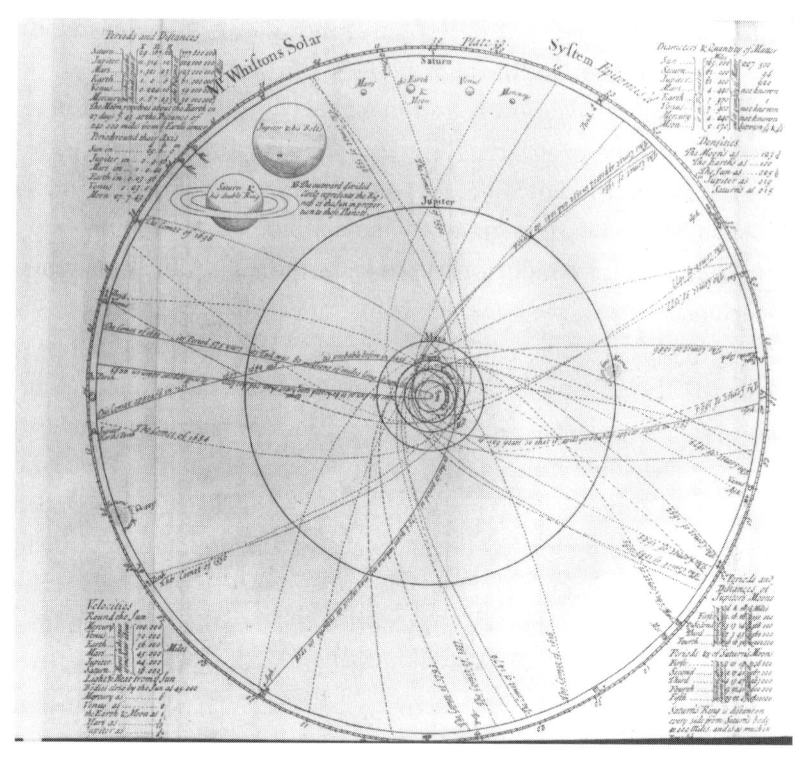

Figure 19 'Mr. Whistons Solar System *Epitomis'd*.' Comets find an essential place in Whiston's conception of the Newtonian universe.

Times of the Restitution of all Things, which God has spoken of by the Mouth of all his holy Prophets since the World began'.[22] It is the link between natural philosophical discovery, Scripture and prophecy that preoccupied him and allowed him thus to reflect on the fate of Galileo at the hands of a mendacious clergy.

Whiston watched for signs in the constellations and in the paths of comets. He would not be distracted by the High-church drone under the dome of St Paul's or in the light of Sacheverell's newly installed stained-glass windows in St Andrew's. Nothing would divert him from the notions that led to his *New Theory*, the intent of which was as far from the accusations of infidelity as all 'Mathematicks, Reason and Philosophy' could reach.[23] As he put it in his *Astronomical Principles of Religion*, which he dedicated to Newton and the Royal Society in 1717, 'Astronomy, and the rest of our certain Mathematick Sciences, do confirm the Accounts of Scripture'.[24]

Here Whiston was not alone; he followed not only Newton but he was also indebted to fellow clergy like Richard Allin of Sidney Sussex College, Cambridge. Their correspondence led to wonderful considerations by which much of the scriptural detail of the post-Noachian world might be explained:

> Noahs drunkeness seems to me a probable argument that there were no vines before ye deluge, & consequently that ye seed of ye vine among those of other Vegetables, came from ye Diluvian Comet. If there had been any vines before ye Deluge, it is not likely that the Antediluivians would have been ignorant of their great use: And if they had not been ignorant thereof, 'tis hard to conceive how Noah should not know it as well as ye rest: And if he had known it, he would not certainly (being so Good a man) have made himself drunk with ye wine proceeding from it, especially, so soon after that terrible Judgment upon a sinful world.[25]

With lunar calculations derived from Newton and the tables of the Astronomer Royal Flamsteed, their views shifted constantly as their calculations became increasingly refined.[26] Whiston, following the discoveries of Newton and Halley, determined that it was a comet that had caused Noah's flood and came to the general proposition that comets when passing through interplanetary spaces 'cause vast Mutations in the Planets, particularly in bringing on them Deluges and Conflagrations... and so seem capable of being Instruments of Divine Vengeance upon the wicked Inhabitants of any of those Worlds'.[27] Inevitably, conclusions drawn from nature ran foul of those who would defend a narrower orthodoxy.

HEAVENLY SIGNS

The political imperative tied to Scripture doctrine was not eased by the bombast from many a High-church pulpit. The high stakes could be measured by theological damnation raining down on Whiston from Convocation, from bishops intent on protecting the power of Bench in the Lords, and in the careers of parish rectors, like Sacheverell reserving pews in Holborn. Whiston was a match for it all. In a draft letter to Archbishop William Wake, Whiston in 1721 recounted two observations: 'One made by an honest Italian, who in ye sincerity of his soul[?] turned Protestant, & came into England, in expectation of finding true religion & piety to flourish here . . . [but found instead] '*No Religion in Italy: No Religion in England: All Politicks, Politicks*'. Given the history of Galileo with which Whiston identified, this was frank, but no more brutal than the second reflection made by one he would not name: '*Happy*, said ye person, *is that Man who is not made a worse Christian by being made a Bishop and Thrice Happy that man who is not made a worse Christian by being made an Archbishop*'.[28] No amount of dissembling by those elevated to pulpits or a Bench would divert Whiston from his purpose.

Whiston was truly convinced that method made the difference. It was essential to follow 'the learned Philosophers and Astronomers of the present Age' who eschewed hypotheses and metaphysics for the mathematical certainty and experimental proofs out of which he would construct his second career. Confidence in religious interpretation would only emerge 'if once the Learned come to be as wise . . . as they are now generally become in those that are Philosophical and Medical and Judicial; if they will imitate the Royal Society, the College of Physicians, or the Judges in Courts of Justice'.[29] Certainly, the fact that Whiston could cite the highly controversial General Scholium of the second edition of Newton's *Principia* (1713) in support of his own scriptural exegesis would only have embarrassed the great autocrat.[30] This was far more than an assertion of the Newtonian assault upon metaphysics and hypotheses; it meant the very marriage of Scripture and natural philosophy that Newton had proposed in 1713.

Just as Whiston sought to explain deluge and conflagration by the intercession of a comet, so he held that the reading of astronomical portents or prophetical spirits could not be simply attributed to charlatans, the frightened and the bewildered. Prophecy and portents could be seen or imagined all too frequently in the night sky. Thus, Whiston's explanations of scripture also incidentally constructed a broad appeal for his astronomy. But, more importantly, meteorological and astronomical phenomena demanded interpretation, especially if dire portents were revealed in the skies at the particularly tense crisis of the Hanoverian succession. Edmond Halley felt compelled to address the politically inspired in 1715, lest crowds be induced to run amok in the streets as at the unhappy prosecution of Sacheverell in 1710, which had then resulted in major disturbances in London with the burning of dissenting meeting houses, attacks on the houses of prominent Whigs and the reading of the Riot Act. Five years before the Jacobite rising, however, the crown had not been immediately at stake. As Alice Walters points out, in the circumstance of open rebellion in 1715, Halley argued that it was therefore imperative, through natural and easy explanation, that the eclipse 'may give no surprize to the People, who would, if unadvertized, be apt to look upon it as Ominous and to Interpret it as portending evill to our Sovereign Lord King George and his Government, which God preserve'.[31] This was a powerful argument for the dissemination of natural and experimental philosophy and for the demonstration of astronomical phenomena, both of which Whiston successfully mastered along with Scripture prophecy. No less the case than after the Civil War, when the Royal Society had come into being, politics, religion and natural philosophy were not easily segregated. Thus, Newtonian natural philosophy and scriptural exegesis were preservatives for public order. Whiston would remark in September 1716, precisely when diplomats on the continent were gathering French promises of armies for the restoration of the Stuarts:

> I dare venture to affirm, with St. Peter, that . . . a more sure Word of Prophecy, is one of the strongest Arguments for the Truth of the Scriptures, of all other whatsoever . . . And I declare I am so far from seeing any Reason *from the present Posture of Affairs in the*

> *World*, to doubt the Completion of those which remain, even for the main, as I have expounded them, that I rather find great Cause to believe, that the Prophetick Scheme begins to clear up apace.[32]

The fate of the Protestant succession rested in the proper understanding of scriptural prophecy.

ARMIES OF THE NIGHT

Both prophets and astronomers contended for believers. In the great political contest brewing throughout the spring and summer of 1715, the caldron of Jacobite emotions periodically rattled the metropolis. The Whigs were clearly unsettled, although their greatest fear lay in the plans of the Pretender and his continental allies. Daniel Defoe meanwhile watched 'Gown Incendiaries' incite Tory crowds and Whig clergymen were on alert. Thus, Whiston's fellow Newtonian, John Harris, hounded one Jacobite sympathizer straight into the arms of Sacheverell in Holborn. In such highly charged circumstances, any discussion of prophetic signs was thus bound to attract the attention of natural philosophers who feared politics lurking in the night sky. Consequently, when groups like the so-called French Prophets revealed themselves in Anne's reign, the whiff of popular enthusiasm – which many associated with the radicals and regicides of the Civil War – provoked deep concern. Here too, Whiston took up the challenge. The prophets were Huguenot refugees from the Cevennes although they soon found themselves with English converts. They made a great storm in London in Anne's reign and, in 1707, in his Boyle lectures on *The Accomplishment of Scripture Prophecies*, Whiston attacked what he believed to be their pretences. The Prophets were not deterred; in fact they had attracted the Swiss mathematician and young friend of Newton, Nicholas Fatio de Duillier. Nevertheless, Whiston persisted in attempting to show them the error of their ways, informing them in 1713 that he thought the source of their inspiration to be both supernatural and evil. The dangers were magnified when the Hanoverian crown hung by a thread in 1715.

It is no coincidence that purveyors of a rational astronomy were perceived as agents of the crown and political stability. Such, of course, had been the message of the Restoration Royal Society after a generation of inspired insurrection and bloody regicide. The Civil War had defined religious enthusiasm as incendiary. Few were better prepared than Halley and Whiston to exploit this notion when the heavens literally darkened immediately after the Hanoverian succession and, in the depth of the Jacobite rebellion, were seemingly as full of portents as the streets with prophets. As Patricia Fara has shown, the Royal Society debated the celestial spectacles but it was also an opportunity for Whiston to take employment in the Censorium of the Whig politician and pamphleteer Richard Steele – to give lectures on astronomy and denounce those who foretold disaster.

Events in the skies over early eighteenth-century London, and their convergence with the uncertain survival of the Hanoverian monarchy, were exceptionally crafted for the revival of prophecy. And the French Prophets took full advantage to read in the heavens the survival and resurgence of the Protestant enemies of papist pretenders. A diary of the revelations and rites of the French Prophets tells part of the tale of the visions Whiston was compelled to vanquish. Comparing the sights to the 'many strang [*sic*] and surprizing Lights' seen before the Civil War, the interpretations of the Prophets resonated with scenes of blood and swords:

> [There] appeared in ye Sky two great Armies wch contained certain Thousands of Men and Horse, these seem'd fiercely to Encounter each other, and ye Battle seem'd long and doubtfull, as if fortune was in Debate wth her self, not knowing to whom she s^{hd} give victory.[33]

The rebellion of marauding armies would not succeed. From the mouths of the Prophets, full of the Protestant cause, came the voice of the Holy Spirit: 'I will destroy them, their detested carcases shall fall like dung upon ye earth. Behold the butchers of the Earth are gathering together, and they shall be slaughtered as oxen fitted for destruction'.[34]

The Pretender's plans collapsed, as the Prophets foretold. But in London the crisis brewed still in attacks on dissenting meeting houses and burial grounds and rumours of the rising of a second Sacheverell among the brew houses of Southwark.[35]

THE PUBLIC ACHIEVEMENT OF NATURAL PHILOSOPHY

When Whiston removed to a London full of troubles, with High-church preachers and Jacobite sympathizers watching warily, he brought with him a biblical certainty and a philosophical practice. His natural philosophical lecturing experience at Trinity enabled him to manufacture a new career joining the new experimental entrepreneurs then emerging in London in the reign of Anne. From 1703 at least, the mathematician and later fellow of Trinity Roger Cotes had been aware of Whiston's early efforts to discern the shape of the earth at creation. By the spring of 1707, Cotes and Whiston were jointly presenting a course of twenty-four lectures in experimental philosophy – following ventures well underway in London, most notably by James Hodgson and by the Jacobite-baiting Reverend John Harris. These provided the prototype for lectures that Whiston undertook with the instrument-maker Francis Hauksbee, Sr in London during 1712 and 1713.[36] Very shortly thereafter, with the crisis building in Convocation over heterodoxy, the Whigs Joseph Addison and Richard Steele arranged for Whiston to give a series of lectures on mathematics at the Whig haunt of Button's Coffee-house in Covent Garden. After the succession, he cemented these connections by his engagement in Steele's Censorium at York Buildings on the Thames.

It was in the Censorium that Whiston truly came into his own. No doubt Steele regarded him as one willing to shatter the pretences of prophets which, if sufficient public disturbances arose, might endanger the crown. In April 1715 Whiston presented a lecture and a demonstration of the eclipse with smoked glass in Covent Garden. Almost exactly a year later he was in the Censorium explaining the aurora. His concern was with both the natural occurrence and the supernatural consequence. And Whiston became more visionary with age, later attributing to the 1716 aurora God's watch on the rise of heresy. Toward the end of his life Whiston came to believe that the 1715 solar eclipse was likely 'a Divine signal for the End of the over-bearing Persecution of two of the ten idolatrous and persecuting King-doms, which arose in the fifth Century, in the Roman Empire, the Britains and the Saxons'.[37] Eclipses, of which there were four from 1715 to 1748, formed a staple of Whiston's lecturing ventures. Before and after the April 1715 solar eclipse, the enterprising quondam pro-fessor earned £120 not only directly through lectures and the sales of books and charts, but also in secondary income from patronage gifts. Whiston himself affirmed that this money 'was a very seasonable and plentiful supply', which maintained him and his family 'for a whole year altogether'.[38] In May 1724, he once again made a handsome penny showing the eclipse to those who paid one-half guinea at a room at the Horse Guards. As one astute commentator observed, Whiston got 'his Bread out of the *Stars*'.[39] For the entrepreneurial, astronomical phenomena had become very lucrative indeed.

Whiston charted his obligations to crown and Christianity by explicating the heavens. He also shamed his enemies by persistence and by reminding them of Galileo and of his own service to the na-tion in a scheme for improved navigation. In 1713 he had complained of his persecutors within the Church while informing them of 'his Friends and his own noble Discovery of the Longitude at Sea, (which they are now laying before the Publick, and for which, if it succeed, the Publick cannot but highly esteem and reward them, and look very meanly on those that endeavour to crush any noble Discoveries, for the Benefit of the Publick)'.[40] This was hardly likely to dissuade his en-emies. Whiston, and particularly his mathematical friend Humphrey Ditton, were behind a proposal for a reward of £20 000 for a method of determining longitude at sea. Moreover, it was a Whig enterprise from start to finish. Whiston and Ditton's notion to fire rockets in the air and to determine longitudinal position by difference of the travel of the flash of light and the sound of the explosion was theoret-ically possible but essentially impractical. Whiston tried the method

Figure 20 Hogarth depicts Whiston in Bedlam [1735]. In the final plate of 'The Rake's Progress' the Rake is thrown into Bedlam where he encounters myriad 'lunaticks'. On the back wall one inmate is sketching Whiston's method for determining longitude by means of firing a shell from a mortar.

by firing mortars from Hampstead Heath in the autumn of 1714, hoping to receive reports from observers like Roger Cotes in Cambridge 60 miles away. Cotes saw nothing and indeed he expected little unless Whiston printed a book of instructions for witnesses on how to draw a meridian and take the bearing of the light.[41]

Hampstead was not sufficient for Whiston's tests. He proposed applying to the government for money to set up a gun at Land's End near the shore to demonstrate his method and claim the reward.[42] In 1717 he was back again firing mortars over Hampstead and at Black Heath near the Royal Observatory witnessed by Flamsteed.[43] In 1721 he secured £470 from the Board of Longitude for a method of determining longitude by the dipping needle, and in 1741 he successfully petitioned the Commissioners for a full £500 to be put to use in mapping magnetic variation near the coasts.[44] In any case, the longitude was a wonderful opportunity for Augustan wits – such as the engraver William Hogarth – who already thought Whiston mad. It certainly did not take long for Whiston's mortar project to be tied to his scriptural speculations. As one rival for the reward put it, 'never an Arrian in this Kingdom, shall seduce me to hold with him in so giant an Error'.[45] Whiston was thus widely seen not only as a Latitudinarian ready to cut a broad swath through the rigours of Anglican doctrine, he was a Longitudinarian as well, firing mortars to shake loose the £20 000 reward.[46]

CONCLUSION: MARTYR OF PHILOSOPHY AND RELIGION

To Whiston, and to his critics, all was of a piece. Compared to his millenarian vision, however, High-church frustration was a myopic rage fixed on their immediate failure and a festering enmity which would not finally be shattered until the Stuart catastrophe of Culloden in 1746. Resentments played themselves throughout the reigns of Anne and the first two Georges. Challenges thrown down by Whiston were no more to be ignored than the demagoguery of Henry Sacheverell or than the Jacobite plots of bishops like Francis Atterbury, both of whom felt Whiston's sting. Whiston, the apostate, became more adamant with age, deeply disappointed but not cowed by the tattered, tradition-laden rigidity of the Established Church. Although Whiston was careful to portray himself as a victim of university and Church, there were successes even in martyrdom. With his textbooks, astronomical

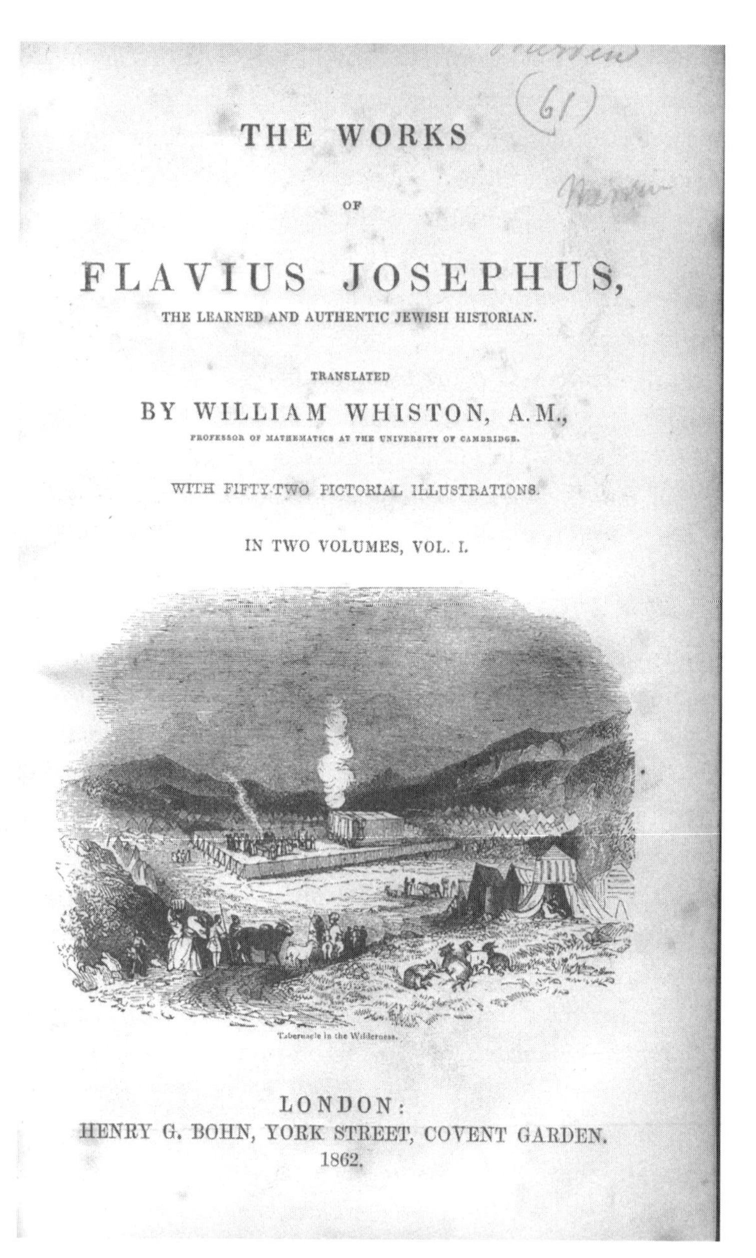
THE WORKS

OF

FLAVIUS JOSEPHUS,

THE LEARNED AND AUTHENTIC JEWISH HISTORIAN.

TRANSLATED

BY WILLIAM WHISTON, A.M.,

PROFESSOR OF MATHEMATICS AT THE UNIVERSITY OF CAMBRIDGE.

WITH FIFTY-TWO PICTORIAL ILLUSTRATIONS.

IN TWO VOLUMES, VOL. I.

Tabernacle in the Wilderness.

LONDON:

HENRY G. BOHN, YORK STREET, COVENT GARDEN.

1862.

Figure 21 The title page to one of countless editions of Whiston's bid to restore primitive Christianity. His 1737 translation of the ancient Jewish historian Josephus not only became the most successful work of any eighteenth-century churchman, but, by the end of the twentieth century, was one of the most-published books in history.

charts and the course of experiments he ran with Francis Hauksbee Jr at Crane Court, Whiston would have had little trouble not only matching, but even substantially exceeding, his £100 Lucasian salary. What is more, annuities beginning in 1727 from Queen Caroline and in 1738 from the Whig statesman Joseph Jekyll gave Whiston greater freedom to pursue the goals of his theological programme. By extending to London the efforts he began at Cambridge to reveal Newton's philosophy to the world, Whiston also extended his natural philosophical career and income. But Whiston was increasingly concerned with publicizing his own reading of Newton's religion as well. Thus, publications for primitive Christianity continued, including his 1737 translation of the ancient Jewish historian Josephus, which not only went on to become the most successful work of any eighteenth-century churchman – high or low – but by the beginning of the twenty-first century had established itself as one of the most-published books in history.

Toward the end of his life Whiston had more to reveal, as in 1742, about meteors and the longitude, along with 'Models and Schemes of the Tabernacle of Moses, and of Solomon's, Zerobabel's, Herod's, and Ezekiel's Temples at Jerusalem'. Firm in his belief that the millennium would come by 1766, Whiston devoted a growing proportion of his time to proclaiming the swiftly approaching apocalyptic denouement. In 1748, he took his ventures to the fashionable diversions of Tunbridge Wells. There the ageing Whiston, a spectacle in his own right, was seen 'showing eclipses, and explaining other phenomena of the stars, and preaching the millennium, and anabaptism . . . to gay people, who, if they have White teeth, hear him with open mouths, though perhaps with shut hearts; and after his lecture is over not a bit wiser'.[47] Whiston received far less than the kind of veneration he believed his due. Upon his death in 1752 in his eighty-fifth year, his obituary recalled his 'sacrificing all his worldly advantages & expectations . . . to the professing & defending such religious sentiments, as, upon the result of the most careful enquiries, appeared to him to be the truth, this characteristick, even those who most opposed his opinions, have ever allowed him'.[48] This was not quite the case. To

Whiston, the censure of his critics was nothing to the greater truth of scripture and nature combined.

Notes

1 Flamsteed to Sharp, 31 March 1702; *The Correspondence of John Flamsteed, the First Astronomer Royal*, Eric G. Forbes, Lesley Murdin and Frances Willmoth (eds) (Bristol, 1997), vol. 2, p. 929.

2 Whiston, *Memoirs of the Life and Writings of Mr. William Whiston*, 2nd edn (London, 1753), vol. 1, p. 32.

3 Whiston, *Memoirs*, vol. 1, p. 38.

4 Whiston, *Sir Isaac Newton's Mathematick Philosophy More Easily Demonstrated* (London, 1716), p. 1.

5 I. Bernard Cohen (ed.) 'Introduction', in Whiston, *Sir Isaac Newton's Mathematick Philosophy More Easily Demonstrated* (New York, 1972), p. vi.

6 Whiston, *Newton's Mathematick Philosophy*, p. 2.

7 Whiston, *Newton's Mathematick Philosophy*, p. 408.

8 Smith, Preface to Cotes, *Hydrostatical and Pneumatical Lectures by Roger Cotes A.M.* (London, 1738), sig. $\mathrm{A3}^{\mathrm{r}}$.

9 Whiston, *Memoirs*, vol. 1, p. 115.

10 Stiles to Whiston, 9 May 1751, Yale College.

11 *The Covent-Garden Journal*, 29 August 1752.

12 W. W. Rouse Ball, *A History of the Study of Mathematics at Cambridge* (Cambridge, 1889), p. 85.

13 Nicholas Rogers, *Whigs and Cities. Popular Politics in the Age of Walpole and Pitt* (Oxford, 1989), p. 367.

14 White Kennett to Arthur Charlett, 29 January (1709[?]), Bodleian Library, Oxford, MS Ballard 7, f. 122.

15 [Whiston], *An Account of Mr. Whiston's Prosecution at, and Banishment from, the University of Cambridge* (London, 1718), pp. 1, 27; William Whiston, *Historical Memoirs of the Life of Dr. Samuel Clarke* (London, 1730), pp. 15, 27.

16 Whiston, *Memoirs of Clarke*, pp. 179–80.

17 William Whiston, *An Account of the Convocation's Proceedings with Relation to Mr. Whiston* (London, 1711), p. 54.

18 Whiston to Robert Nelson, 31 July 1710, Leicestershire Record Office (LRO) Conant MSS DG11/DE.730/2/12–13.

19 [Whiston], *Several Papers Relating to Mr. Whiston's Cause Before the Court of Delegates* (London, 1715), pp. 4, 15, 24.

20 [Whiston], *Several Papers*, pp. 148–52.

21 William Whiston, Jr. to Sarah Barker, 11 February 1719, LRO Conant MSS DG11/DE.730/2/43.

22 Whiston, *Memoirs* (London, 1749), p. 38.

23 Whiston, *An Account of the Convocation's Proceedings with Relation to Mr. Whiston* (London, 1711), p. 60.

24 Whiston, *Astronomical Principles of Religion, Natural and Reveal'd* (London, 1717), pp. 105, 267.

25 Allin to Whiston, 19 March 1700, LRO Conant MSS DG11/DE.730/2/5–6.

26 Allin to Whiston, 29 August 1707, LRO Conant MSS DG11/DE.730/2/10–11.

27 Whiston, *Astronomical Principles of Religion*, pp. 23, 74–5.

28 Whiston to Wake, 18 May 1721, LRO Conant MSS DG11/DE.730/2/48–9.

29 James E. Force, 'Newton's "sleeping argument" and the Newtonian synthesis of science and religion', in Norman J. W. Thrower (ed.), *Standing on the Shoulders of Giants: A Longer View of Newton and Halley* (Berkeley, 1990), pp. 109–27; esp. 115, 120.

30 For example, Whiston, *Astronomical Principles of Religion*, pp. 237–40.

31 Quoted in Alice N. Walters, 'Ephemeral events: English broadsides of early eighteenth-century solar eclipses', *History of Science*, 37 (1999), 11.

32 Whiston, *Astronomical Principles of Religion*, 296. Dated 1 September 1716. Our italics.

33 Osler Library, McGill University. John Lacy diary, ff. 168v–169r.

34 *Ibid.* f. 162 v.

35 *Ibid.* f. 150 (25 February 1716).

36 Whiston, *Memoirs* (1749), 135–6; Roger Cotes to Rev. John Smith, 11 December 1703, Trinity College, Cambridge, MS 4.42, no. 1.

37 Whiston later reflected on the lack of a hypothesis concerning the frequency of the northern lights, noting that there had been more meteors than usual as well as northern lights from 1715 to 1736. As he put it in 1736, where mechanical causes failed, 'we must therefore have recourse to such as are Immechanical, as we have in Cases everywhere'. See British Library copies (873.l.2 [1], [2]) with MS notes by Whiston, of

Whiston, *An Account of a Surprizing Meteor, Seen in the Air, March the 6th, 1715/16, at Night*, 2nd edn (London, 1716), 67, and Whiston, *An Account of a Surprizing Meteor, Seen in the Air March 19. 1718/19 at Night*, 2nd edn (London, 1719), pp. 33–4.

38 Whiston, *Memoirs* (1753), vol. 1, pp. 204–5.

39 *The Craftsman*, 19 July 1729.

40 [Whiston], *Several Papers*, pp. 26–7.

41 Whiston to Cotes, 26 November 1714; no. 9, Trinity College, MS R.4.42, no. 8; Cotes to Whiston, 2 December 1714, no. 11; Whiston to Cotes, 7 April 1715. Also LRO Conant MSS DG11/DE.730/2/28.

42 Whiston to Samuel Barker, 27 January 1715, LRO Conant MSS DG11/DE.730/2/29.

43 Cambridge University Library, RGO 1/37, f. 150r (n.d.). Notes in Flamsteed's hand on circular from Whiston.

44 See Royal Society MSS. Journal Book (C), XII, 219, 22 March 1722; Whiston to the Commissioners of the Longitude, 28–29 September 1730, LRO Conant MSS DG11/DE.730/2/57–8; Whiston to his son[George Whiston], 23 February 1742, LRO Conant MSS DG11/DE.730/2/104.

45 R.B., *Longitude to be Found out with a New Invented Instrument, both by Sea and by Land* (London, 1714), pp. 15, 26–27.

46 See Jeremy Thacker, *The Longitudes Examin'd* (London, 1714), p. 2.

47 Samuel Richardson to Miss Westcomb, *c.* 1748, in Anna Laetitia Barbauld (ed.), *The Correspondence of Samuel Richardson* (London, 1804), vol. 3, pp. 318–19.

48 LRO Conant MSS DG11/DE.730/2/149, transcript from the *General Evening Post*, Tuesday 25 August 1752.

4 Sensible Newtonians: Nicholas Saunderson and John Colson

John Gascoigne

School of History, University of New South Wales, Sydney, Australia

A telling anecdote is recorded by Martin Folkes, Cambridge graduate and president of the Royal Society (1741–53): 'After Sir Isaac printed his *Principia*', Folkes recalled what was said about Newton 'as he passed by the students at Cambridge'. Apparently one witty undergraduate remarked, 'there goes the man who has writt a book that neither he nor any one else understands'. The task of rendering the abstruse *Principia* into the common coin of Enlightenment Britain largely fell to the Lucasian professors who followed Newton: William Whiston, Nicolas Saunderson and John Colson. Though the period when Saunderson and Colson held office as Lucasian professors (1711–39 and 1739–60) was not a heroic one in the annals of Cambridge University, it was, none the less, an era when the mathematical tradition for which the university became so well known was consolidated. It was, moreover, a period when the Newtonian version of the workings of the natural world still had rivals who had to be vanquished, an intellectual contest in which the Lucasian professors played a major role. Just as Cambridge's divinity professors saw it as their function to combat heretical assaults on the Church of England so, too, the university's mathematical professors saw their duty as defending the pure gospel of Newtonianism. The analogy extends to the fact that, just as Anglican orthodoxy required the rejection of deviant forms of Christianity, so the Cambridge professors played an important role in

From Newton to Hawking: A History of Cambridge University's Lucasian Professors of Mathematics, ed. Kevin Knox and Richard Noakes. Published by Cambridge University Press.

establishing what constituted the canonical version of the Newtonian system of the world.[1]

When the blind Nicholas Saunderson was elected to the Lucasian chair in 1711, the university was still digesting the great Newtonian achievement. Newton's formidable mathematical masterpiece took some time to be embraced by the great minds of Europe so it is not surprising that it was a slow and arduous process to put it into a form which undergraduates or even, indeed, college tutors could understand. Making the great peaks of the *Principia Mathematica* accessible to gownsmen was a weighty burden that was put on the shoulders of the early-eighteenth-century Cambridge Lucasian professors since Newton himself, always anxious to avoid controversy, reportedly 'made the *Principia* abstruse to avoid being baited by the little smatterers in mathematics'.[2]

The first substantial Cambridge Newtonian textbooks of natural philosophy were produced by Newton's successor as Lucasian professor, William Whiston, who held the office from 1702 to 1710. But reorientating a curriculum which had, since the high Middle Ages, been dominated by Aristotelian philosophy was, inevitably, a slow process. True, the traditional scholastic curriculum was in rapid decline following the achievements of the Scientific Revolution of the seventeenth century. Moreover, Cartesian philosophy had made rapid progress in late-seventeenth-century Cambridge partly because, as an all-embracing philosophical schema deduced from first principles, it readily filled the void created by the waning of scholasticism. The result, however, was that Newton's early disciples in Cambridge had to fight a battle on two fronts: demolishing the ailing Aristotelianism and combating the influence of Cartesianism which was still vigorous on the continent.

The task was further complicated by the character of Cambridge teaching. The foundation of the Lucasian professorship in 1663 and the lustre of its first two incumbents, Barrow and Newton, indicate that the professorial mode of instruction still played a major role in late-seventeenth-century Cambridge. Professorial teaching was

Figure 22 Portrait of Saunderson. Engraving of Saunderson from the frontispiece to his *Elements of Algebra*, published posthumously in 1740.

particularly valuable in areas such as mathematics or the recent developments in natural philosophy which the college tutors did not feel equipped to teach. The fact that Newton himself did 'of times ... in a manner for want of Hearers, read to the Walls' is, however, an indication that mathematics remained at the margin of the undergraduate curriculum, a situation not improved by the fact that Newton's lectures were often drafts of his *Principia*. This helps to explain the impact of William Whiston, the third Lucasian professor, whose lectures and books provided one of the few means by which Cambridge undergraduates could come to terms with the Newtonian natural philosophy. But at both Oxford and Cambridge (in contrast to the Scottish universities) the tide was flowing away from professorial instruction towards that based on college tutors.[3]

Throughout Cambridge's history there has always been a tussle for dominance between the colleges and the central university. In the Middle Ages the university was largely dominant but from the Reformation onwards the colleges were in the ascendant (though in our own times the balance is beginning to swing back towards the university). In the late seventeenth century the lingering remains of the traditional curriculum with its established round of disputations and other exercises still lent some plausibility to the notion that the central university set a course of studies controlled by the professors. But the eighteenth century was a period that favoured the colleges which, with their wise investments in property, were becoming wealthier. This wealth, coupled with decreasing numbers of students, meant that they could easily take over most undergraduate instruction, thus leaving the professors with little to do.

During Saunderson's tenure, there was still some demand for professorial lectures on natural philosophy, particularly since many of the college tutors did not feel qualified to provide instruction in the field. But, to judge by student guides such as Daniel Waterland's influential *Advice to a Young Student*, college instruction by the 1730s had largely caught up with what the Lucasian professors could offer. This state of affairs helps to account for the faint impression made on the university by the lectures of the fifth Lucasian professor, John Colson (1739–60), for his lack of success in attracting students cannot be entirely put down to his personal inadequacies as a teacher. Nor did Colson's mathematically more distinguished successor, Edward Waring (Lucasian professor, 1760–98), fare much better in attracting students for, as the pedagogue Samuel Parr diplomatically remarked, his 'profound researches were not adapted to any form of communication by lectures'.[4]

The trajectory of the eighteenth-century Lucasian professorship is then something of a downward spiral. At the beginning of the century it could bask in the lustre imparted to it by Barrow and Newton and its impact on the instruction of the university was considerable thanks to the exertions of Whiston and, to a lesser degree, Saunderson. By Colson's time, professorial instruction in mathematics and natural philosophy was both less necessary and less valued as the college tutors assumed control over a curriculum increasingly dominated by mathematics and natural philosophy. To some extent, Colson created a niche for himself by publishing editions or translations of textbooks on mathematics or natural philosophy, while Waring's mathematical publications partially justified his low-key status; but neither had a significant impact within Cambridge. In a university dominated by colleges where the student body (such as it was) was in the hands of tutors, professors often seemed like anachronistic pedants.

But this is to take the long view: when Saunderson was elected following Whiston's expulsion, he had a clear sense of mission. Like Whiston, he sought to win over both the undergraduates and the senior gownsmen to the pure gospel of Newtonianism. He had come to know the 'Great Man' himself and contemporaries attributed his advancement within Cambridge partly to the result of Newton's good offices.

How far Newton involved himself actively in the election for the Lucasian chair is doubtful. Saunderson's chief rival was Christopher Hussey, a Trinity man enjoying the powerful backing of Richard Bentley, the Master of Trinity and a master of academic politicking.

Naturally, Bentley sought to enlist Newton's support: Bentley reckoned that Newton had been indebted to him since 1692 when Bentley had given his celebrated Boyle lectures. For these lectures, *A Confutation of Atheism,* were among the first major works to publicize the *Principia* and to establish it as a resource for defending the existence of God, as well as the established Anglican order. But Newton was not to be cajoled, even by the formidable Bentley, and he responded lukewarmly: 'I have made a resolution not to meddle with this election of a Mathematick Professor any further then in answering Letters & have given this answer to some who have desired a certificate from me'.[5]

None the less, Newton's letter does leave open the possibility that he did write on behalf of Saunderson, and perhaps did so with sufficient warmth to sway the electors. The voting record indicates, however, that Hussey had the support of those in Cambridge most closely associated with Newton, including Bentley and Sir John Ellys, the Master of Caius College. Despite this strong support for Hussey, Saunderson was elected by six votes to four. Bentley's ruthless tactics had made him many enemies in Cambridge and it is possible that his identification with Hussey may have hindered, not assisted, his chosen candidate's prospects. Or, Saunderson's victory might be attributed to his discretion in religious matters: Halley, for example, is supposed to have remarked that 'Whiston was dismissed for having too much religion, and Saunderson preferred for having none'.[6]

For the blind and humbly born Saunderson to gain such a post was a remarkable achievement. Following his birth in Thurlston, Yorkshire in 1682, the contraction of smallpox at age one cost him not only his sight but his eyes. Notwithstanding this handicap, his father, an exciseman, taught him a little arithmetic 'which he took very readily and made surprizing progress'. His interest in mathematics was also stimulated by a Richard West of Undebank, 'a gentleman of learning'. Following schooling at the Penistone grammar school he spent a year or two at the dissenting academy, Christ's College, Attercliffe (Sheffield), despite the fact that the master 'forbad [his pupils] the

Mathematicks, as tending to scepticism & infidelity'. Nevertheless, there must have been some mathematical instruction at Attercliffe since we are told that many students 'made a considerable progress in that branch of Literature', even though Saunderson did not find the traditional curriculum of logic and metaphysics to his taste.[7]

This association of mathematics with infidelity was, however, to prove prophetic since Saunderson appears to have been a Deist with little commitment to revealed religion – in so far as we can gauge his religious views, in an age when, as the Whiston case testified, it was unwise to publicize one's infidelity. That industrious early-nineteenth-century historian of Cambridge, George Dyer, remarked that Saunderson was 'no friend to Divine revelation', though Dyer added that he was said to have taken Holy Communion on his deathbed.[8]

Although probably an exaggeration, Halley's quip concerning the contrasting religious views of Whiston and Saunderson conveys an essential truth about eighteenth-century Cambridge: as an arm of the established Church of England it expected its members not to challenge the central tenets of Anglicanism. The radiance of the Enlightenment, however, was sufficiently strong to stifle attempts at enforcing orthodoxy when unbelief did not take an overtly public form. Saunderson does appear, however, to have tested the limits of Cambridge's tolerance: his entry in the late-eighteenth-century *Biographia Britannica* praises him for his vivacity and honesty, but adds that since he was predominantly remembered for his 'indulgence of women, wine, and profane swearing to such a shocking excess, ... he did more to harm the reputation of mathematics than he did good by his eminent skill in the science'. One wonders, too, what Saunderson's father-in-law, William Dickons, rector of Boxworth, Cambridgeshire, made of the less than devout husband of his daughter, Abigail. Yet, as with Charles Darwin's father, paternal unbelief did not prevent him from placing his son in the Church, for Saunderson's son, John, took orders after graduation. As a clergyman and a fellow of Peterhouse, John also had the leisure to help in the publication of his father's *Algebra*.[9]

After his early introduction to mathematics at school and at Attercliffe, Saunderson became ever more captivated by mathematics through his personal study (which he turned to some practical effect by assisting his father with excise calculations). He aspired to study at Cambridge but this appeared to be a goal denied him by his slender resources. However, in 1707 he made his way there by more indirect means, acting as the personal tutor to Joshua Dunn, the son of a wealthy mercer of Halifax and Attercliffe, who enjoyed the high status of a fellow-commoner at Christ's (which entitled Dunn to dine with the fellows, exemption from undergraduate examinations and higher fees). Thus began a long association by Saunderson with Cambridge University – his high reputation as a private tutor eventually obtaining for him an honorary MA (1711) which, subsequently, was followed by an LL.D. (1728). Christ's College remained his home until 1723 when he moved into a house in Cambridge, where he lived with his wife.

As a private tutor he devoted himself to the enterprise which dominated his Lucasian professorship: winning over the undergraduate body to Newtonianism. In this endeavour he had the support of the then Lucasian professor, William Whiston. Though Saunderson's classes posed some competition to his own, Whiston was ready to embrace a fellow Newtonian proselytizer and, as his willingness to sacrifice his university position indicates, Whiston was not one to put private gain before principle. Predictably, Saunderson's lessons were based on Newton's *Opticks* and *Principia* and his pupils were among those who helped to establish Newtonian natural philosophy as part of the normal fare of undergraduate university disputations: 'We every year heard the Theory of the Tydes, the *Phaenomena* of the Rainbow, the Motions of the whole Planetary System as upheld by Gravity', Richard Davies wrote in an obituary of Saunderson, 'very well defended by such as had profited by his Lectures'.[10]

For Saunderson, being a private tutor meant, inevitably, catering particularly to the wealthier students and especially the fellow-commoners. These gilded youths did not need academic success or even a degree and, consequently, were often less than attentive. Recalling such vexations, Saunderson reportedly exclaimed, 'that if he was to go to hell, his Punishment would be to read lectures in the mathematics to the gentlemen commoners of that university'. Saunderson was not always willing to cast his mathematical pearls before such swine even if it meant pecuniary sacrifice. When the young Horace Walpole became a pupil, Saunderson told him within a fortnight: 'Young man, it is cheating you to take your money: believe me, you never can learn these things; you have no capacity for them'. After a year with another tutor, Walpole acknowledged the justice of Saunderson's assessment and abandoned his struggle with mathematics.[11]

Meanwhile, Saunderson soldiered on with his other pupils and gained an excellent reputation as a tutor. Once elected as Lucasian professor in 1711 he had a more prominent platform on which to continue the work to which he had devoted himself as a private tutor: the promotion of Newtonianism in a university where the increasingly important college tutors were often still mathematically innocent. Around 1730, William Heberden records, 'Newton Euclid and Algebra were only known to those who chose to attend the lectures of Professor Saunderson, for the college lecturers were silent on them'.[12]

As the statutes of the Lucasian chair required, his lectures were deposited in the Cambridge University library and still survive (which, despite the regulation, is not true of those of Colson). They cover most major topics in natural philosophy – hydrostatics, mechanics, optics, sound, the tides, astronomy – all, of course, along Newtonian lines. The number of copies indicates their wide diffusion and they appear to have continued to be used in manuscript well into the late eighteenth century. James Bradley used Saunderson's discussion of hydrostatics in preparing his lectures as Savilian professor of astronomy at Oxford, as did James Wood and Samuel Vince in preparing their late-eighteenth-century natural philosophical textbooks (meant chiefly for Cambridge consumption). Edmund Burke, the celebrated aesthete and opponent of the French Revolution, also commented on the excellence of Saunderson's notes on light and colour.[13]

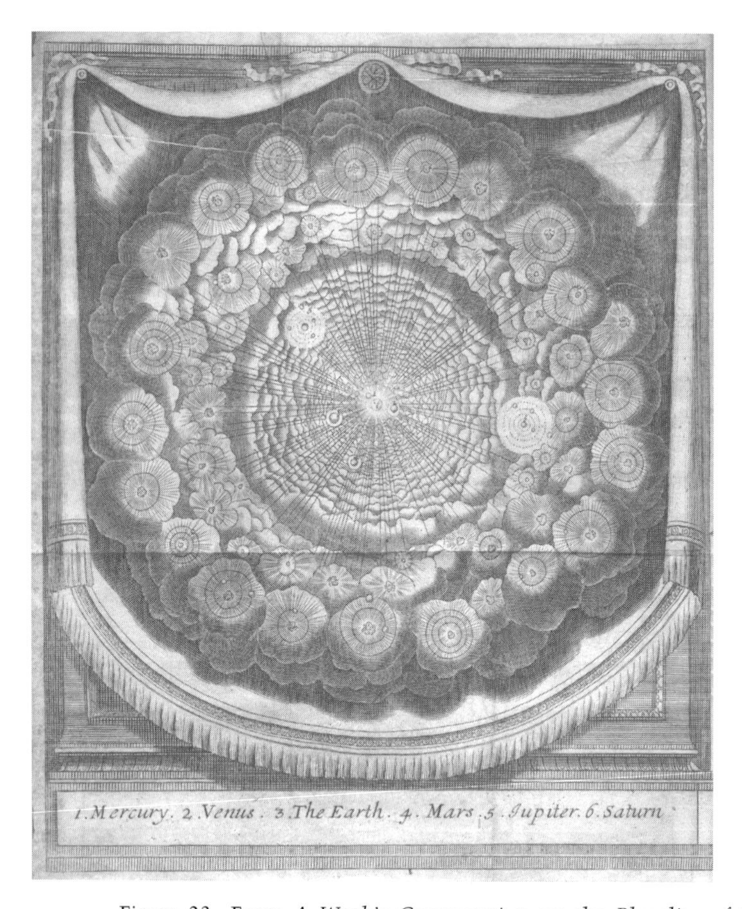

Figure 23 From *A Week's Conversation on the Plurality of Worlds* (London 1738) Fontenelle's solar system of vortices. While Fontenelle suggested to his audience of aristocratic women that the universe was full of Cartesian vortices, Saunderson and Colson combated such misapprehensions with such works as their translation of Professor Donna Maria Agnesi's *The Plan of the Lady's System of Analyticks*.

As befitted a fellow of the Royal Society (a dignity he held from 1719), Saunderson's enterprises drew extensively on the experimental and mathematical heritage of the *Opticks* and the *Prinicipia*, particularly in his discussion of topics such as 'Of the ascent and suspension of liquors in capillary tubes' or 'Of the circulation of the sap'. Along with his advocacy of Newton, Saunderson was determined to vanquish the 'Great Man's' opponents, particularly the Cartesians. Although in the seventeenth century Descartes's philosophy had enjoyed a strong foothold in Cambridge, it had come to be viewed with increasing disfavour because it was seen to endanger the Anglican conception of the Almighty. The particulars of Cartesian natural philosophy were attacked in order to erode the foundations on which Descartes's theologically suspect world system had been constructed. Thus, for example, it is unsurprising to find that Saunderson's lectures on hydrostatics insisted that the Cartesian notion of fluidity – that is, 'a body whose parts are in Continuall motion' – was 'contrary both to sense & Reason'. On the other hand, Saunderson made certain that he demonstrated to his students that the fundamental principles of the Newtonian cosmos were eminently reasonable: for instance, he ensured that pupils understood that the Copernican system was based on pure reason, which also shows us that one of the central tenets of the Scientific Revolution was not entirely to be taken for granted.

There were also opponents of Newton within the citadel of Cambridge. A year after taking up his Lucasian professorship Saunderson commented dismissively to that industrious purveyor of mathematical gossip, William Jones, that 'There has been nothing publish'd here since my last to you, excepting by one Mr Green Fellow of Clare-Hall of this University'. For Saunderson, the best way of dealing with Greene's work was by treating it with contempt. As he remarked to Jones, 'I can find nothing in it, but ill manners & elaborate nonsense from one end to the other'. The book that Saunderson so unflatteringly described was Robert Greene's *Principles of Natural Philosophy* (1712) which expatiated at length on the inadequacies of the 'Corpuscular System, or the Philosophy of Homogeneous Matter' which he saw as a revival of Epicureanism. Though Greene was cautiously polite in his treatment of Newton, his position inevitably conflicted with Newtonian matter theory. In particular, Greene's work challenged Newton's supposition in *Opticks* that 'it seems probable to me that God in the beginning formed matter in solid, massy, hard, impenetrable, movable particles'. The anti-Newtonianism of Greene became even more evident in his vast *magnum opus*, *The Principles of the Philosophy of the Expansive and Contractive Forces* (1727), which characterized as theologically suspect not only 'the Principles of a Similar and Homogeneous Matter' but also characteristically Newtonian concepts, including absolute space and time and action through a vacuum. Certainly, Newton's disciples had no doubt that Greene's work was intended as an attack on the 'Great Man'. Cotes reported to Newton in 1711 that Greene was publishing a book 'wherein I am informed he undertakes to overthrow the Principles of Your Philosophy'.[14]

Saunderson plainly thought that Greene and other opponents of Newton's philosophy could best be combated by ridicule and vigorous promotion of the self-evident truths of Newtonianism. Given the demands on his time (which must have been compounded by his blindness), Saunderson did this principally through his teaching. However, he made sure that his work was informed by the most recent developments in the Newtonian world. Evidently, he closely followed the progress of Roger Cotes's second edition of the *Principia*, reporting to William Jones in 1712 that 'Sr Is. Newton is much more intent upon his principia than formerly, & writes almost every post about it, so that we are in great hopes to have it in our hands[?] in a very little time'. News that Jones was contemplating an edition of David Gregory's '*Notes upon Sir Isaac Newton's Principia* prompted an enquiry as to whether 'tis an explanation of the whole or only of some parts of the Principia'. But the letter to Jones about this ultimately abortive project also revealed the constant demands on Saunderson's time since he apologized to Jones that 'I have had so much business upon my hands, ever since I had the happiness to be acquainted with you, as has hindered me from carrying on a correspondence with you'.[15]

Thanks in part to Saunderson's exertions during his term of office, the university abandoned the decaying remnants of the traditional scholastic curriculum in favour of mathematics and Newtonian natural philosophy. Such a transformation was not solely the work of the Lucasian professors. Along with the contribution of Whiston and Saunderson was that of the first two Plumian professors of astronomy and experimental philosophy, Roger Cotes (1706–16) and Robert Smith (1716–68), who also promoted Newtonian natural philosophy, especially in its more experimental manifestations. College tutors such as John Rowning of Magdalene, author of the avowedly Newtonian *A Compendious System of Natural Philosophy* (1735, with several reprintings up to 1772) also helped to channel Cambridge's academic energies into the mathematical current for which it became known.

By the 1730s contemporaries remarked on the dominant and, to some, oppressive mathematical atmosphere within the university. Those of a literary bent like Horace Walpole, whom Saunderson had so quickly and perceptively dismissed, bemoaned Cambridge's mathematical myopia. Hence his lament in 1735 that Cambridge was 'o'errun with rusticity and mathematics' – a view echoed by the poet, Christopher Smart, in his Tripos verses at his graduation in 1742.[16] But such protests did little to stem the mathematical tide which, from the middle of the century, was increasingly institutionalized in the Senate House Examination which replaced the traditional scholastic disputations. Indeed, mathematics so dominated the examination that it eventually took the name 'the Mathematical Tripos'. Thus from 1753 the published Tripos lists were, for the first time, divided into wranglers and senior and junior optimes – categories which were eventually to develop into the familiar first class and the two subdivisions of the second class.

Beyond the cloistered world of Cambridge, Saunderson's reputation was limited by the fact that he published little during his lifetime, even if his membership of the Royal Society and the Board of Longitude (on which he served ex officio from 1714) did give him some contact with a larger scientific world. Although manuscripts of his lectures and treatises were circulated widely, he did not follow the common practice of publishing them – perhaps because he was blind. To some extent this deficiency was remedied posthumously. In 1740, the year after his death, his *The Elements of Algebra, in Ten Books: To Which is Prefixed, an Account of the Author's Life and Character, Collected from his Oldest and Most Intimate Acquaintance* was published at the initiative of Richard Davies, physician and fellow of Queens' College. An abridged edition was published in 1756 by the Reverend John Hellins of Trinity College, and in that form four more editions were printed, the last of which appeared in 1792. Hellins paid tribute to 'The Excellence of Professor Saunderson's *Elements of Algebra* [which] is universally acknowledged'. He also added, with a remark that underlines the increasing dominance of the college tutors, that since much of it was intended for the 'Advantage and Amusement to proficients in the general Science of the Mathematics, than of necessary Use to Students in Algebra; some of the principal Tutors in the University of *Cambridge*' were anxious for an abridged, more pedagogically focused work.[17]

Saunderson's work indicates his sympathy for analytical methods which perhaps better suited him because of his blindness. This meant something of a departure from the dominant Newtonian tradition. For, though Newton himself greatly advanced the use of analysis by his actual practice, he retained the traditional veneration for geometry over analysis – hence his remark that 'Algebra is the Analysis of the Bunglers in Mathematicks'. Something of the stir created by Saunderson's approach is evident in the mathematical arguments it prompted between his students and those of the London-based mathematician and military engineer, Benjamin Robins, FRS, 1727 and Copley medallist, 1747. But, for Saunderson, whether algebra or geometry, mathematics in all its forms still retained its mystical aura of the Pythagoreian–Platonic tradition. He regarded mathematics as the one sure guide to truth: 'It is in Mathematics only that truth, that is, rational truth, appears most conspicuous, and shines in her strongest lustre . . . whosoever to the utmost of his finite capacity would see truth as it has actually existed in the mind of God from all eternity, he must study Mathematics more than Metaphysics'.[18] It is a passage that underlines the fact that, for all the rumours of his infidelity, Saunderson remained a Deist, if not an orthodox Christian.

Along with the *Algebra*, another posthumous work of Saunderson was his lengthy but revealingly entitled *The Method of Fluxions Applied to a Select Number of Useful Problems: Together with the Demonstration of Mr. Cotes's Forms of Fluents in the Second Part of his Logometria; The Analysis of the Problems in his Scholium Generale and an Explanation of the Principal Propositions of Sir Isaac Newton's Philosophy* (1756). It was a book which served to fuel the growing mathematical blaze within the university, being particularly useful for those aspiring to high honours in the Senate House Examination. Despite this blaze Saunderson was the first Cantabrigian to provide through his lectures a systematic introduction to the fluxional calculus. While the subject of the first part of the book was devoted to the methods of fluxions, the second and third were focused on Cotes's integrals and an analysis of some propositions of the *Principia* (where he remains true to Newton's geometrical approach).[19] Its anonymous editor made the claim that it may 'be reckoned the best for Students in the Universities, of any yet published'. His commentary on Cotes's *Logometria* also testified to 'his intimate Acquaintance with that extraordinary Person'.

Such works, then, enhanced Saunderson's already well-established reputation within Cambridge as a teacher, although they did little to add his name to the larger annals of mathematics. But Saunderson was known throughout Britain, and even beyond, for his extraordinary achievement as an individual in gaining so hotly contested a position as the Lucasian professorship despite his blindness. What particularly fascinated the European philosophers was the extent to which Saunderson could be regarded as a test-case for the Lockean view that all ideas were the outcome of experience. For, as someone who was blind, Saunderson was denied many common experiences (though, of course, Saunderson was blind from infancy, not blind from birth, and may have retained some early experiences – a point not taken up in such discussions). Ten years after Saunderson's

36 LETTRE

7 8 5 6 8
8 4 3 5 8
8 9 4 6 4
9 4 0 3 0

Il eſt Auteur d'un Ouvrage très-parfait dans ſon genre. Ce ſont des élémens d'algébre où l'on n'apperçoit qu'il étoit aveugle qu'à la ſingularité de certaines démonſtrations qu'un homme qui voit n'eût peut-être pas rencontrées : c'eſt à lui qu'appartient la diviſion du cube en ſix piramides égales qui ont leurs ſommets au centre du cube, & pour baſes, chacune une de ces faces. On s'en ſert pour démontrer d'une maniere très-ſimple, que toute piramide eſt le tiers d'un priſme de même baſe & de même hauteur.

Il fut entraîné par ſon goût à l'étude des Mathématiques, & déterminé par la médiocrité de ſa fortune & les conſeils de ſes amis, à en faire des leçons publiques. Ils ne douterent point qu'il ne réuſsît au de-là de ſes eſpérances, par la facilité prodigieuſe qu'il avoit à ſe faire entendre. En effet, Saounderſon parloit à ſes Eléves comme s'ils euſſent été privés de la vue ; mais un Aveugle qui s'exprime clairement pour des Aveugles, doit gagner

Figure 24 Saunderson's calculating board. According to Diderot in his *Lettre sur les Avaeugles* (*Letter concerning the Blind*), Saunderson was able to speak to his students as if he had sight.

death Diderot referred to him in his *Letter on the Blind* (1749) when discussing whether a person born blind would be able to grasp geometrical concepts such as the difference between a sphere and a cube. Saunderson's manifest success as a mathematician led Diderot to conclude that, indeed, the blind could grasp such concepts, though an essentially Lockean epistemology was saved by attributing the formation of ideas to experience based on the sense of touch rather than sight.

Diderot even provided a detailed (though somewhat inaccurate) account of the device that Saunderson used to study geometry by means of touch – a device described at length by his successor as Lucasian professor, John Colson, in a section of his posthumous *Elements of Algebra* entitled 'Dr Saunderson's palpable arithmetic decypher'd'. For, as Colson put it, Saunderson devised a form of '*Abacus* or Calculating Table, and with which he could readily perform any Arithmetical Operations, by the Sense of Feeling only'. The device took the form of a sort of wooden grid with a hole at each point of intersection which was capable of accommodating a pin. By using such pins Saunderson could perform complex calculations and 'could place and displace his Pins . . . with incredible Nimbleness and Facility, much to the Pleasure and Surprize of all the Beholders'. The device could be used both for arithmetical calculations and to describe 'upon it very neat and perfect Geometrical Figures'. Saunderson also devised his own 'Species of Mathematical Symbols' which could take the place of visible signs by again substituting 'Feeling in the place of Seeing' – symbols he derived from listening to others read mathematical works to him.[20]

Diderot also pointed to other ways in which, in Lockean fashion, Saunderson built up ideas from the building blocks of sensory perception. Saunderson, for example, like other blind people, was particularly sensitive to any slight change in his surroundings, so that he could tell if clouds were passing over the sun and hence when was an appropriate moment for astronomical observations. Saunderson, as Diderot pithily put it, 'saw with his skin'. He also remarked on his finely developed sense of hearing so that he could tell who was approaching by the sound of their step. It was a point also dwelt upon by his Cambridge contemporaries who remarked on his acute ear, evident in his abilities as a flautist. It was related that 'if ever he had walked over a Pavement in Courts, Piazzas, &c, which reflected a Sound, and was afterwards conducted thither again, he could exactly tell whereabouts in the Walk he was placed, merely by the Note it sounded'.[21] By such means Saunderson appears to have constructed a fairly complete picture of the world, enabling him to join in conversation in a way that made contemporaries almost forget about his blindness.

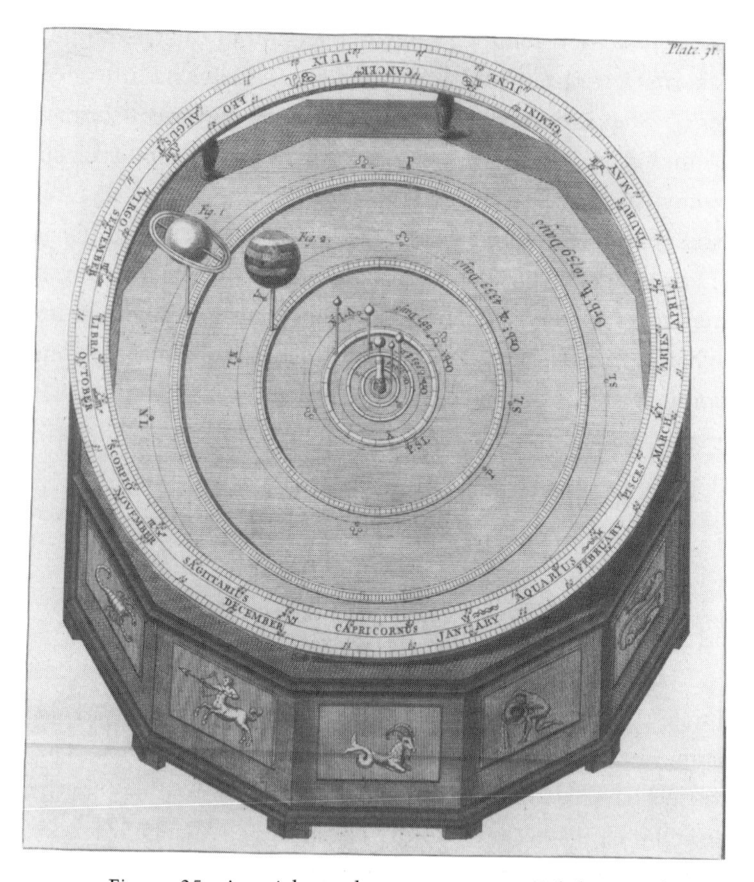

Figure 25 An eighteenth-century orrery. While Saunderson's board helped the visually impaired make complex calculations, other devices, like the orrery, gave eighteenth-century professors and lecturers the ability to enable the sighted to better comprehend the Newtonian universe.

For Diderot, Saunderson was a fascinating case study not only of the way in which the blind could grasp complex mathematical concepts but also as a prosecution witness in Diderot's continuing assault against the claims of religion. For, in Diderot's view, the weaknesses of the argument from design should be particularly evident to someone incapable of seeing the alleged evidences of such design – even if, ironically, the early popularization of Newton's work had been closely associated with clerical popularizers of the design argument. Manufacturing an entirely imaginary conversation, Diderot portrayed Saunderson on his deathbed undermining the arguments advanced by a typical clerical physico-theologian. In the first place, as one that was blind, proofs based on the visible splendours of the universe had no meaning to Saunderson; but even if evidence of design could be conveyed through the sense of touch, this did not necessarily lead to the notion of a divine intelligence. Saunderson's own experience indicated that human beings were too willing to invoke transcendent explanations for what they could not understand:

> People have been drawn to see me from all over England merely because they could not conceive how I was able to do geometry; you must admit that those people did not have very exact notions as to what is or is not possible. We encounter some phenomenon that is, in our opinion, beyond the powers of man, and immediately we say, 'It is the work of God.' Our vanity will admit no lesser explanation.

Saunderson also served as an instance of the problem of evil with his supposed declaration 'Look well at me, Mr. Holmes. I have no eyes. What did we do to God, you and I, that one of us should possess those organs and the other be deprived of them?' Taking the theme of Saunderson's blindness as a instance of the imperfection of the world, Diderot also alluded to the concept of the plurality of worlds – a concept which Saunderson had expounded in his *Elements*. 'How many maimed and botched worlds have melted away', the blind natural philosopher asks, 're-formed themselves, and are perhaps dispersing again at any given moment?'. Despite such scepticism towards divine intelligence, benevolence and design, Diderot portrays Saunderson as finally espousing some form of deistic belief, dying with the cry, 'O God of Clarke and Newton, pity me now'.[22]

In his *A Philosophical Enquiry into the Origin of our Ideas of the Sublime and Beautiful*, Burke argued that Saunderson also served as an important test-case in the relationship between ideas and experience – in particular, ideas relating to colour. What particularly struck Burke was that the blind Saunderson 'gave excellent lectures upon light and colours; and this man taught others the theory of those ideas which they had, and which he himself undoubtedly had not'. Burke then assumed in Lockean fashion that it was impossible to have an idea of something like a colour unless one could build on experience which was not available to a blind person. Saunderson must then, argued Burke, have merely been using the words for colours without possessing the ideas: 'But it is probable, that the words red, blue, green, answered to him as well as the ideas of the colours themselves . . . He did nothing but what we do every day in common discourse'.[23]

Dr Johnson, too, was fascinated by the question of whether Saunderson could discern colours by touch but concluded in a conversation with Boswell, 'that to be sure a difference in the surface makes the difference of colours; but that difference is so fine, that it is not sensible to the touch'.[24] For Johnson, as for Burke or Diderot, Saunderson became an instance of the inseparable link between ideas and experience. All these figures in their different ways used the example of Saunderson to shore up the Lockean tenet that ideas are the product of experience – a tenet that underlay the Enlightenment's confidence that it was possible for humankind to move forward without the inherited burden of innate ideas.

When Saunderson died in 1739 there was no well-established Cambridge candidate for the Lucasian chair, which is itself an indication of the declining status of the professorship. The competition was therefore between the Huguenot refugee, Abraham De Moivre, and John Colson. For all De Moivre's mathematical achievement – for he was the author of pioneering works in the field of probability – he was too old and infirm to be a credible candidate. As the antiquary, William Cole, remarked, De Moivre was 'brought down to Cambridge and created an MA when he was almost as much fit for his coffin; he was a mere skeleton, nothing but skin and bone'. The post therefore went to John Colson, a decision which, according to one contemporary source,

Cambridge was later to regret: although he was 'a plain, honest man of great industry . . . the university was much disappointed in their expectations of a Professor that was to give credit to it by his lectures'.[25]

As a pedagogue, Colson arrived with good credentials, coming from a family of mathematical instructors. His namesake (fl. 1671–1709) and probable Kinsman ran a mathematical school in Wapping, London, intended particularly for those destined for the sea. Hence Colson's involvement in producing an edition of Sturmy's *Mariner's Magazine* and his role as an examiner in the skills of navigation on behalf of Trinity House. John Colson senior also took a close interest in astronomy, contributing to the Royal Society an observation of the solar eclipse of 1 June 1676. Such connections led to his friendship with Halley and a rather tense relationship with Flamsteed, who took him to task for spreading the story that Newton's account of the moon's motions was based on Halley's observations. But Flamsteed evidently continued to hold a high opinion of Colson's pedagogical abilities since he referred on to him Jacob Bruce who came to London in 1698–9 as part of Peter the Great's entourage. Bruce studied under Colson for six months and returned to Russia with a collection of scientific works which included accounts of the Newtonian system. Subsequently Bruce was involved in the foundation of a Russian navigation school and the St Petersburg Academy of Sciences.[26]

Another mathematical member of the Colson clan was Nathaniel (fl. 1674), whose *Mariner's New Kalendar* included an advertisement for John Colson's school as a place where 'are taught these Mathematical Sciences, viz: Arithmetic, Geometry, Algebra, Trigonometry, Navigation, Astronomy, Dialling, Surveying, Gauging, Fortification and Gunnery, the Use of the Globes and other parts of the Mathematics'. Yet another mathematical Colson was Lancelot (fl. 1668–87), whose astronomical observations appear in the correspondence of Newton and Flamsteed.[27]

In contrast to these London-based Colsons, John Colson, the future Lucasian professor, was born at Lichfield, the son of the vicar-choral of the cathedral. He then matriculated at Christ Church, Oxford in 1699, though he left without taking a degree. At this time, Oxford was a centre for the dissemination of Newton's work, thanks to the energetic activity of the two Episcopalian exiles from Scotland, David Gregory and John Keill, and it is possible that they may have helped to stimulate further the young Colson's interest in the mathematical sciences. It is also possible that after Oxford Colson taught at his namesake's school in London. If he did so, this experience, together with his family connections, help to explain why he was appointed in 1709 as the first master of the new mathematical school founded at Rochester by Sir Joseph Williamson – a post that brought a house and a salary of one hundred pounds a year. Certainly, the goals of this new school – to educate boys 'towards the Mathematicks and all other Things which may fitt and encourage them for the Sea Service' – were very similar to those of the school run by John Colson senior in London.[28]

In a society offering few opportunities for mathematical careers, the Mastership was a much sought-after appointment. Among his rivals was the hapless Christopher Hussey, who was subsequently defeated by Saunderson in the election for the Lucasian chair. As a fellow of Trinity, Hussey was able to enlist Newton's aid through the good offices of the college's Master, the redoubtable Richard Bentley, who intended to 'get of him [Newton] a Character of Mr. Hussey'. Consequently, after interviewing Hussey Newton wrote a testimonial about his mathematical abilities, describing Hussey as 'very well qualified'. But, as Bentley had feared, the post went to Colson – something which, predictably, Bentley attributed to the fact 'y^{t} Interest rather than Merit will prevail in ye Election, & one Coleson has ye best friends'. Bentley believed that these 'friends' of Colson swayed the seventeen electors, who were local dignitaries such as the mayor, the recorder and the 'eldest Resident Prebendary' (who may have had some clerical connection with Colson's cathedral-based father or with Colson's godfather, the ecclesiastical historian, John Strype). Perhaps, too, the Royal Society had some say in Colson's appointment since the Rochester school was patronized by Sir Joseph Williamson, the former President of the Royal Society from 1677 to 1680. Certainly it did not hurt that Colson's Kinsman had been friends with the likes of Halley and Flamsteed. True to family tradition, once Colson was appointed he shared his good fortune with his kinsfolk: he appointed his sister, Mary, as his housekeeper and his brother, Francis, as an usher.[29]

Colson must at some stage have taken orders, for after 1724 he was able to augment his salary as Master of the school with the living of Chalk near Gravesend and when he died in 1760 he was rector of Lockington, Yorkshire. His theological interests are evident, too, in his translation (along with Samuel D'Oyley) from the French of the massive *An Historical, Critical, Geographical, Chronological, and Etymological Dictionary of the Holy Bible, in Three Volumes.*

Even before assuming his position at Rochester, Colson established his credentials at the Royal Society with a mathematical paper in the *Philosophical Transactions*, 'The universal resolution of cubic and biquadratic equations, as well analytical as geometrical and mechanical' (1707), and he was duly elected a fellow in 1713. The only other original mathematical papers that he published were an 'Account of negativo-affirmative arithmetick' (1726) and 'The construction and use of spherical maps' (1736), which also appeared in the *Philosophical Transactions*.

On 23 April 1728 Colson was formally entered as student at Emmanuel College, Cambridge, and shortly afterwards he was one of the seventy-one persons granted a degree when George II visited the university. At this stage Colson's links with Cambridge appear to have been rather tenuous. His place on the King's list probably derived from the good offices of the mathematician, Robert Smith, a fellow of Trinity and from 1716 Plumian professor of astronomy and experimental philosophy. Certainly, Smith had good connections at court, being master of mechanics to George II, as well as being mathematical preceptor to William, Duke of Cumberland. It was largely thanks to Smith's influence that Colson eventually obtained the Lucasian chair.

Colson continued to hold the post of Master at the Rochester school until a year after he was made Lucasian professor in 1739. By

Figure 26 Portrait of John Colson. From the collection of the Cambridge antiquarian, William Cole.

1737, however, he was resident in London, for it was there that the young David Garrick (who, like Colson, was from Lichfield) boarded with him while being instructed in 'the mathematics, philosophy and human learning'. On his trip to London Garrick was accompanied by

Samuel Johnson, who is said to have modelled his character Gelidus in the *Rambler* (no. 24) on Colson.[30]

It was in this period that Colson published the work for which he is chiefly remembered and which probably did much to secure the Lucasian chair: *Sir Isaac Newton's Method of Fluxions; Translated from the Author's Latin Original not yet Made Publick. To Which is Subjoined a Perpetual Comment on the Whole Work* (London, 1736 and 1737). Its dedication to William Jones made apparent Colson's debt to the mathematician, friend of Newton, and vice-president of the Royal Society: he profusely thanked Jones for 'the free access you have always allow'd me, to your Copious Collection of whatever is choice and excellent in the Mathematicks'. The gratitude was merited, since Colson's edition was based on Jones's transcript of Newton's tract.[31] The work was, however, considerably more than a simple translation, for Colson added to it some two hundred pages of explanatory notes. He also endeavoured to use the work to place the calculus of fluxions on a sounder mathematical (and, in particular, geometrical) footing.

The footings of the Newtonian fluxional calculus had been undermined not only from continental supporters of the Leibnizian differential calculus, but also from within the British Isles, notably from Bishop Berkeley in his *The Analyst* (1734) and *A Defence of Free-thinking in Mathematics* (1735). For Berkeley, the analytical foundations of fluxional calculus, with its reliance on the concept of infinitesimals, was obscure. It was philosophically and even theologically corrupt: 'if we remove the veil and look underneath, if, laying aside the expressions', Berkeley grumbled, 'we set ourselves attentively to consider the things themselves which are supposed to be expressed or marked thereby, we shall discover much emptiness, darkness and confusion; nay, if I mistake not, direct impossibilities and contradictions'.[32]

Within Cambridge there were those who regarded such a critique as an attack on the whole Newtonian heritage with which the university was now identified. The sense of outrage in Cambridge – a

Figure 27 In Newton's *Method of Fluxions*, John Colson's 'perpetual commentary' responded to Berkeley's vehement attack on 'infidel mathematicians' by showing that the calculus was a rational and tangible means of expressing real motions in space. To explain the second derivative, Colson has the reader, presumably a Cambridge undergraduate, imagine an attempt to pot two ducks with one shot while the Greek masters look upon the action. Colson presumes that this '*sensibiles sensibilium velocitatum mensure*' becomes 'not only the object of the Understanding, and of the Imagination ... but even of sense too'.

university which had prided itself on welding the defence of Christianity to the promotion of Newtonian natural philosophy – was all the more violent since the bishop had linked infinitesimals and heterodoxy. This Cantabrigian outrage is encapsulated in the title of one major response to Berkeley: James Jurin's *Geometry no Friend to Infidelity: Or, a Defence of Sir Isaac Newton and the British Mathematicians* (1734), who wrote under the revealing pseudonym, *Philalethes Cantabrigensis*. A year later Jurin's defence was followed by his *The Minute Mathematician: Or, The Free-thinker no Just-thinker*. Jurin, a former fellow of Trinity and, at Bentley's insistence, an early popularizer of Newton's work, had been, like Colson, a mathematical educator, having obtained a position at Christ's Hospital School with the support of Bentley before becoming master of Newcastle grammar school.

For Jurin, Berkeley's work was unpatriotic both because it was a critique of the 'British mathematicians' and because it undermined the mathematical achievement of Britain's great cultural hero, Isaac Newton. For, as he wrote in his first pamphlet, Berkeley's 'avowed design' is 'to lessen the reputation and authority of Sir *Isaac Newton* and his followers, by showing that they are not such *great masters of reason, as they are generally presumed to be*'. More dangerously, in Jurin's estimation, Berkeley implied that science could lead to infidelity when the energies of many within the Established Church and (particularly within Cambridge) had been devoted to demonstrating the opposite: 'I am afraid,' Jurin asserted, 'it would be a great stumbling block to men of weak heads, if they were made to believe, that the justest and closest reasoners were generally infidels.' There is even a whiff of anticlericalism in Jurin's response to Berkeley's assault, as when, in the second pamphlet, he contrasted the intellectual freedom of the mathematicians with the clergy who are bound by creeds. Discussing his hostility to Berkeley's clerical supporters, he remarked that, 'I am a Layman. If the Clergy obtain more power, I shall have less liberty: if they will have more wealth, I am one of those must pay to it.'[33]

By contrast, Colson was less hostile towards 'the very learned and ingenious Author of the Discourse call'd *The Analyst*'. He sought to counter Berkeley's critique by placing the fluxional calculus more securely on a geometrical rather than an analytical foundation. By doing so, he helped to provide that form of calculus with a more kinematical approach which corresponded more directly with the behaviour of a body in motion. As he himself put it, it was his aim to use moving diagrams to exhibit 'Fluxions and Fluents Geometrically and Mechanically... so as to make them the objects of Sense and ocular Demonstration'. Thanks to Colson, then, fluxional calculus thereafter had a much more overt geometrical foundation which served to distance it from the use of infinitesimals and thus distinguish it more clearly than in the past from the differential calculus. But, for Colson, such an enterprise was not an innovation but rather the recovery of the true Newtonian heritage: 'For it was now become highly necessary, that at last the great Sir *Isaac* himself should interpose, should produce his genuine Method of Fluxions, and bring it to the test of all impartial and considerated Mathematicians'.[34]

When Colson was duly rewarded with the Lucasian professorship for his defence of the Newtonian mathematical heritage, he also gained with it the first Taylor lectureship in mathematics at Sidney Sussex College. As Lucasian professor, Colson, who appears to have had considerable facility with languages, chiefly contributed to the scholarly world by publishing scientific translations – for his lectures made little impact on the university. But, as we have seen, the need for such lectures had greatly diminished now that Newtonian natural philosophy and mathematics had become part of the staple of college instruction. Fittingly, Colson's first professorial publication paid tribute to Saunderson, his predecessor as Lucasian professor, for prefixed to the posthumous 1740 edition of Saunderson's *Elements of Algebra* was Colson's dissertation on 'Dr Saunderson's palpable arithmetic decypherd'. Evidently, too, his good opinion of Saunderson was reciprocated, since in his *Algebra* Saunderson refers to 'The learned Mr. John Colson, a gentleman whose great genius and known abilities in these sciences I shall always have in the highest admiration and esteem'.[35]

In the often parochial world of English and, *a fortiori*, Cambridge science, Colson's translations exposed students to continental natural philosophy. For, in 1744, he published a translation from the Latin of the great Dutch experimentalist and Newtonian popularizer, Peter van Musschenbroek's *The Elements of Natural Philosophy Chiefly Intended for the Use of Students in Universities* (though with a preface which, in good English Newtonian style, dissented from the Dutchman's reliance on Leibniz). This was followed in 1752 by a translation from the French of the *Lectures in Experimental Philosophy* by Abbé Nollet, who was particularly well-known for his work on electricity. This in some ways was a brave publication in Newtonian Cambridge since Nollet, a thorough-going experimentalist, was agnostic in his allegiance to any particular system of natural philosophy. Hence Nollet's claim in the preface to the work that 'I do not list myself here, under the standard of any of the celebrated philosophers. 'Tis not the Physicks of *Des Cartes*, nor of *Newton*, nor of *Leibniz*, that I propose particularly to follow . . . I take nothing upon their bare Word, except what I find to be confirmed by Experience.'[36]

In old age Colson even set to work to learn Italian to translate the mathematical work of Donna Maria Agnesi, that rare phenomenon in the eighteenth century: a female professor of mathematics (at the university of Bologna). One of his intentions in doing so was to make the study of mathematics more accessible to women, a goal reflected in the title of his manuscript draft, *The Plan of the Lady's System of Analyticks*, and the detailed exposition of contents it included. Such a goal of ensuring that English women did not fall behind those of Italy in the study of mathematics was an unusual one in male-dominated Cambridge and perhaps reflects the fact that Colson brought to the post of Lucasian professor a long family tradition of mathematical instruction which could go beyond the immediate needs of the university. More predictably, it was also a work which could appeal to Colson as a demonstration of the value of fluxions.[37] However, he died before it could appear and it was eventually published posthumously in 1801 by the Reverend John Hellins, FRS (the editor of Saunderson's abridged *Algebra*) at the expense of Baron Maseres.

Such was the contribution of Colson to the consolidation of natural philosophy and mathematics at Cambridge. But, in many ways, the task had already been achieved by the time of Colson's election and the role of the Lucasian professor had become somewhat superfluous as such studies had become institutionalized in the college teaching and university examinations. Much of the period when Colson held office (1739–60) was one when the great initial wave of enthusiasm for Newton's work had subsided into a dutiful orthodoxy as Cambridge drilled its students in the accepted canon of Newtonianism. It is significant, for example, as Geoffrey Cantor points out, that only two new textbooks of natural philosophy were published at Cambridge between 1750 and 1780 as compared with eleven in the period 1720 and 1750.[38]

When Colson died in 1760, then, the role of the Lucasian professor had shrunk from one of advocacy and catalysis for mathematical (and especially Newtonian) studies to that of echoing the instruction that undergraduates routinely received from the tutors. The revival of the chair as an active focus of mathematical activity required a more pluralist intellectual climate in which the confidence that Isaac Newton, Cambridge's most famous son, had revealed all – or very nearly all – was challenged. It also required a re-evaluation of the role of professors whereby their role as educators could be better integrated into the round of undergraduate teaching and their role in promoting their subject through research could receive greater recognition. Such changes, however, were to be the work of the next century when the Lucasian chair and the university as a whole were energized by the reforming impulse of the age.

Notes

1 Anecdote of Martin Folkes in King's College, Cambridge, Keynes MS 130.
2 Anecdote of John Durham in King's College, Cambridge, Keynes MS 130.

3 Anecdote of Humphrey Newton in King's College, Cambridge, Keynes MS 130.

4 Samuel Parr, *A Spital Sermon . . . to Which are Added Notes* (London, 1801), p. 121.

5 Newton to Bentley, 20 October 1709, *The Correspondence of Isaac Newton*, H. W. Turnbull, J. F. Scott, A. R. Hall and Laura Tilling (eds) (7 vols, Cambridge, 1959–77), vol. VII, p. 479.

6 J. Tattersall, 'Nicholas Saunderson: "The blind Lucasian professor"', *Historia Mathematica*, 19 (1992), 356–70 on 366.

7 Saunderson's early arithmetic in Add. MS 4223, f. 194 (biographical anecdotes from the Birch collection); maths in Sheffield in J. W. Ashley Smith, *The Birth of Modern Education: The Contribution of the Dissenting Academies 1660–1800* (London, 1954), p. 109; Saunderson's logic in Add. MS 4223, f. 194v.

8 G. Dyer, *The Privileges of the University of Cambridge* (London, 1824), vol. II, pp. 142–3.

9 Saunderson's debauchery cited in Tattersall, 'Nicholas Saunderson', p. 362; John Saunderson's work in J. Peile, *Biographical Register of Christ's College, 1505–1905* (2 vols, Cambridge, 1910–13), vol. II, p. 243.

10 Whiston's support in *Gentleman's Magazine*, 24 (1754), 373. Davies' reminiscence in N. Saunderson, *The Elements of Algebra, in Ten Books: To Which is Prefixed, an Account of the Author's Life and Character, Collected from his Oldest and most Intimate Acquaintance* (Cambridge, 1740), p. vi.

11 Saunderson on gentleman commoners in *Gentleman's Magazine*, 24 (1754), 373; Walpole's study in 'The life and works of Thomas Gray', *Quarterly Review*, 94 (1853–4), 1–48 on 8.

12 C. Wordsworth, *Scholæ Academicæ* (Cambridge, 1877) p. 66.

13 Saunderson's lectures are held in Cambridge University Library, Add. 589. Other copies are Add. 6312, 2977. (There is also a set of student notes from his lectures by John Mickleburgh (professor of chemistry, 1718–56) in Gonville and Caius College Library, MS 723/74a.) Outside Cambridge there are copies at Bodleian, MS Rigaud 3-4, British Library, MS Add. Eg. 834 and University College, London, MS Add. 243; Burke on colour in Tattersall, 'Nicholas Saunderson', 363.

14 Saunderson's response to Greene in Newton, *Correspondence*, vol. V, p. 247, Saunderson to Jones, 16 March [1711/12]; Newton on matter in H. Thayer, *Newton's Philosophy of Nature. Selections from his Writings*

(New York, 1974), p. 175; Cotes on Greene in Newton, *Correspondence,* vol. V, p. 166; Greene on Newton in R. Greene, *The Principles of the Philosophy of the Expansive and Contractive Forces* (Cambridge, 1727), preface, pp. 17, 41–7.

15 Saunderson to Jones, 16 March [1711/12], Newton, *Correspondence,* vol. v, p. 247; Saunderson to Jones, 4 February 1713–4, S. Rigaud, *Correspondence of Scientific Men of the Seventeenth Century* (2 vols, Oxford, 1841), vol. I, p. *264.

16 W. S. Lewis (ed.) *The Correspondence of Horace Walpole*, vols, 13–14 (New Haven, CT., 1948), p. 94 and A. Sherbo, *Christopher Smart: Scholar of the University* (East Lansing, MI., 1967), p. 27.

17 N. Saunderson, *Select Parts of Saunderson's Elements of Algebra for the Use of Students at the Universities,* 5th edn. Revised and corrected by the Rev. John Hellins of Trinity College, Cambridge, p. iii.

18 G. Hiscock, *David Gregory, Isaac Newton and their Circle* (Oxford, 1937), p. 42; Saunderson's approach to algebra in N. Guicciardini, *The Development of Newtonian Calculus in Britain 1700–1800* (Cambridge, 1989), p. 24; Saunderson, *Elements of Algebra,* vol. II, p. 740.

19 Guicciardini, *The Development,* p. 24.

20 *Lettre sur les Aveugles, à l'Usage de ceux qui Voient* in *Oeuvres Philosophiques de Mr. D[iderot]* (Amsterdam, 1772), vol. II, pp. 73, 29–36. On Diderot's inaccuracies see M. Kessler, 'A puzzle concerning Diderot's presentation of Saunderson's Palpable arithmetic', *Diderot Studies,* 1981 (20), 159–73; Saunderson, *Elements of Algebra,* pp. xx–xxv.

21 Diderot, *Lettre,* p. 46; Saunderson, *Elements of Algebra,* p. xiii.

22 Diderot, *Lettre,* 48–55. Translation from L. Crocker (ed.), *Diderot's Selected Writings* (New York, 1966), 20–4.

23 E. Burke, *A Philosophical Enquiry into the Origin of our Ideas of the Sublime and Beautiful,* J. T. Boulton (ed.) (London, 1958), p. 169.

24 J. Boswell, *Life of Johnson* cited in K. MacLean, *John Locke and English Literature of the Eighteenth Century* (New Haven, 1936), p. 107.

25 Cole on De Moivre, BL, Add. 5866, 200. My thanks to Dr A. Capern for transcriptions of this. Colson's achievements in *History and Antiquities of Rochester* (1772), vol. VIII, p. 220 cited in BL, Add. 5866, f. 3.

26 Colson on navigation in E. G. R. Taylor, *The Mathematical Practitioners of Tudor and Stuart England* (Cambridge, 1954), p. 264; Flamsteed to Colson, 10 October 1698, Newton, *Correspondence,* vol. IV, pp. 284–5;

Jacob Bruce in V. Boss, 'Russia's first Newtonian', *Archives Internationales de l'Histore des Sciences,* 15 (1962), 233–65.

27 John Colson's curriculum in Taylor, *Mathematical Practitioners,* p. 264; Flamsteed to Halley, 17 February 1680/1; Newton to Flamsteed, 16 April 1681, [12 January 1684/5], Newton *Correspondence*, vol. II, pp. 339–40, 365.

28 R. V. and J. Wallis, *Biobibliography of British Mathematics and its Applications. Part II, 1701–1760* (Newcastle-upon-Tyne, 1986), p. 29; Colson's school in Taylor, *Mathematical Practitioners*, 530.

29 Bentley on Hussey in Bentley to Cotes [21 May 1709], J. Edleston, *Correspondence of Sir Isaac Newton and Professor Cotes* (London, 1850), p. 1; Newton on Hussey in Richard Westfall, *Never at Rest. A Biography of Isaac Newton* (Cambridge, 1980), p. 700. Westfall surmises that the interview was for a post at the Christ's Hospital School but Cotes in a letter to Newton refers to the occasion on which Hussey 'formerly waited upon You for Your recommendation of him to a Mathematical School at Rochester'. Cotes to Newton 25 October 1711, Newton, *Correspondence,* vol. V, p. 202; Bentley on election in Edleston, *Correspondence*, p. 2; Colson's family in Add. 5866, f. 1.

30 T. Davies, *Memoirs of the Life of David Garrick* (2 vols, Dublin, 1780), vol. I, p. 9; Samuel Johnson on Colson in Taylor, *Mathematical Practitioners*, 115.

31 J. Colson, *Sir Isaac Newton's Method of Fluxions* (London, 1736–7), p. iv; D. Whiteside (ed.), *The Mathematical Papers of Isaac Newton* (Cambridge, 1967–81), vol. VIII, p. xxiii.

32 H. Pycior, *Symbols, Impossible Numbers, and Geometric Entanglements. British Algebra Through the Commentaries on Newton's Universal Arithmetick* (Cambridge, 1997), p. 233.

33 [J. Jurin], *Geometry no Friend to Infidelity* (London, 1734), pp. 9–10; [J. Jurin], *The Minute Mathematician* (London, 1735), pp. 2–3, 9.

34 Colson on Berkeley in Pycior, *Symbols*, p. 301; Colson, *Method of Fluxions*, vol. x, p. 234: cited in Guicciardini, *The Development*, p. 57; N. Guicciardini, 'Flowing ducks and vanishing quantities' in S. Rossi (ed.), *Science and Imagination in XVIII-century British Culture* (Milan, 1987), pp. 231–5 on 231.

35 Saunderson, *Elements of Algebra*, vol. II, p. 720.

36 Jean Antoine Nollet, *Lectures in Experimental Philosophy*, translated by J. Colson, p. xvi, (London, 1752).

37 Colson's draft manuscript in Cambridge University Library, MSS Ee.2.36-38; W. Johnson, 'Contributions to improving the teaching of calculus in early nineteenth century England', *Notes and Records of the Royal Society*, 49 (1995), 93–103 on 100.

38 G. Cantor, *Optics after Newton. Theories of Light in Britain and Ireland, 1704–1840* (Manchester, 1983), p. 5.

5 The negative side of nothing: Edward Waring, Isaac Milner and Newtonian values

Kevin C. Knox

Division of Humanities and Social Sciences, California Institute of Technology, Pasadena, CA, USA

> In order to prove that a guinea and a feather would descend *in vacuo* in the same time, [Milner] made use of a glass tube hermetically sealed, in which the guinea and the feather were enclosed; it so happened that in several attempts the guinea had the advantage: he then managed to place the guinea above the feather. At the end he exclaimed 'How beautifully this experiment has succeeded!' for if you observed attentively, you would perceive that the feather was down sooner than the guinea.
>
> Henry Gunning, *Reminiscences of Cambridge*.[1]

Nothing beats a good conspiracy theory to capture an extensive audience. The British readership of the 1790s was particularly smitten with conspiratorial literature and Georgian publishing houses loved to disseminate stories of malevolent intrigue. Take John Robison's 1797 *Proofs of a Conspiracy against all the Religions and Governments of Europe, Carried on in the Secret Meetings of Free Masons, Illuminati, and Reading Societies*: this classic in the genre was gobbled up by gallophobic readers, going through five editions in its first two years. In the wake of the French Revolution and terror, Robison hit a sensitive nerve amongst Britons dedicated to established religion and its scientific ally, the Newtonian philosophy.

Robison, the University of Edinburgh's professor of natural philosophy, contended that continental masonic meeting houses had been annexed by the revolutionary 'Order of the Illuminati'. 'The Lodges', he told his readers, 'had become the haunt of many projectors

From Newton to Hawking: A History of Cambridge University's Lucasian Professors of Mathematics, ed. Kevin C. Knox and Richard Noakes. Published by Cambridge University Press.

and fanatics, both in science, in religion, and in politics.' Robison reckoned that the illuminati were supplanting the traditional, benign values of freemasonry with those of 'Jacobinism', a general tendency to embrace irreligion and anarchy. According to the distinguished professor, the Jacobin distemper was now poised to contaminate Britain. Without brisk intervention, Robison warned, the toxic writings of Voltaire, d'Holbach, d'Alembert and Diderot would topple the British constitution.

Part of the 'proof' that Robison tendered to convince readers of the conspiracy consisted in recent developments in English natural philosophy. His principal target was the great chemical philosopher and rational dissenter, Joseph Priestley. Robison alleged that Priestley had purposely misinterpreted the writings of Cambridge's Isaac Newton and David Hartley in his bid to promote a theory of matter and mind which would lead to atheism, regicide and anarchy, the trademarks of the French Revolution. Now, with the works of Priestley in hand, the 'illuminated' masons of Britain were primed to subvert the benevolent state.

Recent historians have tended to dismiss Robison's sensational allegations as the paranoid ravings of an opium-addicted lunatic. How serious – or real – the Jacobin threat truly was is still debated, but to contemporary Britons Robison's conjectures were anything but preposterous. Countless eminent writers did battle with the Jacobin 'scourge'. In Cambridge, Robison's diatribe was but one of dozens of anti-Jacobin tracts that could be found on booksellers' shelves. *Proofs of a Conspiracy* struck a familiar chord for dons, especially Isaac Milner, the university's seventh Lucasian professor and a close associate of the Edinburgh philosopher. A decade after the appearance of the *Proofs* Milner reminisced about his own efforts to quash the conspiracy. Invoking his part in the expulsion of one of Priestley's close collaborators, William Frend, Milner boasted that he had not

> been merely an indolent Spectator of what has passed during these late critical years. – Your Lord^{sp}. will remember that in the year 1793 when I was Vicechan^{cr}. of this Univers^{y}. I had a very difficult

> business to conduct respecting the trial of M^{r}. Frend of Jesus Coll[ege]: but Governm^{t}. are not aware how much depended upon the successful management of that affair. There then flourished in this place a School of Jacobinism, with M^{r}. Frend for its Leader, which has never been able to hold up its head since his Banishment. – It was crushed effectually; & at a most critical moment. – The difficulties I had to struggle with on that occasion are only known to myself.

Never one to underestimate his influence in the arcane arena of university politicking, the mathematics professor triumphantly declared: 'Since this event Jacobinism is scarcely heard of'.[2]

This chapter places the Lucasian professors within the contexts of scientific and political controversy in late-Georgian England. While the furore over American independence or the resurgence of millennial prophets would also be apt, the dread of Jacobinism is an appropriate theme around which to discuss the careers of the late-Georgian Lucasian professors since it draws together natural philosophy, mathematical practice, revolutionary politicking and the tensions between Anglicans and dissenters. Indeed, the interrelationship of these entities engrossed Isaac Milner and his predecessor, the introverted Edward Waring. Working in a period of great political ferment and dramatic transition in the sciences, the university's mathematical professors needed to come to grips with the significant changes in which enlightened men viewed the relationship between natural and political worlds.

Although the Lucasian professors could not have predicted it, theirs was the last generation of the natural philosopher (the next generation of academic scholars would see the advent of the professional 'scientist'). Nevertheless, during the tenure of these two Lucasian professors (1760–98; 1798–1820), their sanctuary of higher learning remained a place where research degrees were unknown and specialization discouraged. Theirs was a world in which a scholar might find himself the chair of chemistry one year and Regius professor of divinity the next (as Richard Watson did). Theirs was a world in which

Figure 28 Unlike Milner and Waring, who as undergraduates had to differentiate themselves from *hoi polloi* by distinguishing themselves on the Tripos, it was impossible to mistake the fellow commoners and noble undergraduates in their resplendent gowns.

religion continued to permeate most academic, and many social, activities. Late-Georgian Cambridge also remained an exotic space of resplendent garments, honour, status, gentlemanly pursuits and demeaning customs. With the first appearance of female undergraduates still a century away, women fitted into the university's intricate hierarchy only marginally. Amidst all of this, Cambridge remained a world of inscrutable ritual: 'Aegrotats', 'Apostles', 'gulphing it', 'huddling', 'jobatios' and 'Wranglers' were as familiar to the eighteenth-century Cantabrigian as they are foreign to us. The freshman was swiftly encultured; but it took decades to become a savvy player in the *microcosmographia academica*.

Neither Waring nor Milner questioned the values that underwrote this site of privilege, chauvinism and ceremony. Both, however, toiled hard to protect this domain and to ensure these values were disseminated throughout the empire. And, as mathematicians and natural philosophers, both were convinced that the sustenance of these moral values hinged upon a proper understanding of Isaac Newton's conception of the natural world. Their adversaries, by contrast, argued that this was actually a valueless project: the infinitesimals of the calculus, the application of negative numbers and the statements about nature derived from this use were signs that the university's edifice rested on nothing of substance. The story of Edward Waring and Isaac Milner, then, is primarily the ongoing struggle by dons to be Newton's true representatives, and to use this representative status to ward off attempts at radical reform, both within and without the university.

To set the stage for a discussion of the major campaigns that Milner and Waring waged in their careers, it is valuable first to see how Waring and Milner themselves have been represented by their contemporaries and by past historians. I then examine the professors' lectures and mathematical works in relation to the goals of other enlightened natural philosophers. In the final section, I focus upon two great controversies in which the professors were embroiled: the first involves the spirited endeavours of radical materialists such as Joseph Priestley and William Frend to foist reform upon the university; the second controversy, also involving the intersection of mathematical and biblical dispute, centres on the complex relationship between the formations of the Cambridge Bible Society and Babbage's celebrated Analytical Society. Both episodes illuminate the priority that the professors placed upon the role of spirit in their world and the ways in which the mathematical sciences could be deployed to sustain the traditional values of Cambridge. The episodes also help us to form a social history of late-Georgian mathematics that unravels an unexpected substantial continuity between the Georgian and the Victorian university.

'A HISTORY LEADING NOWHERE'?

If the Lucasian professors of the first half of the eighteenth century have been remembered as something less than heroic, their late-Georgian counterparts have been all but forgotten. Indeed, in his influential history of mathematics at the university, W. W. Rouse Ball tried to persuade his readers that one need not worry if this period remained permanently buried: 'You will not blame me I think, for making my account of the Cambridge mathematical school at this time little else than a list of names', he declared: 'Its history leads nowhere and hence it is not necessary to discuss it.'[3]

Rouse Ball's hasty dismissal echoed the complaints of earlier mathematicians who had been exasperated by the mathematical regime of the late-Georgian university. Some of these criticisms emanated from those who had experienced this regime first-hand. Charles Babbage, for instance, reminisced that upon entering Cambridge as a freshman in 1811, he found himself in an intellectual wasteland. He contrasted his 'energetic mind' with the 'disguised ignorance' of the mathematical tutors. He grumbled that the *Principia* had become a 'mill stone around the necks' of Cambridge gownsmen. Meanwhile, John Robison's successor as professor of natural philosophy at Edinburgh voiced similar complaints in his widely read review of Laplace's *Mécanique Céleste*. Although admitting that 'teaching the doctrines of Locke and Newton is sufficient to cover a multitude of sins', the professor none the less castigated his Cantabrigian colleagues: 'Newton is taught there in the way least conducive to solid mathematical advancement.'

These criticisms of the university's mathematical researches reflect a larger, pervasive sentiment concerning late-Georgian Cambridge. Eighteenth-century critics such as Samuel Johnson enjoyed mocking the 'stagnant conversation and private scandal' of the university's 'genuine idlers'. For Johnson, the college fellow became the epitome of languor and inebriation. Commentators since have derided the 'dull potations', 'torpidity' and even the 'supine slumber' of the dons. They suggest that the only thing which could rouse the drunken dons was the hotly contested mêlées to acquire sinecures. Indeed, in filling vacant professorships, expertise in manipulating the intricate networks of patronage, not academic excellence, seemed the order of the day: upon his election to the chair of chemistry, Richard Watson admitted that, 'At the time this honour was conferred upon me, I knew nothing at all of Chemistry, had never read a syllable of the subject; nor seen a single experiment on it.' With such self-incriminating evidence, it is compelling to follow the conclusion of one famous historian: 'The Professors of the eighteenth century have incurred the indignant scorn of posterity, and for the most part they deserve it.'

Neither Edward Waring (1734–98) nor Isaac Milner (1750–1820) have been immune from this 'indignant scorn', although, perhaps, criticisms of Waring have been slightly less scathing. Rouse Ball dismissed Waring's major mathematical works as 'defective in classification and arrangement'. 'Except [William] Emerson', another critic moaned, 'there is scarcely any writer whose works are so revolting as those of Waring.' Another detractor looked to 'a certain want of order and method in his mind' to explain Waring's shortcomings and grumbled that the mathematical characters in his manuscripts 'were often utterly inexplicable'. His moral character was also questioned, Waring being deemed 'one of the strongest compounds of vanity and modesty which the human character exhibits, the former [being] his predominant feature'. Modesty *qua* shyness often prevailed, however. In a letter to William Frend, Thomas Jones admitted that Waring was a 'scrupulously honest man', but then described the professor's painful timidity:

> He was shy before strangers & his address was embarrassed & awkward... He came to dine in our Hall at our festival on Trinity Sunday; it happened that there was no dish very near him which he liked; choosing to not give 'trouble' he did not eat a bite the whole time.[4]

Spending his later years in virtual isolation (although with his wife Mary) on his estate in Pontesbury, he became increasingly gloomy,

Figure 29 Portrait of Edward Waring.

withdrawn, and possibly mad. Bishop Richard Watson, never one to let his devotions cramp his lifestyle, speculated that Waring's excessive piety incapacitated the Lucasian Professor towards the end of his life. Waring, he said, had ended up 'sunk into a deep religious melancholy approaching to insanity'.

While Isaac Milner's piety has been mocked, no one has yet declared shyness as his foible. Rather, he was described by his Victorian biographer as 'the life of the party'. As one colleague reminisced, Milner's 'public dinners were very merry, but the *private* ones were quite uproarious'. True to Cambridge form, 'the bottle circulated very freely'. Despite his conviviality, and despite a hagiographic biography by his niece, Milner's reputation has not fared well in posterity. Criticisms of him have tended to focus on three shortcomings: his indolence, his religious hypocrisy and his superciliousness. According to Hugh Trevor Roper, Milner was the 'oracle of evangelical complacency'. Roper's verdict is severe: 'This noisy Christian . . . sounds a dreadful bore.'[5] Invoking Milner's evangelicalism, Gilbert Wakefield stressed that 'his theological conceptions were always, I confess, one of the inscrutabilities of mystery; a *heterogeneous* composition of *deistical* levity and *methodistical* superstition: disparaging the ceremonies of religion, and performing them with slovenly precipitation'. Ignoring such benevolent activities as his fight to abolish the slave trade, others have described Milner as 'an unappealing character', 'a thief', 'a glutton', 'awkward, arrogant, and contemptuous', as well as being 'indolent in all affairs which did not interest him'. Echoing a famous quip about William Whewell, the sardonic James Stephen recounted how Milner's

> sinecures were a burden, beneath which the most buoyant spirit could scarcely have moved with freedom . . . From such toils he might have broken away, if the wily courtesan had not thrown around him the more seductive bondage of social and colloquial popularity. The keen sarcasm, that 'science is his forte – omniscience, his foible,' could never have been aimed at any of the giants of Cambridge with more truth, or with greater effect, than at the former president of Queen's.[6]

No doubt Milner was a noticeable character, especially given his immense girth (one contemporary observed that he was 'the most enormous man it was ever my fate to see in a drawing room!'). Yet, while much of his notoriety came from tales of his parties, his corpulence and – as Thomas De Quincy noted in the *Confessions of an Opium Eater* – his 'habit', his intellectual endowment was renowned. One Victorian critic acknowledged that 'he had looked into innumerable books, had dipped into most subjects, whether of vulgar or of learned inquiry, and talked with shrewdness, animation, and intrepidity, on them all'. Even his long-time adversary, Gilbert Wakefield, esteemed Milner to be 'endowed with one of the most vigorous and penetrating minds I know'. Another foe admitted, 'the University, perhaps, never produced a man of more eminent abilities'.

Waring, too, from an early age was accounted a 'marvel', and, by his own account, he reckoned his contributions substantial: 'To every one of [the mathematical] sciences I have been able to make some additions, and in the whole, if I am not mistaken in enumerating them, somewhere between three and four hundred new propositions of one kind or other, considerably more than have been given by any English writer; and in novelty and difficulty not inferior'. Most of these additions were in the realm of cubic and biquadratic equations, including new techniques for combining independent equations. His 'formulas' and his 'problem' are still known eponymously: 'Waring's formulas' revealed fresh relations between equations' roots and coefficients. His 'problem' was actually introduced as the theorem that every integer is the sum of four squares, nine cubes, nineteen biquadrates, 'and so on'. Increasingly, the theorem became a 'problem' as future mathematicians failed to provide its proof. Only in 1909 was a general proof adduced, but it continues to test the ingenuity of number theorists and spawn novel mathematical researches.

For his contributions to algebra Waring was awarded the Royal Society's prestigious Copley medal. Waring himself proudly noted that d'Alembert and Le Grange had praised his *Meditationes Algebraicæ* 'as a book full of excellent and interesting discoveries in Algebra'. It is likely that the less analytically inclined Britons, on the other hand, found his work too obscure, his *Meditations* deemed by one Englishman 'one of the most abstruse books written on the abstrusest parts of

Algebra'. For this reason he depended on allies such as the Astronomer Royal, Nevil Maskelyne, to expound his virtues. After the caustic denunciation of Georgian mathematics by Joseph Jérôme Lefrançais de Lalande, Maskelyne countered the French astronomer's judgement: 'It is strange that the author should be ignorant of or inattentive to such eminent British Analysts such as Lyons, Emerson, Landen, Waring; who all flourished in the period he speaks of; the last of whom is still living, being the author of some of the greatest discoveries in Algebra, algebraic curved lines, infinite series, increments and fluxions.' More pithily, Dugald Stewart called him 'one of the greatest analysts that England has produced'.[7]

In many ways Waring and Milner followed a typical career path in what has been described as Cambridge's newly invented tradition of mathematical meritocracy. Milner was perhaps the luckier of the two to find himself at Cambridge. His father, ruined by the Jacobite rebellion of 1745, had died penniless; it seemed only his older brother, Joseph, would receive a university education, while weaving would become Isaac's lifelong trade. However, with substantial aid from his brother and another patron, Milner was unshackled from his apprenticeship and given the opportunity to enter Cambridge. Waring, on the other hand, was the son of a wealthy Shropshire farmer, and as such had fewer financial worries.

Notwithstanding their different backgrounds, both Waring and Milner matriculated as humble sizars, Waring at Magdalene in 1753, and Milner at Queens' in 1770. Milner found the experience of being a sizar – serving dinners to wealthier undergraduates and ringing the chapel bell – particularly humiliating, and pledged to eradicate the position 'when [he] got into power'. Given their unenviable status as sizars, Waring and Milner realized that 'hard reading' and a conquest of the Senate House Examination could be the panacea for their worldly woes. An outstanding performance in this gruelling four-day test of mathematical skill would guarantee their future within the university system of fellowships, tutorships, masterships and college livings. The world of ecclesiastical preferment, though more fickle, would also be open to them. Neither candidate was disappointed: both Waring and Milner became senior wranglers, Waring (in 1757) accounted as a mathematical 'prodigy' and Milner (in 1774) described as *'incomparabilis'*. With such esteem were their intellectual achievements regarded that they were both invited to join the snobbish Hyson Club, a university society dedicated to 'drinking China tea and engaging in rational conversation'.

After his triumph at the Senate House, Waring was immediately offered a fellowship at Magdalene College. Fortune smiled upon him when only a year later John Colson died, leaving the Lucasian chair vacant. Yet Waring was so young that a special royal mandate was needed to grant him an MA, a requisite degree for anyone gunning for a professorship. He also had to contend with two other strong candidates. On top of these hurdles, several college masters exhibited disdain for the young upstart during the weeks before the vote. Typically, Waring also faced attacks from fellows who favoured candidates from their own colleges. In an ill-disguised promotion of William Ludlam of St John's, William Powell lashed out at Waring's *Miscellaneæ Analyticæ*. Asking why Waring had troubled gownsmen with 'the least entertaining and least useful branch of Mathematics, the intricacies of modern Algebra', Powell advised him to 'destroy the present impression of his book'. Meanwhile, the Duke of Newcastle, the university's Chancellor and chief string-puller, was being wooed by Baron Francis Maseres in his bid for the professorship. Despite the Baron's stratagem, and despite the venom of his detractors, Waring narrowly won the election.

Other elections in which Waring was a candidate were less fraught. Members of the metropolis's Royal Society gave him the nod in 1763; fellows of Göttingen and Bologna elected him to their royal societies soon thereafter. He also became a commissioner on Greenwich's Board of Longitude during the committee's most critical years. Back in Cambridge, he frequently moderated exams in the university's Senate House: there, 'his examinations for the Smith's prizes were considered the most severe test of mathematical skill in Europe'. In a move that seems peculiar to us but was perhaps representative of the eighteenth-century don, Waring took the degree of MD in 1767 and began practising physic in London and Cambridge. This was the least successful of his endeavours, short-sightedness and a nonexistent bedside manner being cited for his failure as a physician.

Isaac Milner's professional advance at once evinces the rapidity with which a wrangler might rise, and the heterogeneous composition of a wrangler's career. With his undergraduate laurels, he secured a fellowship at Queens' College in 1776 and was hired by the Astronomer Royal as a 'computer', making endless calculations. Milner's semi-menial computational labour was short-lived, however. After the publication of three impressive papers in the *Philosophical Transactions*, he was duly elected a fellow of the Royal Society (1779). Meanwhile, after taking orders, he was presented with the parish of St Botolph's in Cambridge. Then, in rapid-fire succession he became a college tutor, lecturer of chemistry, and, in 1783, Jacksonian professor of natural philosophy. As the chair's founder stipulated, the Jacksonian professor was expected to make 'further discoveries' in natural philosophy which would 'best tend to set forth the Glory of the Almighty God, and promote the welfare of mankind'.[8] For Milner, this included joining forces with Joseph Banks and Humphry Davy in a quest to cure gout, as well as teaming up with the university gardener to investigate the 'inoculation of plants'. He invented a luminous water clock and an ingenious oil lamp, and, as he put it, 'constantly dabbled' in his private laboratory. Like other Oxbridge professors, he served on the Board of Longitude and consulted on several large engineering projects. In 1789 he developed an important process to fabricate nitrous acid, a key ingredient in the manufacture of gunpowder. Ironically, it was the French Jacobins who first utilized the process on a large scale after England blockaded their nitre supply.

University administration and the evangelical cause remained central to Milner's life. In 1788 he stood unopposed for the presidency of Queens', while he served as the university's Vice-chancellor in 1793 and 1810. He often meddled in the affairs of the University Press and stuck his nose into town politics. He became a doctor of divinity in 1789 while an intimate friendship with William Wilberforce and a timely dispatch to William Pitt helped him to secure the position of Dean of Carlisle in 1791. Although he never produced a book on mathematics, he published several volumes on religious subjects, including the substantial *History of the Church of Christ*, co-written with his brother. Even if the last paper that Milner published on a mathematical subject was in 1778, he none the less coveted the chair of Isaac Newton. In 1798, after waiting impatiently for the death of Edward Waring, he finally gained the professorship.

NEWTON AND 'NOBODY'S FRIENDS'

By the time that Edward Waring secured the Lucasian professorship in 1760, the post was unequivocally 'Newton's chair'. Albeit imprecisely, the works of Newton defined the eighteenth-century university. Professors and proctors, bursars and beddells, tutors and tuted seemed univocal in their adulation of Sir Isaac Newton. Such was his esteem for the 'Great Man' that, upon graduating as senior wrangler and Smith's prizeman, Isaac Milner was 'tempted to commit his first act of extravagance'. As his niece recalled, 'In the pride of his heart, he ordered from a jeweller a rather splendid seal, bearing a finely-executed head of Sir Isaac Newton.'

Milner's infatuation with Grantham's favourite son was not unique. Representations of and tributes to Newton flourished in mid-century Cambridge as a medley of statues, portraits, medals, stained-glass windows and poems were hewn, painted, struck, glazed and composed. Almost unanimously, the divines, tutors and professors of Cambridge opined that studying Newton's natural philosophy and fluxions helped undergraduates develop mental and moral discipline which would aid them in their future, whether at the law courts, on the estate, at the stock exchange, in the pulpit or in parliament. When prodded about this commitment to Newton's *œuvre*, Milner retorted: 'I have often contended, the best answer that we could give to persons who sometimes accuse the resident members of the University of Cambridge of employing their time too much in mathematics and natural philosophy, was to inform them, that our lectures on these subjects were, indirectly at least, subservient to the cause of religion; for that we endeavoured, not only to fix in the minds of young students the most important truths, but also to habituate them to reason.' In a letter to William Pitt, he evoked the Prime Minister's mathematical expertise: 'I never yet heard that he found the habit and accuracy & method, which in those subjects is indispensable, to be . . . an Incumberance or Obstruction to sound reasoning in practical matters.'[9]

Yet the university's tutors and professors were forced to toil hard to make their own unique interpretation of Newton's works seem self-evident. Much to their consternation, there were countless interpretations of Newton's philosophy circulating throughout Europe, most of which were antithetical to interpretations by Newton's proponents in Cambridge. For instance, William Ludlam dissented from his fellow Cantabrigians in wanting to modify Newton's second law. Perhaps still smarting from being passed over in the last election for the Lucasian professorship, Ludlam dared to cast doubt on Newton's conception of force. In a paper presented to the Royal Society (but not printed in the *Transactions*) he justified his dissent: '*Nullius in verba* is the motto and maxim of the Society, and this the author hopes may be his apology for questioning the Logic of so great a master as NEWTON.' Such intestine discord did not sit well with the Cambridge Newtonians, however. In the 1770s Milner published a series of papers in the *Transactions* to quell disputes about Newton's three laws. He reckoned that recent quarrels had arisen from 'the violence of prejudice and party-spirit', not from any inconsistencies in Newton's reasonings. Culprits included Leibnizians, French rationalists and British engineers, all of whom had tried to test Newtonian mechanics with unsuitable experiments. In exposing the errors of artisans and continental philosophers, Milner also revealed the depth of his loyalty to his master: as he warned Jean D'Alembert, 'when we

Figure 30 A 'Frontispiss' by William Hogarth. This 1763 engraving accompanied a pamphlet by the FRS Gregory Sharpe, written, but never published, against the Hutchinsonians. Hutchinson is depicted as a devilish witch, who relieves her/himself upon the work of Isaac Newton, while the Hutchinsonians are represented as rats. While some of the living vermin storm Newton's telescope and *Principia*, another has died by feeding on the volume of Hutchinson.

venture to differ from Sir Isaac Newton in these matters, it is with the utmost difficulty that we can arrive at certainty'.

And, despite heroic labours by the likes of James Jurin, Samuel Clarke, John Colson and Nicholas Saunderson (whom John Gascoigne has discussed in the previous chapter), overtly anti-Newtonian sentiments continued to bedevil gownsmen. Cartesians, who continued to linger in academia, chortled at the notion of empty space. So too did Hutchinsonians. Some wits – like William Hogarth – found Hutchinsonianism philosophy ripe for satire, but many intelligent readers considered it a legitimate philosophy. Deriving their name from the author of *Moses's Principia*, the Hutchinsonians claimed they did not intend to eradicate entirely the Newtonian philosophy, only to make it conform with scripture. Branded 'Nobody's Friends' in their stronghold of Oxford, these High-church Tories pointed to William Whiston and Samuel Clarke to show that Newtonianism was the philosophy of vain heretics. They believed that Newton, in neglecting the Bible, had erroneously introduced the malevolent concepts of the void, force and action-at-a-distance. Hutchinsonians countered with a plenum in which the physical contact of material bodies was the only thing that could effect change. They posited the existence of a triune æther, modifications of which were the universal agents of fire, light and air (or spirit). Unlike gross bodies, these subtle agents corresponded to the Holy Trinity and were responsible for the formation and preservation of the material world. It was a philosophy in which many things stood above man's corrupted reason.

To battle anti-Newtonians and to show that they were the rightful heirs of Newton's philosophy, the Cambridge Newtonians produced an assortment of textbooks, sermons, pamphlets and demonstration devices. The objective in these works was fourfold: to remove ambiguities in the *Principia* and the *Opticks*; to show that by donning the spectacles of the university Newtonians one could see clearly the perpetual activity of a benevolent deity and the role of immaterial substances within His system; to remind Britons (and especially Hutchinsonians) that Newton's philosophy was perfectly consistent with revelation as understood by Anglicans; and to stymie French Deists who seemed to be using Newton's philosophy to undermine the traditional social order.

In their endeavours to make these truths transparent to Britons and foreigners alike, Waring and Milner were accompanied by a number of industrious dons. Like Milner, many of these Cambridge Newtonians had left their humble stations as bricklayers, dyers, weavers and water carriers in order to brave the 'meritocratic' system of Cambridge. One of these victorious wranglers, Thomas Parkinson, produced two influential texts, his 1785 *System of Mechanics* and his 1789 *System of Mechanics and Hydrostatics*, that detailed how Newtonian mechanics was based on laws which were experimentally demonstrable, incontestable and incompatible with materialist philosophies. Samuel Vince, Plumian professor of astronomy and experimental philosophy, used celestial motions and mathematical analysis in three weighty books – *Observations on Gravitation*, *Observations on Deism* and *A Confutation of Atheism* – to affirm the existence of a '*superintending* Providence'. These publications chimed in marvellously with the natural theology of the great Christ's College don, William Paley, who produced a number of works that argued for the existence of a wise Creator from a minute examination of animals and fauna. Yet, whereas Paley used a natural theology of pigeons and plants to express his distaste for French deist philosophy, Vince appealed to the heavens and the existence of immaterial forces. For Vince, the study of attraction and repulsion was a critical enquiry, 'particularly . . . when many of the most eminent Philosophers upon the Continent have been endeavouring to account for all the operations of nature upon merely mechanical principles, with a view to exclude the Deity from any concern in the government of the system'.

For their part, the Lucasian professors produced the homophonously entitled *Essay on the Principles of Human Knowledge* and *Essay on Human Liberty*. Like their colleagues, they cringed at active-matter philosophies which denied the existence of immaterial substances. Such philosophies, in their estimation, limited God's power, mocked the Holy Trinity and snatched free will from the individual. A natural world without spirit was ineluctably *un*-Christian. Attributing activity to matter effaced the fundamental differences between good and evil, jeopardized the concept of salvation and paved the way for heresies such as Socinianism, Deism, predestination and transsubstantiation. For this reason, Waring used his *Essay on Human Knowledge* to show why his training in mathematics and Lockean philosophy had convinced him that 'Freedom of will can arise from no mechanism.' He scoffed at those who proposed that a philosophical system of self-moving matter might be incorporated into any benign form of Christianity: 'Many writers have defended the materiality of the soul; but we must from its properties . . . conclude the substratum of the soul to be different from that of meer matter.'

Similarly, Milner used his *Essay on Human Liberty* to elucidate how Cambridge's expertise in mathematics and experiment might be used to steer between the twin evils of Calvinism and Arminianism, both of which, he assumed, had arisen when theologians had misconstrued their natural environs. Being 'naturally led to observe a remarkable difference between the operations of matter and of the mind', Milner outlined why this affected morality:

> Immaterial substances are essentially different from material ones; and, as far as we can judge from experience, seem to be possessed of certain active principles, which are best described by saying that they are the direct contrary of inertness which is allowed to be essential to matter . . . The essence of virtue and vice, whether we mistake in our opinions or not, never depends on the external motives, but always on the internal disposition of the agent.

How did a university professor ascertain such vital truths? 'You must be sensible that there is no part of this reasoning', Milner explained, 'but what is founded on the common and established principles of Experimental Philosophy.'[10]

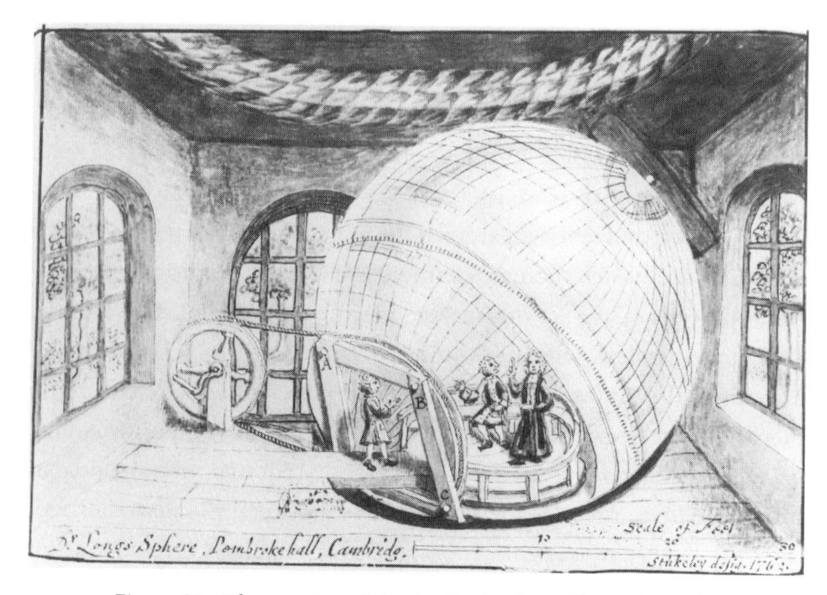

Figure 31 The rotating globe, in Pembroke College, drawn by William Stukeley, one of Newton's greatest advocates. The globe, with iron ribs and tin plates, was designed in the late 1750s by Roger Long, first Lowndean professor of astronomy. Experiencing this Newtonian universe was intended to 'lift up their [the students'] thoughts to the adorable author and Lord of all things'. It was dismantled in the 1870s, and sold for scrap metal.

Experiments and demonstration devices therefore attained a crucial niche within Cambridge's scientific milieux. At Pembroke College, generations of students were encouraged to enter a grand, rotating celestial globe so that they might better comprehend what Waring called the 'supreme Architect's infinite machine'. Samuel Vince accumulated a vast array of experimental equipment – ranging from 'Thunder houses' and Leyden jars to 'whirling tables' and 'Magic Pictures' – to inspire awe in his students. George Atwood devised a remarkable machine of weights and pulleys that gave students the opportunity to see passive material bodies being moved by means of demonstrable forces. While Waring never lectured ('The profound researches of Dr. Waring,' Samuel Parr tactfully mused, 'were not adapted to any form of communication by lectures.'), Milner, as

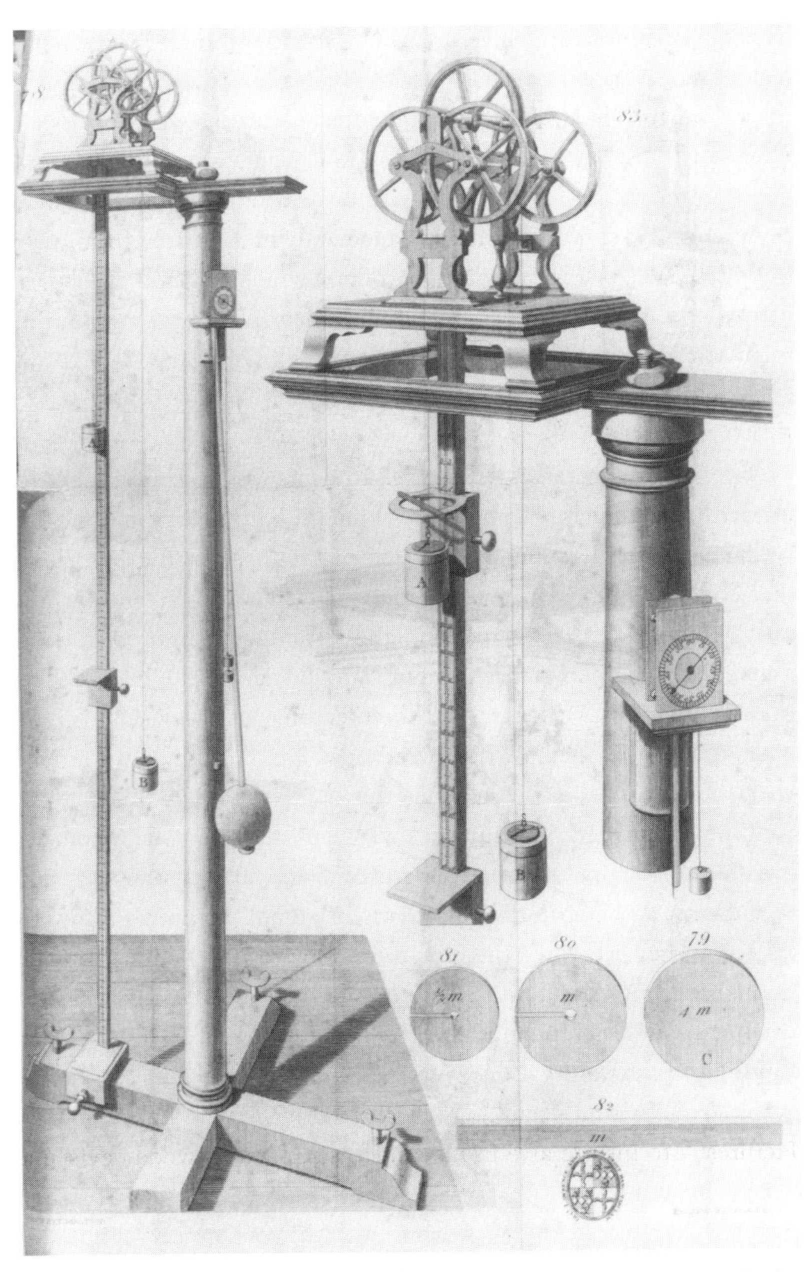

Figure 32 Atwood's machine. Developed by George Atwood, the instrument helped dons demonstrate to younger gownsmen many of the principles of rational, Newtonian mechanics.

Jacksonian professor, gained a reputation as a 'first-rate showman': 'Dr. Milner was always considered a very capital lecturer', reminisced one former student, 'The Chemical lectures were always well attended; . . . the audience was always in a high state of interest and entertainment.' Milner understood the value of 'astonishing' entertainment. On a small scale, his spectacular demonstrations mimicked omnipotence. As the Magdalene College wrangler, Samuel Cooper argued, the astute philosophers of Cambridge, 'like Newton with the prismatic glass', had caught a 'glimpse of the DIVINE MIND'.[11]

ZERO TOLERANCE

The wonderful spectacles with which students were mesmerized were not without problems. Some critics found it hard to differentiate between university lectures and more base forms of scientific diversions that could be found at the town's fairs and theatres. Given that the university lecturers used many of the same demonstration devices as popular showmen this sentiment is understandable. Undoubtedly, the dons would have denied that their lectures were devised to compete with the 'learned pigs' of Stourbridge fair, but they did contend with the likes of the 'anti-combustible Mr. Gyngell' and 'Sieur REA Sen', who delighted his audiences with a wide variety of 'Astonishing Philosophical, Mathematical, and Mechanical AMUSEMENTS', not to mention the 'unparalleled enchanted lemon'. Although a mystical lemon never entered his repertoire, George Atwood was known as the university 'conjuror'. After considering Richard Watson's fiery chemical demonstrations, T. J. Mathias reworked Pope's couplet about Newton's vanity:

> Eager, they press, they shout, they stare, they gape,
> And view a Watson, as we view an Ape!

Another dissatisfied student condemned Milner's optical lectures as 'nothing more than Magic Lantern shows', and further complained that, as Jacksonian professor, Milner's chemical lectures were more 'contrived to amuse' than enlighten.

Joseph Priestley, arch-enemy of both Edinburgh's John Robison and the coterie of Cambridge Newtonians, also worried over the distinctions between showmen and philosophers proper. Priestley, Britain's foremost Unitarian and chemical philosopher, contrasted the use of scientific spectacle with the work of genuine natural philosophers who painstakingly derived systems of nature from a perpetual round of tedious experiments. When all was said and done, he decided that Oxbridge scholars belonged to the former group. Lazy and over-wooed by spectacle, gownsmen clung to superficial, antiquated scientific beliefs. Writing to William Pitt in 1787, Priestley damningly reported: 'your Universities resemble pools of stagnant water secured by dams and moulds, and [are] offensive to the neighbourhood'.

Priestley advocated a system of nature derived from a belief that Cambridge Newtonians found repugnant: a universe of active matter that did not need to be guided by immaterial substances. In contrast to the stagnation of Cambridge, Priestley claimed that the universe was comprised of a conglomeration of forces, including electric, chemical and *thinking* powers. Especially revolting to the Cambridge Newtonians was the fact that, in developing his philosophy, Priestley made extensive use of the work of David Hartley, who had been the protégé of Nicholas Saunderson. In his 1749 *Observations on Man*, Hartley used Newton's æther to develop his own 'associationist' conception of mind which attributed thinking to undulations in subtle matter, not to an incorporeal substance. Combining Hartley's speculations with years of experimentation on electrical fluids and 'airs', Priestley determined that any philosophy that posited immaterial substances was heretical. In direct opposition to statements by Waring and Milner, Priestley enunciated that 'the doctrine of a separate immaterial soul . . . is a notion borrowed from heathen philosophy'. Moreover, in denying the divinity of Christ, he rejected crucial Anglican doctrines such as original sin, redemption and the efficacy of the sacraments.

Waring, Milner and the Cambridge Newtonians were also harangued by several champions of Priestley within the university, the most ardent being William Frend, mathematics tutor at Jesus College. Frend concurred with Priestley's censure of the conservative establishment, and combined Priestley's matter theory with his own etymological and mathematical critique of university life. According to the young mathematician, the analogue to the corrupt language of Anglicanism could be located in the mathematically based writings of Cambridge Newtonians. Frend – 'the last of the learned anti-Newtonians,' as Augustus De Morgan deemed him – took up where Bishop Berkeley and Baron Maseres had left off. Although Berkeley had been a High-church Tory, Frend agreed with the bishop's condemnation of infinitesimals as the 'the ghosts of departed quantities'. Maseres, on the other hand, was a Unitarian who had attacked the 'nonsensical unintelligible jargon' of Newtonians during his ill-fated campaign to secure the Lucasian professorship. He was especially repulsed by negative numbers: 'It is by the introduction of needless difficulties and mysteries into algebra (which for the most part take their rise from the supposition of the existence of quantities *less than nothing*)', Maseres grumbled, 'that algebra . . . has been rendered disgusting.'

Frend's condemnation of negative numbers was more severe: 'negative numbers are abortions, and impossible [negative] roots are unnatural crimes'. Similarly, he ridiculed the unintelligibility of Newton's method of fluxions. As a Unitarian, he suspected that the use of infinitesimals and negative numbers was an integral part of a corrupt alliance between the university and the Church of England: 'Athanasian [i.e. Trinitarian] theology and Cambridge mathematicks are very much on par', he groaned. It was a conspiracy against plain reason, driven by mystery and the desire of Cambridge dons to retain their privileged status. 'There is . . . no danger to modern [orthodox] theology from the mathematical knowledge taught at Cambridge', Frend told readers of *The Gentleman's Magazine*,

> for the mathematicians who have swallowed the notions, that a quantity may be less than nothing; and that an infinitely great

> quantity multiplied into an infinitely small quantity is something; and that a quantity less than nothing multiplied into another quantity less than nothing is also something; and a hundred other articles of the same kind; are not likely to be shocked at the dreams of Athanasius, or the tenets of [Anglican priests].[12]

The negative roots that Waring offered as real solutions to cubic and biquadratic equations were exemplary of the nonsense that corrupted the university's mathematicians and divines. As Frend later told Augustus De Morgan, he could do little else than hope that 'the figment $\surd - 1$ will keep its hold among the Mathematicians not much longer than the Trinity does among the Theologians.'

Meanwhile, to eradicate the 'idle inventions' and 'ghost stories' that plagued the Church of England, Frend joined Priestley's team, who were preparing a new translation of the Bible. If they thought they could reform the Anglican Church by expunging references to immaterial substances in the Bible, they were mistaken. Even before the French Revolution, Priestley and his followers were attacked from all quarters. An English Catholic challenged Priestley's exclusion of spirit: 'why do you not, as a *metaphysician*, aim to rise above the *visible world of matter*, where you may discover the existence and reality of other beings, whose ethereal forms cannot be confined in a tub of water, or a basin of quick-silver; nor be extracted by friction from a globe of glass; nor infine be analysed by all the powers of chymistry?'[13] An irate Anglican philosopher complained, 'Sir Isaac Newton ... never pretended to any knowledge about the essence of God or his mode of existence. But, perhaps, your vanity and presumption may be occasioned by your improvements in experiments, and the great facility with which you analyze an ounce of air.' In Cambridge, the wrangler Richard Omerod was only one of dozens to attack Priestley's 'fanciful and vague interpretations' of both scripture and chemical phenomena. In response to the 1788 publication of *Vindiciæ Priestleianæ: An Address to the Students of Oxford and Cambridge*, the Oxonian and Hutchinsonian George Horne fired a tremendous volley at Priestley:

'You have given the world much fixed air: let it have some fixed principles.'[14] Pointing at the mischief brewing in Priestley's 'theological laboratory', Horne declared, 'we already see the ditch filled with bodies, all of his own killing'.

After the French Revolution began, things got really tough for Unitarian reformers. They were now identified as 'Jacobins', power-hungry and bloodthirsty. Edmund Burke famously evoked the 'wild gases' and 'insurrectionary nitre' of 'alchemical legislators' in his censure of Priestley. And, according to John Robison's *Proofs of a Conspiracy*, Priestley had reduced the workings of the mind to the 'quiverings of some fiery marsh *miasma*'. By hijacking the 'unmeaning jargon of Dr. Hartley,' Robison reckoned that Priestley was 'preparing the minds of his readers for Atheism'. 'For, if intelligence and design be nothing but a certain modification of *vibratiuncloer* or undulations of any kind,' reasoned the Edinburgh professor, 'what is supreme intelligence, but a mere extensive and refined undulation, pervading and mixing with all others?'

In Cambridge, the university Newtonians determined that Frend's philosophizing was equally as nefarious. The masters were determined to show their patrons in Whitehall that they too were 'foes to infidels and Jacobins'. In 1793, after Frend published a radical pamphlet which suggested that the French regicide was not problematic, twenty-seven senior members of the university visited Milner in his lodge at Queens'. They demanded that as Vice-chancellor he needed to put an end to Frend's radical pamphleteering. The 'Twenty-seven' (whom Frend referred to as 'the cubicks', at once deriding their trinitarianism and Waring's cubic equations) brought the mathematician to trial in the Senate House, alleging that he had transgressed the same ancient statute that had resulted in the expulsion of William Whiston in 1710. Milner served as the trial's judge. Despite passionate support from Samuel Taylor Coleridge, one of Frend's rebellious students, 'it was apparent that the vice-chancellor was determined to convict'. Frend, Milner concluded, had 'defamed the public liturgy of the established church . . . and ridiculed the most sacred offices of

Figure 33 Portrait of Isaac Milner.

religion'. His expulsion from the university was appropriate punishment.

Before leaving the courtroom Frend could do little else than berate Milner: 'From one whose early years were employed in labourious occupation of mechanick life, the manners of a gentleman and the taste of a scholar are not to be expected.' Frend was no nobleman, but none the less believed that Milner's coarseness and desire for monetary riches had compromised his academic integrity. After his banishment, Frend continued to take great pleasure in ridiculing

Cambridge Newtonians. For one fashionable London magazine he composed a parody of his famous trial in which both Waring and Milner were lampooned. Milner is portrayed as a place-seeking ogre, willing to make 'something out of nothing' to impress his patrons. Waring, his partner-in-crime, is also depicted as a vulgar toady: he hawks and spits and blows his nose before exclaiming that 'matter may be divided into parts infinitely smaller than the infinitely smallest part of the infinitesimal of nothing', and that to deny this would be 'a most cruel innovation and usurpation, tending to destroy all just subordination in the world, making universities superfluous, levelling vice-chancellors, doctors, and proctors . . . to the mean and contemptible state of butchers and brick-layers'. Frend's point is obvious: as former labourers, the Cambridge Newtonians were financially dependent upon the establishment, and as such were prepared to perpetuate meaningless scientific stories that served the interests of the state. As Frend imagines Waring to boast triumphantly in the courtroom, 'the science of *nothing* is taught in the best manner in the university'.[15]

With the death of Edward Waring in 1798, Frend continued to hound Milner. He even declared himself a candidate for the Lucasian professorship, contending that, as president of Queens' College, Milner was ineligible for Newton's chair. Milner, however, held all the trump cards. The surviving ballot for the professorship reveals his influence in Cambridge: no one but Milner was on it. Yet if Milner reckoned his days of dispute were behind him, he was mistaken. That perpetual irritant to both Waring and Milner, William Ludlam dogged the Milner brothers with a series of pamphlets which contended that they misunderstood scripture metaphors since their understanding of the Word was too 'mechanical'. Shrewdly, Milner waited for Ludlam to die before publishing his response.

Another religious controversy in which Milner was embroiled flared up in 1812. The dispute, which received great attention in the press, involved none other than William Frend's High-church cousin and the Lady Margaret professor of divinity, Richard Marsh. The feud was precipitated by the formation of the Cambridge auxiliary branch of the British and Foreign Bible Society. In other parts of the empire the Society had great success distributing free Bibles to 'infidels' and the 'poorer sort'. In Cambridge, however, the society was controversial since the Bibles were distributed unaccompanied by the Anglican Book of Common Prayer. Some Anglicans, like Marsh, presumed that without the guidance of the prayerbook, readers would formulate dangerous misinterpretations of scripture. Marsh announced exactly this in a sermon at St Paul's cathedral.

Since the Society comprised low-church evangelicals, Milner soon became its vocal champion. While Society members believed that 'Scripture was so perfect that any comment [upon it] was unnecessary', Milner suggested that Marsh had created his own dangerous hypothesis: 'that the Bible alone has a tendency to make men dissenters'. In addition to this seemingly anti-Anglican hypothesis, Milner evoked Marsh's appalling track record of wild conjectures concerning the composition of the New Testament. In his *Strictures on Marsh*, he summed up the divinity professor's wont to use hypotheses and probabilities in theology, 'the queen of the sciences':

> What a notion indeed, will men form of Cambridge academics, when they observe the use which they make of their attainments in mathematics and natural philosophy, subjects on which they are by many persons supposed to bestow too much time and attention; when they observe that they transfer into religious inquiries of the greatest moment, not the safe, solid, and sound principles of the Newtonian philosophy, but rather principles which resemble the dangerous and fanciful levities of Des Cartes! . . . Des Cartes was not without genius and invention, but, was vain, bold, and precipitate; extremely injudicious, and fond of novelty . . . [16]

It may seem surprising that a mathematician-philosopher would decry 'novelty'; but Milner worried that the 'extravagant hypotheses' in religion and science would feed off each other and produce monstrous systems. For this reason, when Milner had first become Lucasian professor, he pledged to the celebrated evangelical, William

Wilberforce, that he would eradicate 'knick-knackery' from mathematics. Although this pledge was never realized in the form of a mathematical treatise, Milner never stopped fretting about the superfluities in the sciences – whether the science of algebra or the science of theology – in Cambridge.

Charles Babbage did not share Milner's anxiety about Cambridge academics. Rather, he found the Bible Society debate amusing, and 'thought it . . . a good subject for a parody'. Ferociously satirizing both of the warring tribes, Babbage transformed the central issues of the debate into a mathematical allegory. He fantasized that he and his undergraduate cohorts could change Cambridge's dogged adherence to Newtonian fluxions. Imitating a Bible Society poster, Babbage sketched an advertisement for an Analytical Society dedicated to distributing the pure gospel of continental mathematics as found in Silvestre Lacroix's 1802 *Traité Élémentaire de Calcul Différentiel et de Calcul Intégral.* As members of the Bible Society maintained that scripture was so flawless that any comment was superfluous, he claimed that 'the work of Lacroix was so perfect that any comment was unnecessary'. Convinced in the efficacy of Leibnizian mathematical notation over the dots of Newtonian fluxions, he punned on the strengths of '*D*-ism' over the 'dot-age' of Cambridge dons.

Babbage's friend, Michael Slegg, believed that the scheme was 'too good to be lost'. With John Herschel, George Peacock and three others, they drew up a 'Plan for the Society': as missionaries of analysis, the society would oppose the *de rigueur* of Cambridge's undergraduate curriculum. 'Geometry, & Geometrical demonstration', for the Society's members, were 'contrary to *its* ultimate objects.' They were especially enamoured with the power of French analytical mathematics and wanted to 're-import the exotic' methods. Despite waning enthusiasm from all but Babbage and Herschel, the Analytical Society published its *Memoirs for the Year 1813*, less than two years after its formation. Promising in its preface to 'aid to that spirit of enquiry, which seems lately to have awakened in the minds of countrymen, and which will no longer suffer them to receive discoveries in science second hand', the *Memoirs* contained but three papers: one by Babbage on continued products, and two by Herschel on trigonometrical series and finite differences.

The next year Herschel fashioned himself and Babbage as 'the ringleaders, if not the only actors in this literary assault upon the peace and quietness of the world'. He and Babbage had their own crusade and were happy to be branded 'infidels'. And along with mathematical 'infidelity', Babbage, like William Frend before him, also displayed great enmity towards the Anglican Church. His disputation 'Acts' in his second year seemed so heretical that he only managed to obtain an ordinary degree, which later became a substantial obstacle in his bid for his Lucasian professorship. Nevertheless, in 1817, George Peacock managed to import French mathematics into the Senate House Examinations. As Peacock triumphantly reported to John Herschel, 'The introduction of *d*'s into the papers excited much remark. Wood, Vince, Lax, & Milner were very angry & threatened to protest against [the infiltration of] French mathematics.'[17] To the original members of the Analytical Society, it seemed that the values of the eighteenth century could finally be left behind. The story seems complete when, in 1828, only eight years after Milner's death, Babbage was named the university's tenth Lucasian professor.

CONCLUSION

Contrary to W. W. Rouse Ball's assessment, the history of late-Georgian mathematics at Cambridge is rich and does indeed lead somewhere. The world of the eighteenth-century Lucasian professors was not simply a somnolent one against which heroic, rebellious characters like Charles Babbage would inevitably fight. Although the formalist mathematics of Babbage's Analytical Society was a means of rejecting the established institutions of social management and spiritual guidance, recent historians have argued that the reforms of Babbage and Herschel have been overemphasized. The late-Georgian dons were not ignorant of French mathematics, as Babbage charged. Edward Waring switched indifferently between the *dot*age of

Newtonian fluxions and the *de*ism of Leibnizian calculus. His colleague, Robert Woodhouse, published his *Principles of Analytical Calculation* in 1803, a full decade before Babbage's attempt to 'import the exotic'. Milner himself bequeathed a state-of-the-art mathematical library to his college in 1820, replete with the works of Biot, Laplace, Lagrange and Lacroix. And, though Babbage envisaged himself as a heretical outcast during his undergraduate years, he did not irrevocably repulse Milner: 'Mr. Babbage's talents and attainments as a person of profound knowledge and extensive acquirements in difficult branches of mathematics and general science', Milner gushed in a letter of reference for the young mathematician, 'place him in the very highest rank of philosophers of the present time.'[18]

Neither Edward Waring nor Isaac Milner revolutionized the way that mathematicians or natural philosophers conceived the natural world, but their careers as Lucasian professors reflect the complex contexts in which mathematics was practised, evaluated and taught in the Georgian university. Even if Milner admitted the power of the new analytical mathematics, he had good reason to be furious about the intrusion of Leibnizian *d*s on the Tripos exams. Milner's generation of Cambridge mathematicians did not reject French '*D*-ism' because they were unable to recognize its analytic efficacy, nor because they felt it was their patriotic duty to stick unfalteringly with Newton's fluxions. Rather, like Newton himself, they understood fluents and fluxions to be an intricate part of Newton's entire theological and philosophical corpus. As part of the conception of preexisting unbounded space and absolute flowing time, and argued to be consistent with the 'Geometry of the Ancients', the method of fluxions was intimately bound to Britons' understanding of God's omnipresence and dominion in the universe. For these late-Georgian mathematicians, it was hard to disagree with Newton's own claim that, in contrast to Leibnizian differences, his 'Method [was] derived immediately from Nature her self'.

Yet, although both Waring and Milner justified the privileged place of mathematics at the university in terms of its subservience

eighteenth-century Cambridge see John Gascoigne, *Cambridge in the Age of the Enlightenment* (Cambridge, 1989); D. A. Winstanley, *The University of Cambridge in the Eighteenth Century* (Cambridge, 1922).

4 T. Jones to William Frend, 20 May 1799. Cambridge University Library MS Add. 7886.138.

5 Hugh Trevor- Roper, 'The second reformation', *Times Literary Supplement* (12 February, 1993), 4.

6 James Stephen, *Ecclesiastical Studies* (London, 1849), p. 233. See also Mary Milner, *Life of Isaac Milner* (London, 1842).

7 For Waring's work, see Edward Waring, *Meditationes Algebraicæ* (Cambridge, 1770). See also Ian Stewart, 'The Waring experience', *Nature*, 323 (1986), 624–5.

8 The stipulations of the Jacksonian professorship are reprinted in John Willis Clark, *Endowments of the University of Cambridge* (Cambridge, 1904), pp. 206–14.

9 Isaac Milner, *Strictures on Some of the Publications of the Reverend Herbert Marsh, D.D.* (London, 1813), p. 230. Milner's comments on Pitt in Isaac Milner to William Pitt, 7 November, 1791. CUL MS Add. 6958.1016.

10 See Isaac Milner, *An Essay on Human Liberty* (London, 1824), pp. 5–6, 10, 20–21. See also Edward Waring, *An Essay on the Principles of Human Knowledge* (Cambridge, 1794), pp. 48–9.

11 For Gunning's experiences with Milner see *Reminiscences*, pp. 236–8. See also L. J. M. Coleby, 'Isaac Milner and the Jacksonian chair of natural philosophy', *Annals of Science*, 10 (1954), 234–57.

12 Frend, 'Farther remarks on St. John', *The Gentleman's Magazine* 70 (1800), 92–6. See also his 'A letter to the Vice-Chancellor' (Cambridge, 1798), 1–13.

13 See Joseph Berington, 'Remarks on Dr. Priestley's *Experiments on Air*', in *Letters on Materialism and Hartley's Theory of the Human Mind, Addressed to Dr. Priestley, F.R.S.* (London, 1776), pp. 220–1.

14 George Horne, *A Letter to the Revered Doctor Priestley by an Undergraduate* (Oxford, 1787), p. 3.

15 William Frend, *An Account of the Proceedings in the University of Cambridge* (Cambridge, 1793).

16 Milner, *Strictures*, p. 212.

17 Peacock to Herschel, 4 March 1817, Royal Society MS HS.13.247. For the Analytical see Menachem Fisch, '"The emergency which has arrived":

the problematic history of nineteenth-century algebra – a programmatic outline', *British Journal for the History of Science*, 27 (1994), 247–76; David Bloor, 'Hamilton and Peacock on the essence of algebra', in H. Bos, H. Merhtens and H. Schneider (eds), *The Social History of Nineteenth-century Mathematics* (Boston, 1981), pp. 202–33.

18 Milner's letter for Babbage cited in Mary Milner, p. 699. Babbage's memory is found in *Passages from a Life of a Philosopher* (London, reprint, 1992), p. 21.

19 Augustus De Morgan, *Newton: His Friend: and his Niece* (London, 1885), pp. 150–1; Charles Kelsall, *Phantasm of an University* (London, 1814), p. 33. For Newtonianism's meanings see Richard Yeo, 'Genius, method, and morality: images of Newton in Britain, 1760–1860', *Science in Context* 2, 2 (1988), 257–84.

6 Paper and brass: the Lucasian professorship 1820–39

Simon Schaffer

University of Cambridge, Cambridge, UK

> We are reading mechanics, which rather amuse me, and would much more, if in this our University simplicity were the fashion. As it is, we have to wade through darkness visible.
>
> William Makepeace Thackeray,
> Trinity College Cambridge, 1 November 1829[1]

In the years after Waterloo many Britons took the new opportunities of wealth and peace to escape south from their dark island for a better climate, the monuments of classicism and Catholicism, and the traces of more recent secular martyrs, such as the atheist Shelley and the philhellene Byron. In spring 1828 one of Byron's near-contemporaries from Trinity College Cambridge, the wealthy mathematician and convinced secularist Charles Babbage reached Rome in flight from London to recover from the effects of the deaths of his son and wife the previous summer. Accompanied by a workman turned valet, he had already visited northern European workshops, arsenals, instrument shops and mines. Never cowed by obstacles, Babbage first planned a journey to China, but war in Greece, where Byron had recently met his end, blocked the road east. So he took rooms on the Piazza del Popolo, toured the eternal city and travelled further south to Naples, the temples at Pozzuoli, hot springs at Ischia and explosions at Vesuvius. While working on the economic uses of volcanic heat, Babbage kept up contacts with his surviving family and friends, especially the astronomer John Herschel, another distinguished Cambridge graduate

From Newton to Hawking: A History of Cambridge University's Lucasian Professors of Mathematics, ed. Kevin C. Knox and Richard Noakes. Published by Cambridge University Press.

with whom he had previously scouted French museums and Italian alps. The able Herschel was now charged with managing Babbage's ambitious and troubled project to build a calculating engine for making astronomical tables in the British capital. Then, at the start of April 1828, the Roman papers reported to Babbage's apparent astonishment his election as Cambridge's new Lucasian professor of mathematics. His mother confirmed the report, as did his campaign manager Herschel. There was, however, a cloud on Babbage's horizon. Herschel reported sadly that the calculating engine 'gets on very slowly, and more is done on paper than in brass'.[2]

Babbage had firm views about the right relation between paper and brass. He reckoned paperwork should manage brass machinery. Rational mathematics must dictate to artisans the right construction of efficient engines and careful statistical surveys and graphical methods should govern the entire factory system of industrial Britain. But he also held that paperwork could be performed by brass. His new

Figure 34 A portion of Babbage's Difference Engine. In this portion of the addition mechanism, brass cogwheels rotate cylinders surrounded by paper on which are displayed the inputs and outputs of mechanical calculation. The machine was completed by Babbage's son, Henry, after the project to build the machine was abandoned in 1842.

calculating machines were designed to execute intelligent functions till then the prerogative of mind. Abstract mathematics, indeed the entire work of reason, he argued, was best understood as a system of machine-like operations subject to predictable laws. These claims had major political significance. Any institution whose legitimacy relied on the apparently occult or tacit skills of its members, whether artisan workshops, aristocratic courts or England's two ancient universities, was to be opened up to public scrutiny, its operations analysed in the name of mechanical efficiency.

Babbage, privately wealthy and untroubled by self-doubt, easily hitched his campaigns to other bourgeois demands for reform in the final years of the Hanoverian regime: campaigns for disestablishment of the Anglican church, extension of parliamentary suffrage and the transfer of political power to the economically productive sector of society. Machinery's powers could discipline the workforce and challenge old corruption. Four years after his election to the Lucasian chair, Professor Babbage put forward these principles in the first poll for the newly reformed House of Commons. His political allies urged mitigated change lest worse befall the landed establishment at the hands of truly dangerous plebeians. Babbage recorded the telling fact that, whereas in Italy, for example, above three times more workers were engaged in agriculture than in all other enterprises, in Britain barely one-third of the national workforce laboured on the land. Under successive Tory administrations, the postwar years witnessed unprecedented levels of immiseration and class conflict. The newly vast industrial cities saw explosive conflicts between the urban workforce and the armed forces of the state; in the countryside, protests against grain prices and levels of rent turned into widespread revolt.

In response, in newly founded mechanics' institutes, pamphlets and journals pouring from the automated presses, and agencies designed to rationalize welfare and economic provision, reformers set out to diffuse useful knowledge and social discipline among the working classes to convince potential rebels of the values of gradual progress. Temporary alliances were forged with Whig leaders and

Figure 35 Charles Babbage, Lucasian professor 1826–1828, drawn by William Brockendon [1840].

popular agitators in the name of prudent change of the established order. Conservatives retorted that there was nothing to choose between reform and revolution. It was pointless, or foolish, to give the tools of knowledge and power to ignorant subversives. There was no need to compromise with agencies they judged to be threats to the values of tradition.

These polemics mattered much within Cambridge, where mathematical doctrine was entangled with theological and political issues of social hierarchy, proper knowledge and spiritual welfare. Babbage recalled his professorial elevation as 'an instance of forgiveness unparalleled in history'.[3] There was a lot to forgive. A Cambridge pupil in the wartime years of 1810–13, he ignored the established curriculum for his own scientific pursuits, otherwise played chess and whist, hunted ghosts and skipped chapel for plush boat rides downriver to King's Lynn. It was characteristic that when he and his dissident colleagues launched an undergraduate campaign to change the university's mathematical creed, they would make fun of the slogans of Isaac Milner's evangelical Bible-thumpers. Babbage left Cambridge without an honours degree, joined Herschel and his friends in London in 1820 in a new Astronomical Society which was viewed with loathing by the eminent gentlemen of the Board of Longitude and the Royal Society, and then teamed up with London engineers and financiers.

Given these circumstances, the Lucasian chair was an unpromising site from which to launch such initiatives for reform. Milner's immediate successor as Lucasian professor, the forty-seven-year-old Caius fellow Robert Woodhouse, senior wrangler in 1795, had in his late twenties adopted a boldly reformist strategy to overhaul Cambridge teaching by redefining the calculus as a series of purely mechanical operations. Babbage and his allies found helpful resources in Woodhouse's early claims that algebraic reasoning was based on rational labour, not inspired intuition. But conservative dons rejected Woodhouse's alien innovations, forcing him to issue a series of more prudent textbooks which better fitted the university's curriculum. He refused to join new philosophical societies lest he offend the powers that be, and in the end even resisted reforms of bastions of state patronage such as the notoriously inefficient Board of Longitude. This belated conformity helped him to secure the chair of mathematics. When Woodhouse resigned the chair to head the new university observatory, he was swiftly replaced in 1822 by Thomas Turton, another senior wrangler. Turton was a Yorkshire Tory who devoted his time as fellow and tutor to defences of Cambridge classical scholarship and attacks on foreign readings of the Scriptures. Foe of mechanics' institutes and so-called useful knowledge, Turton left the Lucasian professorship for a better-paid living in Huntingdonshire; soon after he was awarded the university's divinity chair, lucrative deanships and eventually the see of Ely. Dean Turton was a prolific composer of orthodox hymns and enemy of the admission to Cambridge of those outside the Anglican communion. Babbage competed for the Lucasian chair unsuccessfully against Woodhouse in 1820, and again after Turton's resignation in 1826. With these distressing precedents, his election in 1828 might indeed be seen as a remarkably forgiving Cambridge victory for the reformers, belatedly bringing modern mechanical values to courts and colleges. Yet it was in fact a temporary, though disappointed, signal that his machine philosophy could be used for orthodox purposes. His prodigal career has distracted attention from the equally important concerns with engineering, mechanics and economics among the slightly less colourful, certainly more effective, leaders of the Cambridge mathematical programme.

The question was not whether, but how, the machine systems of industrial society would be treated by Cambridge instructors. One protagonist of such systems was the professional astronomer and heroic product of the university's strenuous discipline, George Airy. He was sternly devoted to rigorous training in the mixed mathematical sciences of celestial mechanics, geometrical optics and the theory of machines. Turton's successor as Lucasian professor between 1826 and 1828, then head of the university's observatory and from 1835 Astronomer Royal, he was a conscientious lecturer and very close ally of the network of competent engineers and instrument-makers which flourished in industrial Britain. He used their products to teach rational mechanics and reorganize the machinery of data production. Airy held that 'an observation is a lump of ore, requiring for its production, when the proper machinery is provided, nothing more than the commonest labour, and without value until it has been smelted'. Under the division of astronomical labour, his observatories were 'quietly contributing to the punctuality of business through a large portion of this busy country'.[4] During the early 1820s Airy emulated Babbage's scientific projects, his Whig sympathies prompting moderate support for the older mathematician's reforms. But Airy saw how to integrate the philosophy of machinery within the university regime without subverting its traditional order, and was richly rewarded by dons and politicians for his efforts. In 1826 he beat Babbage to the Lucasian chair with the support of the Trinity mathematics tutors, George Peacock and, decisively, the prodigiously energetic and highly conservative philosopher and educator William Whewell.

The proverbially omniscient Whewell – a Lancaster carpenter's son who rose to Cambridge eminence as Vice-chancellor, professor and legislator of the university's scientific curriculum – had his own view of the appropriate relationships between on the one hand Cambridge, and on the other the permanent establishment and the progressive forces in British society. By re-equipping the academy's curriculum and the nation's constitution, Whewell judged, it would be possible to preserve the virtuous system of technical instruction and spiritual exercise, and avoid the excesses of radicals who would destroy established principles in morality, mathematics and metaphysics. When Turton urged the exclusion of nonconformists from Cambridge, Whewell backed him: 'I think the fellowships a necessary support to the established church and I think the church a necessary part of our social system'. Turton repaid the compliment, praising Whewell's textbook defences of disciplinary permanence and the alliance of mathematics with established religion.[5]

One challenge to these men came from changes in the sciences. Whewell wanted a firm distinction between permanent disciplines, formed on the deductive pattern of Newtonian dynamics and safely taught to vulnerable students, and progressive fields, whose shifting principles might confuse, if they did not inflame, mere undergraduates. In the first decades of the nineteenth century, there was a European revolution which ended the clerical order of natural philosophy, established new disciplines such as mathematical physics, political economy and comparative anatomy and institutions such as teaching laboratories, public museums and national scientific associations. In the ingenious political geography mapped by Babbage and his allies, a sclerotic Anglican regime was fatally sundered both from the rational sciences cultivated beyond the Channel and the explosive industrial forces of urban Britain. Babbage held that 'the amount of patient thought, of repeated experiment, of happy exertion of genius, by which our manufactures have been created and carried to their present excellence is scarcely to be imagined'. But Cambridge dons, subjects of the victorious British state, did not necessarily suppose they were lagging behind foreign savants or out of touch with the triumphs of grimy engineers. Whewell argued that 'England has constantly impelled the progress of thought and of institutions in Europe, while at the same time she has held back from the extravagances and atrocities to which the progressive impulse has urged more unbalanced nations'.[6]

A master of analytical mechanics and complex engineering, Airy struck the right balance between progress and permanence, as Whewell defined it. The two friends sometimes bickered on issues such as religious nonconformity and oceanic tides, but Whewell used Airy's wave optics to teach true religion, while Airy adopted Whewell's project for re-energized professorial teaching in mixed mathematics. The Tory premier Robert Peel awarded Whewell the mastership of Trinity, and gave Airy a civil list pension and the management of Royal Greenwich Observatory. Both did their best to halt Babbage's projects and challenged the claims of his analysis. Whewell attacked his materialism, his mathematics and his political economy; Airy's advice to Peel undermined Babbage's claims to technical expertise and ensured the failure of the calculating engines. By the end of the 1830s, traditional academic order had been preserved by the ingenious machinations of these dons, who reinvented their world to save it from destruction. This was no mere reimposition of clerical order. Central to all these programmes was the pervasive figure of the rationally designed machine, simultaneously an emblem for the right kind of disciplined routine, an exemplar for mixed mathematical analysis and the central element of the new industrial system.

THE INVENTION OF A TRADITION

This chapter uses the intertwined early careers of the two Lucasian professors Airy and Babbage to explore the reinvention of discipline and the central role of the rational machine between 1820 and 1839. It contrasts the standard story of these developments which has been rather more preoccupied by mathematical notation than mechanized production. Thanks to the narrative skills of Babbage and his nineteenth-century Cambridge-trained avatars such as Augustus De Morgan and Walter William Rouse Ball, it has seemed as if the extraordinary achievements of Victorian mathematical physics depended on a sudden decision to abandon the indigenous notation of fluxions and instead adopt the foreign sign-system of differentiation. Within the European scientific revolution, we are told of a more parochial transformation, the 'Analytical Revolution', which from 1811 brought up-to-date French mathematics to Britain via Cambridge. This revolution was one reason why the university was, as one wry Scot soon put it, 'the Holy City of Mathematics'.[7] The undergraduates Babbage, Herschel and Peacock, in alliance with wealthy scholars such as Edward Bromhead, then notoriously helped form a short-lived Analytical Society, adopting the algebra of the Paris professor Joseph-Louis Lagrange. Dotted notations hallowed by Newton were abandoned for the symbolism of differential ratios. In 1816 the young analysts published a modified translation of a short calculus textbook issued in 1802 by Lagrange's colleague Sylvestre Lacroix. By 1820, the three mathematicians had also issued a much more successful collection of analytic examples for use by students, and aided the introduction of new notation and methods in the Cambridge examination system. It follows that this Analytical Revolution was the precondition of subsequent Cambridge-sponsored triumphs in the fields of fluid mechanics, optics and electromagnetism by James Clerk Maxwell, William Thomson, George Stokes and Joseph Larmor.

Not all of this story rings true. Though Cambridge was indeed an important mathematical institution, research in pure mathematics was never its main concern, nor was it a lonely oasis in a British wasteland. Furthermore, although Newtonian fluxions were banished from the university's highest examination after 1820, this did not dictate a revolution in Cambridge sciences. As in the epoch of Milner and Frend, changes in mathematical technique were intimately bound up with theologically sensitive conflicts of doctrine. The Analytical Society's algebraic definition of the calculus assumed that for every value of a real variable x, all functions $f(x+h)$ could be serially expanded in h with successive differentials defined as the coefficients of this series' terms. In comparison with Cambridge's traditional stories about vanishing quantities and intuitions about the behaviour of flowing variables, the analysts reckoned this a more natural and correct method, closely associated with Lagrange's attempts to reduce such dynamical problems to analytical statics. A general science of analytical operations would master subordinate disciplines such as mechanics and astronomy. Since these operations embodied the highest functions of mind, the analysts also seemed committed to a decidedly materialist account of mental labour. Orthodox churchmen would have none of this kind of reductionism while university tutors loathed the potential subordination of the disciplinary order of mixed mathematics to this higher analysis. In any case, the analysts' putatively radical approach had already been widely abandoned in Paris itself, where Augustin Cauchy showed that Lagrange's assumptions were false and that rigorous definitions of limits provided better means of grounding calculus.

Nor was early-nineteenth-century mathematical physics a monolithic body of doctrine to be transmitted *en bloc* to francophile scholars. In revolutionary Paris the masterly astronomer Pierre-Simon Laplace and his disciples, such as Jean-Baptiste Biot and Siméon-Denis Poisson, promoted a complex programme of reduction of all natural philosophy to central forces between material particles in empty space. They dominated characteristic institutions: the Paris Observatoire, the Bureau des Longitudes, the Ecole Polytechnique. The defeat of the Napoleonic regime which patronized these institutions was accompanied by a legitimacy crisis for their physics. In Restoration Paris, savants mounted a powerful critique of the Laplacian programme using the undulatory optics of Augustin Fresnel, Jean-Baptiste Fourier's new analytical theory of heat and investigations into electromagnetism by André-Marie Ampère and Dominique-Jean-François Arago. Cambridge mathematicians well knew the protagonists and places of these debates, but were not therefore slavishly compelled to imitate them.

In 1825, after visits to Paris, Herschel and Babbage staged important London versions of electromagnetic trials by Arago and Ampère, and the following year, both Babbage and Airy were in Paris studying Fresnel's new wave optics. Babbage's therapeutic expedition of 1827–8, during which he was made Lucasian professor, was but one among many similar tours of scientific Europe conducted by his colleagues. Whewell learned at first hand that the French were permanently on the point of further political revolution; Babbage's visits confirmed his admiration of Bonapartism. In their travels, different men of science took different approaches to the efficacy and danger of a host of techniques and projects. Even the original promoters of the analytical programme differed widely in their aims, thus in their use of Parisian mathematics. Babbage argued for pure algebraic analysis and the mechanization of reason. He found resources in a version of Laplacian materialism which turned mind into machinery. He used Lagrange's model of algebraic analysis, but attacked him for preserving intuitive notions of limits in his account of the differential, and instead designed a solely operational model of mathematical work. Peacock found less problem with the traces of intuitive notions of the limit-value in Lagrange's theory of the differential, so considerably and successfully modified the original aims of Babbage's overhaul of the calculus. Whewell, no ally of the analysts, was responsible for the final abandonment of fluxional notation in the Cambridge examination system, but never accepted Lagrangian analysis as a basis for mathematical training. In successive textbooks issued from 1819, he used the physics of Biot and Poisson as the basis for a reinvigorated programme in mixed mathematics within the university, but then backed Fresnel's attack on their particulate theory of light. He even held that Newtonian geometry would have let the French analysts make swifter progress than their newfangled mathematics. The story of Cambridge mathematics' unique centrality, a decisive shift in its symbolism and its automatic francophile modernization mistakes the dons for a homogeneous gang of somnambulists.

The university was indeed a 'holy city', committed to liberal education and the established church. Senior scholars were wary of the reforms and the industrial changes, both of which much questioned these commitments. Many established institutions experienced crises of legitimacy in those years – the monarchy and Church, the institutions of public sciences such as the Royal Society and the Board of Longitude, the older universities whose role would be challenged by new colleges in London and elsewhere. Appeal to institutional precedent therefore played a vital if mutable role in interpreting innovation. Half a century later, in his inaugural lecture as Cambridge's new professor of experimental physics, Whewell's former pupil Clerk Maxwell would half-jokingly describe a 'law of evolution' by which the university, 'while maintaining the strictest continuity between the successive phases of its history, adapts itself with more or less promptness to the requirements of the times'. Continuity and adaptation were indeed the watchwords of the institution in the Age of Reform. For Whewell, the Lucasian professorship was 'Barrow's, Newton's and Milner's' chair. At the same time, Babbage wrote suavely to the Master of Trinity: 'the names of Barrow and Newton have conferred on the Lucasian chair a value far beyond any which mere pecuniary advantage would bestow'. He would later gratefully recall the award of 'the Chair of Newton' as 'the only honour I ever received in my own country'.[8] Whatever their role in Cambridge's troubles, their attitudes to social change, public ingratitude and bad pay, Whewell, Babbage and their colleagues described shifts in the order of mathematical learning as threats of revolution and defences of inheritance. Novel projects such as the operations of algebraic analysis and the mechanics of working machinery might be adopted, modified or rejected to the extent that Cambridge savants could interpret them as conforming to institutional tradition. A crucial factor in the adoption or rejection of such programmes was the power of the academic establishment. This is why the question of professorial authority was so fraught in the years after 1820.

THE MECHANICS OF THE HOLY CITY

Inside the Holy City the Lucasian professors occupied ambiguous positions. Election was governed by the college heads under the chairmanship of the Vice-chancellor. None of the four incumbents between 1820 and 1839 saw the post as a career grade. Neither lecturing nor, save to an exiguous extent, examination was required of them. The Senate House Examination, taken at the start of an undergraduate's fourth Lent term, stayed at the head of the *cursus honorum*. It was divided between book-work on the mathematical canon and more advanced problems for those who would become wranglers. High success relied on the burgeoning system of private coaching, an indispensable mechanism for the increasing number who would become wranglers. Coaching did not sit easily with the formal requirements that candidates take annual college examinations and participate in university-wide preliminary public disputations, the Acts. Babbage's defence of an astonishingly heretical proposition in the Acts of Lent 1813 made him a mere 'poll man', with an ordinary rather than an honours degree. This failing might have been forgiven, but was not forgotten. Fifteen years later, Whewell still wondered whether the fact that Babbage was 'one of the $\pi o\lambda\lambda o\acute{\iota}$' [*polloi*, i.e. poll men] would disqualify him from the Lucasian chair.

Compare the spectacular career of Airy, who matriculated at Trinity in Michaelmas 1819. In his first year he was given a copy of the analysts' version of Lacroix by one of its editors, George Peacock. Just after reading Whewell's new mechanics textbook, he then triumphed in the college examination in May 1820. Lectured by Peacock, and provided by Whewell with notes on optics, geodesy and astronomy, Airy succeeded again in May 1821. In early 1822 Airy kept his first Act in the Schools, defending orthodox propositions from the section on perturbation theory in Newton's *Principia*, the geometry of the rainbow from James Wood's conservative *Elements of Optics* and the notion of obligation from William Paley's utilitarian *Moral and Political Philosophy* (using the twentieth edition of 1814). Whewell reckoned that the Acts were 'well adapted to try the clearness and soundness of the mathematical ideas of the men'. On the basis of these tests, candidates would be sorted into classes to guide the level of questions later faced in the Senate House Examination itself. Airy sat this examination in January 1823, graduating as senior wrangler: 'The hours were sharp, the season was a cold one, and altogether it was a severe time.' His sole significant personal encounter with any of the professoriate came two weeks later, when Professor Turton and his colleagues examined him for the Smith's prize, a trial of more wide-ranging topics, which he won. 'The form of examination which I have described was complicated and perhaps troublesome', Airy recalled, 'but I believe that it was very efficient.'[9]

To reinforce their system's efficiency, Whewell, Airy and their allies wanted to break the private coaches' power over instruction, and lobbied for better professorial facilities. They knew the rank that a chair could grant. Whewell held the mineralogy chair from 1828 to 1832, and was only beaten to the Lowndean geometry chair in 1836 by the better-patronized Peacock. Whewell imagined such posts might let him publicize his metaphysics, plans for university reform and natural philosophy. Yet undergraduate satires penned by his unabashed students, such as William Makepeace Thackeray, well showed that unless entertaining or novel, lectures scarcely affected scholars' interests. Being a professor did not yet give great influence, but tutoring, examining and producing textbooks could. As author of a series of textbooks in mixed mathematics, Whewell saw himself as 'a worthy successor to James Wood', the astonishingly reactionary Master of St John's whose texts in optics, algebra and mechanics long dominated undergraduate reading. By rejecting Lagrangian attempts to reduce dynamics to statics, and insisting on the supremacy of mixed mathematics, the new textbooks restored Cambridge tradition by reforming it. The conservative metaphysics which Whewell and his allies invented for public consumption in the 1830s reinforced this 'permanent' order of disciplines with textbook dynamics as the ideal of securely deductive knowledge to which other sciences might eventually aspire. Their metaphysics also celebrated a stately process of historical emergence, and left to a wrangler élite the decidedly extracurricular task of making new advances in the sciences. 'Universities and colleges have for their office not to run a race with the spirit of the age', Whewell urged, 'but to connect ages as they roll on by giving permanence to that which is often lost sight of in the turmoil of more bustling scenes.'[10]

Such a conservative vision of the university sat well with the career options of its undergraduates, more than half of whom, unlike Babbage and Airy, would enter holy orders. This proportion rose during the first half of the nineteenth century. Each shift of the undergraduate system reinforced the importance of the Senate House Examination and the strenuous schooling of the professional coaches. So, though a Classical Tripos was started in 1824, entry into the new programme was contingent upon receiving honours in the Mathematical Tripos, as the Senate House Examination was thenceforward known. The scandalously superficial Acts were abolished as preliminary classification, but the mathematics tests were tailored to a wide spectrum of abilities, and the length of the Tripos examination doubled to six days by 1838. As moderator, Whewell wrote of 'manufacturing wranglers', which he judged a 'very laborious' duty. His Edinburgh critic, the philosopher William Hamilton, argued in early 1836 that in Cambridge mathematics 'the routine of demonstration, in the gymnastic of mind, may, indeed, be compared to the routine of the treadmill in the gymnastic of body ... both equally educate to a mechanical continuity of attention'. Whewell acknowledged that the pure disciplines of Lagrangian analysis might well be only fit 'for men of business', but Cambridge mixed mathematics, in contrast to the mechanical rituals of algebraic computation, was a vital 'part of the gymnastics of education'.[11]

Cambridge intellectual gymnastics relied on hierarchies of church and property which reforms were designed to salvage, not subvert. New institutions appeared, including the Cambridge Philosophical Society in 1819, and the University Observatory in 1823. But plans to house the scientific professors in King's College Old Court in the early 1830s were shipwrecked. Attendance at professorial lectures fell from the 1830s. The Smith's prize, which the Lucasian professors helped to administer, did offer a limited means of giving new matter to the wranglers. To aid such innovation, Airy started printing his Smith's prize question sheets during his brief tenure of the chair in 1827, while his successor, Babbage, used the questions to promulgate his own project in analytical mathematics. 'Explain the difference between the nature of the reasoning employed in Geometry and in Analysis', Babbage demanded in 1833. Babbage reckoned that the Smith's prize was a 'court of appeal' from the Senate House Examination. Yet, though energetic and conscientious as examiner, Babbage never delivered a single lecture, and elicited much fury when he failed to print his examination papers. Advanced research remained the prerogative of gentlemanly specialists.[12]

Babbage loudly claimed in 1830 that the natural sciences were not yet a genuine career in Britain. But he could afford the luxury of vocation. His annual allowance as student from his banker father was already more than three times that of the Lucasian professor, rising to £450 by 1820. He could buy Lacroix's calculus for seven guineas before going up to Cambridge in 1810, pay a similar sum for eight volumes of the journal of the Ecole Polytechnique in late 1811, and obtain a complete chemical apparatus for himself and Herschel in spring 1812. When the allowance was threatened by a breach with his father after his precipitate marriage, Babbage began competing for university chairs. He told Herschel that he would have become a clergyman when he 'had expectations of possessing my father's fortune', since this would give him intellectual leisure. Without the cash 'it would be madness to enter a profession where I have no hopes of getting on'. His father's death in early 1827 left him an estate of £100 000 and the standing to pursue his machinations with or without a chair. In the tumultuous year of 1832, Babbage joked that after parliamentary reform he expected to become 'the secularized Bishop of Winchester'. There were some comparably independent figures among the Cambridge mathematicians, such as the Whig astronomer Richard Sheepshanks, heir to considerable income from the Halifax wool trade, who left the University Observatory £10 000 for research in meteorology and geomagnetism.

Sheepshanks's friend, Airy, knew all about the importance of regular income. After his father, an Ipswich exciseman, was sacked for financial impropriety, Airy carefully monitored his budget and paid for his own equipment. He spent his early years as Lucasian professor and Observatory director negotiating a better salary. In early 1827 he

> had resigned my assistant tutorship of £150 per annum and gained only the Lucasian Professorship of £99 per annum. I had a great aversion to entering the Church. My prospects in law or other professions might have been good if I could have waited, but then I must have been in a state of starvation for many years. I had now in some measure taken science as my line, though not irrevocably.

So the professorial career was dependent on facts of family, economic and social life. For example, William Hopkins, seventh wrangler in 1827, could not as a married man obtain a college fellowship, so instead became Cambridge's principal private coach, manufacturing senior wranglers for cash. He failed to win the Lucasian chair when Babbage resigned in 1839. Privately wealthy bourgeois savants, such as Babbage, could back a radical overhaul of the entire academic system but those whose status most depended on the system's good order, such as the Tory evangelical Milner, the Whig bureaucrat Airy or the conservative moralist Whewell, were distinctly judicious custodians of the reorganized curriculum and its institutions. The conflicts between these groups intimately affected, and were illustrated by, the Lucasian succession.[13]

ASTRONOMICAL AMBITIONS, 1820–3

In April 1820, Milner's Lucasian chair was vacant. The young London analysts Herschel and Babbage decided to stand. Before taking the fast coach to Cambridge Herschel told Babbage that 'from what I know of the bigwigs of the University, who will settle this point their own way, you will have less chance even of success than myself, and that, I well know, is none whatsoever'. Herschel had already resigned a fellowship at his college, St John's, to continue his father William's heroic astronomical labours in vast surveys of nebulae and double stars, and his own experimental work in optics and chemistry. Pessimism about Cambridge success was understandable, and justified. Herschel later recalled that its curriculum was managed by 'souls fenced with the tough bull-hide of Vince and Wood'. Wood's self-described successor, Whewell, had just published a Cambridge mechanics textbook which Herschel judged out of touch with 'the taste of the age' because unsympathetic to the analysts' version of Lagrange. Babbage and Herschel were still negotiating with Peacock to complete their joint collection of analytical examples for Cambridge use. When it appeared in Michaelmas 1820, to some applause, the original purist motif of algebraic analysis had been much modified by Peacock's provision of worked cases which compromised with limit notions of differentiation. The more aggressive Babbage had already failed to win chairs at the East India Company's college in 1816 and Edinburgh in autumn 1819, despite prestigious backing from professors at the Royal Military schools and from his new friends in Paris, including Lacroix, Laplace and Biot. Herschel explained the rationale for a Cambridge campaign: 'I only offer myself that it may become a notorious fact that I want a Cambridge professorship in a mathematical or physical department, and so help myself at the next vacancy, by keeping in their sight. Come up, and we will canvass together.'[14]

On 1 May 1820, barely a month later, Babbage and Herschel both had papers read at the new Cambridge Philosophical Society, an organization their friend Edward Bromhead described as the successor to the Analytical. While Herschel offered the Society a development of Biot's work on the behaviour of crystals and polarized light, Babbage analysed the operational symbolism of the calculus of functions and used difference equations to pursue this calculus. The invention of new mathematical notation could thus be mechanized. This was part of a project for a philosophy of analysis which would define the use of signs in mathematics. Since Babbage supposed all reasoning resembled the mechanics of mathematical computation, his theory of operations with signs would unlock the mechanisms of human thought. It was 'the great object of my life', already projected in the 1813 preface to the Analytical Society's *Memoirs* whose recent achievements had been well-received among Parisian mathematicians, including Cauchy. Soon this cognitive and semiotic project was industrialized. In summer 1819, a Paris printer gave Babbage invaluable copies of the trigonometric tables managed by the engineer Gaspard François de Prony for the republican government. When the new London Astronomical Society tried to seize the Nautical Almanac from the moribund Board of Longitude, Babbage launched a machine to manufacture polynomial tables using Prony's method of finite differences. In the halcyon days of the Analytical Society, Bromhead had told Babbage that 'with regard to the machinery you talk of having invented, I should call it more important than any Analytic Discovery, even the root of an equation of n dimensions'.

This difference engine relied on the techniques Babbage offered the Cambridge Philosophical Society. While Herschel developed the theory of finite differences for their examples book, Babbage's engine generated polynomial tables as close approximations for the logarithmic and trigonometric functions astronomers needed. Within two years he had a device manufacturing six-figure numbers and third-order polynomials at forty numbers per minute. Whewell carried letters from Herschel to Laplace in Paris, and gave Babbage's account of his calculating engine to Prony and Cauchy. Babbage's timing was typically theatrical. On the same day his new paper was read in Cambridge, the college heads unanimously elected Robert Woodhouse as the new Lucasian professor.[15]

Woodhouse had achieved Cambridge glory as senior wrangler and first Smith's prizeman in 1795, and subsequently moderated the Senate House Examination in 1799–1800 and 1803–4. He agreed with previous critics of Cambridge mathematics, such as George Berkeley and William Frend, that the doctrines of infinitesimals and of negative and imaginary quantities as then taught were misty nonsense. Learning from Parisian mathematicians and contemporary English algebraists, he wanted to expel vague intuitions from mathematical work, defining calculus as a system of 'mechanical operations'. He even believed Lagrange's series definition of differentials to be faulty, since it tacitly used unjustifiable intuitions about limits. None of this operationalism persuaded Cambridge tutors. Woodhouse's manifesto of 1803, *Principles of Analytical Calculation*, was never used for Senate House book-work. Accordingly, like other reformers, he insinuated what he could not impose. In 1809 he released a more successful trigonometry textbook, using differential notation with fluxional equivalents in footnotes, and in 1818 followed with a textbook of Laplacian celestial mechanics. Younger analysts saw the promise and compromise of his career. Babbage read his *Principles* before coming up to Cambridge, and was encouraged by Bromhead to produce a 'compact and tangible' version because 'Woodhouse's book is too controversial'. They emulated Woodhouse's prudence both in their edition of Lacroix and their examination papers. Babbage's early papers on functional calculus were sent to Woodhouse by his friends for approval. When Herschel and Babbage met Laplace in Paris in autumn 1819, they were struck by his admiration of 'vous deux', then realized the French astronomer was in fact referring to 'Woodhouse'. The analysts rather often managed to confuse themselves with Woodhouse, a prophet whose vision of the future of analysis they had made real. Cambridge chroniclers record Woodhouse's role in 'the revolution in our mathematical studies', and his early attacks on the inconsistencies of Lagrangian calculus. For Babbage and his friends, these might be understood as mere expressions of Woodhouse's purism about algebraic operations. For more traditional dons, rather, they were signs of the new professor's rejection of French, and radical, excesses. Whewell perceptively told Herschel that 'Woodhouse is known to have no liking for the ultra-analysts'. Bromhead sadly told Babbage that because of Woodhouse's fear of the ruling élite in metropolitan sciences, he had 'left the Cambridge Philosophical Society in the lurch'. A don so concerned with institutional proprieties was a fit successor to Milner's chair.[16]

A key element in Woodhouse's elevation, and the appearance in the Senate House Examination of his mathematical texts of 1809–18, was his intelligent decision to tie algebraic analysis to physical astronomy. Celestial mechanics was the exemplary discipline in the core Cambridge curriculum. Many of its protagonists started working in Laplace's wake. Like Milner, Babbage also sought work as a computer at the Royal Observatory. Then he lectured on Laplacian astronomy at the Royal Institution in 1815. He showed how the ideal methods of Laplacian celestial mechanics exhibited the correct deductive hierarchy of 'grand and general laws' and 'series of observations' applicable to all sciences. In September 1820 Airy bought Woodhouse's physical astronomy textbook. It was 'an epoch in [his] mathematical knowledge', for it demanded a study of the logic of differential calculus, and introduced the young mathematician to his 'desired land' – Laplacian perturbation theory. The task of the more judicious Cambridge dons was to tease apart the useful from the subversive aspects of Laplace's programme, because they were highly suspicious of Laplacian reduction of spirit and mind to matter in motion.

These more dangerous aspects of Laplacian natural philosophy and mechanical analysis were already apparent in Babbage's remarks in an 1816 paper on the calculus of functions, which he then reiterated in the major essays on the philosophy of analysis and the mechanics of thought he directed at the Cambridge Philosophical Society in the early 1820s. Whewell's Tory friends moaned that 'teaching people to occupy themselves with the theory of differences and the knowledge of formulae' would turn students' minds into 'the same sort of thing as Babbage's calculating machine . . . the approximation of the operations of mind and of machinery is a very fit occupation for the Mechanic School'.[17] Babbage instead teamed up with London engineers, military mathematicians and one of their astronomy students, Thomas Colby, who was head of the Ordnance Survey from 1820. This military survey office, established three decades earlier in the face of considerable resistance from the Royal Society, was undergoing rapid expansion throughout the postwar British empire, notably in India and Ireland. Babbage manufactured mathematical tables for the Irish Survey and dedicated them to Colby, who supported Babbage's plans for calculating machines and backed his abortive plans to take over the corrupt Royal Society. This military and astronomical milieu played a decisive role in the reorganization of physical sciences in Britain. Instrument-makers such as Edward Troughton and George Dollond collaborated on the Survey with savants such as Airy, Babbage and Herschel. Combinations of precision machinery, strenuous measurement and grand cosmological principles could give Cambridge mixed mathematics opportunities to prove its universal application.

The nodes of this astronomy network were the observatories of Europe and its colonies. Late Hanoverian Cambridge's condition in this respect was uneven. A small observatory and lecture-room atop Trinity gatehouse was demolished in 1797. St John's maintained an observatory efficiently managed from 1791 by one of the fellows, Thomas Catton, endowed with a fine Dollond refractor by its Master, the proud James Wood. It seemed urgent to the younger reformers to establish in Cambridge well-funded provision for measures of optical refraction and stellar parallax and give students the chance to make observations. Between late 1817 and spring 1820, a university syndicate including Peacock and Herschel agreed to build an observatory and hire a paid observer. Wood reckoned his own college had a fine observatory and found allies among the professors and college heads to resist Peacock and Herschel's plans. Links with the entrepreneurs and artisans of the capital proved decisive. Colby's military surveyors fixed the new observatory's position. Babbage and Herschel collected funds from the Astronomical Society. Peacock told Babbage to get observatory plans from Troughton, a council member of the Astronomical Society and supplier of survey devices for Colby's work in Ireland and India. By then, as Sheepshanks told Whewell, observatory plans were 'so far advanced that to recede or limit the plan may be impossible, even for the Heads'. The conservative establishment was mollified. Land was bought from Wood's own college at a generous price. In January 1822 Woodhouse's prudent conformity and astronomical competence earned him election as Plumian professor and observatory director. A week later, the Heads replaced him as Lucasian professor with Thomas Turton, mathematically inert and utterly reliable.

Institutions such as the Observatory were symbols of, and sites for further work in, precision measurement and systematic management. In the six years that he ran the Observatory, the increasingly enfeebled Woodhouse oversaw the installation of a ten-foot transit instrument made by Dollond, and worked out its major sources of error. His manifesto as observatory director, published at the front of a new edition of his astronomy textbook in December 1822, was characteristically judicious. Woodhouse thought the science 'had reached a kind of maximum state of excellence'. No further revolutions were to be expected in precision celestial mechanics. Permanent public observatories brought 'philosophical and intellectual' benefits. They became sanctuaries of what Woodhouse called 'moral control' and the 'zealous discharge of duty'.[18] The view that celestial mechanics and meridional astronomy were the definitively permanent sciences became Cambridge orthodoxy. Remarkable innovations such as Herschel's nebular cosmology did not shake this institutionalized confidence. Celestial mechanics was a model for each field of natural philosophy, as moral exemplar, mathematical system and mechanical enterprise. This was the programme which Airy implemented as Lucasian professor from late 1826 and as Woodhouse's successor at the Observatory eighteen months later.

Figure 36 Caricature of George Airy (1801–92), Lucasian professor 1826–8, later Plumian professor, director of Cambridge Observatory, and from 1835 Astronomer Royal at Greenwich. As Astronomer Royal, Airy found himself caricatured in *Vanity Fair*, one of the most widely circulated periodicals of the day.

AIRY'S TRACTS AND BABBAGE'S SIGNS, 1823–6

Airy's work after his triumph in the Senate House Examination of January 1823 exemplified the combination of mechanics, machinery and mathematics. With Peacock's support he gave papers on refraction and achromatic lenses at the Cambridge Philosophical Society and worked in London with Herschel and other astronomers on telescope design and Josef von Fraunhofer's methods for measuring refrangibility and dispersion. Airy also developed a fine taste for practical mechanics. His closest Ipswich connexions included the eminent civil engineer, William Cubitt, designer of canals and iron bridges, and a candidate architect of the University Observatory. Cubitt gave the young Airy his first copy of the Nautical Almanac. Airy's early work was in many ways a dialogue, imitative and critical, with Babbage's general analytical science of machinery. When an undergraduate he copied Babbage's plans for calculating machines. As Babbage launched factory tours in summer 1823 as fact-finding missions for his calculating engine project, so in the summers of 1824 and 1825, while making notes on Laplacian orbital theory, Airy visited suspension bridges, aqueducts, mines and cotton mills. In summer 1826, they both toured France, whence Airy, his Whig sympathies temporarily in evidence, smuggled a souvenir copy of the Marseillaise. Both studied Fresnel's new work in wave optics and the researches by Ampère and Arago on electromagnetism. Babbage gave Airy introductions to the Laplace circle, whose dogmas of dynamic reduction to interparticulate forces were now under threat from these younger physicists.

With his patron Whewell, Airy staged dramatic but abortive experiments down a Cornish tin-mine using chronometers and pendulums requisitioned from the London instrument-makers to measure terrestrial density. In autumn 1826 he then gave the Cambridge Philosophical Society lectures on Cornish steam-engines and on Babbage's novel mechanical notation to analyse the behaviour of pendulums and clock escapements. Precision pendulum trials were a basis of new imperial length standards, and became exemplary of novel physical work in the 1820s and 1830s. Much of Airy's metrological enterprise was also tied to his search for well-funded employment, a search made the more urgent by his betrothal to a Derbyshire chaplain's daughter and, in stark contrast to Babbage, his lack of private income.

Soon Airy's financial situation was meliorated by his election to a Trinity fellowship and appointment as assistant mathematics tutor. From this comparative eminence, he began to exercise a more forthright role in Cambridge mathematical society. Like Babbage, he attacked the inefficient regime at the Board of Longitude and its inaccurate astronomical tables. He privately performed a demolition job on the version of Whewell's optical lectures which Coddington had published, 'a work which deserved severe censure'. He also politely corrected the differential calculus in Whewell's 1823 *Treatise of Dynamics*, and explained significantly that the observable behaviour of Atwood's machine, first demonstrated at the Trinity observatory fifty years earlier, was a much better way of teaching ideas about velocity than 'the limit of ds/dt, which is rather difficult to get'.[19] Mechanics was best approached through the analysis of machines. In a burst of creative energy during Michaelmas 1825, Airy pulled together his essays on geodesy and celestial mechanics, and reworked an old undergraduate paper on orbital mechanics inspired by his reading in 1820 of Woodhouse's astronomy textbook. By combining the axioms of Whewell's dynamics, notably the orbital equation for physical astronomy, and the applied analysis of Woodhouse's astronomy and his studies of the calculus of variations, Airy produced a remarkably advanced text designed to introduce aspiring wranglers via variational calculus to the principles of rational and celestial mechanics. These *Mathematical Tracts*, with Turton's somewhat redundant blessing, were published in Cambridge in Easter 1826. They formed a manifesto for the mixed mathematical sciences Airy wished to promote.

In the same month that Airy's *Tracts* appeared in Cambridge, Babbage presented the Royal Society with his own manifesto for the mechanics of analytical technique. His paper on 'a method of expressing by signs the action of machinery' appeared at the apogee of his calculating engine project, which had just been awarded the first gold medal of the Astronomical Society and promised substantial government funds. He hired the brilliant engineer Joseph Clement to build the machine under his direction. Babbage designed a system of mechanical notation as an analytical technology for surveillance and calculation. He used it, vainly, to direct the engine project at a distance. His sign-system distinguished between logical operations performed by the engine and the numerical values that it stored. Thus his machine semiotics had close links with his more cloistered projects in pure analysis. Babbage's 1826 manifesto appealed to 'the advantage which results from conveying information or reasoning to the mind by means of signs instead of words', evident in 'the vast and complicated results which Analysis enables us to investigate'. His Cambridge mathematical essays of the early 1820s, such as the study 'of the influence of general signs in analytical reasoning', advanced from the same definition of analysis as a language of signs to a purely operational account of the calculus of functions.[20] This was the semiotics Airy immediately borrowed to analyse the behaviour of mine-shaft pendulums and chronometers at the Cambridge Philosophical Society later in 1826 and which Babbage used to break the reclusive control his engineers exercised over the work through their own vocabularies of mechanical drawing.

Designed to 'see at a glance what every moving piece in the machinery was doing at each instant of time', Babbage's notation was a technology of universal management and knowledge production. He explained to his friend, the Polytechnicien mathematical engineer Charles Dupin, who toured industrial Britain in 1824 and 1826, that the mechanical notation was 'a system of signs which like those of Analysis enables us to look with different degrees of generality at the same object', and which 'having been practically employed in the workshop ... is more interesting to the philosopher as being one among those systems of signs by which man aids his reasoning power'. Reasoning power and the analysis of machines also mattered much beyond college walls. In 1825, after mass protests in the manufacturing districts, the Tory government was compelled to repeal legislation against trade unions. Then throughout spring 1826 fiscal collapse prompted Lancashire workers to organize systematic attacks on the detested factory machines. It was vainly explained in the mechanics' institutes that the machine-breaking riots were caused by plebeian ignorance of 'useful knowledge' in the principles of political economy, which could be corrected by the graphics of machinery. Babbage underlined a telling passage from a parliamentary report on machine-breaking: 'any violence used by the workmen against the property of their masters is almost sure to be injurious to themselves'.

On his factory tours, Babbage sought machinists who used his sign language. 'I have met with a very intelligent clockmaker who not only understood the mechanical notation but suggested an improvement', he told Herschel later in 1826. 'I think it likely to get into general use in the description of complicated machinery.' In a draft of the 1826 paper he stressed the advantages of machine semiotics because 'of all our senses that of sight conveys intelligence most rapidly to the mind'. The same year, Dupin produced a major mechanics textbook for artisans and factory foremen to 'render their conduct more moral while impressing upon their minds the habits of reason and order that are the surest foundations of public peace'.[21] But the factory system did not look peaceful and rational. Nor did the public institutions of science. Babbage reckoned his machine language, with its analytical foundation and its recipe for industrial harmony, surely deserved an honour from the Royal Society. He was furious, if unsurprised, when he was passed over as medallist by the Society for more orthodox savants in autumn 1826. Not long afterwards Turton gave up the Lucasian professorship for the rewards of the church. Both Babbage and Airy decided to compete for the vacant chair.

THE CONTEST OF 1826

The Lucasian election of late 1826 was a decisive moment in the campaign to maintain Cambridge tradition by reforming it. Both radicals and ultra-conservatives lost. Unchanged heritage was represented by

William French, second wrangler in 1811, a figure of unimpeachable theological probity and no evident mathematical interests, and an obvious replacement for Turton. As Master of Jesus College, French was one of the electors to the Lucasian chair with the promised support of most of the other Heads. As in the Observatory campaign, so in Michaelmas 1826, an attempt to assert unchanged collegiate authority was countered by a broad alliance of otherwise conservative and reformist dons, who expected but deplored French's election. Bromhead, Peacock and Whewell all supposed this outcome would render the chair contemptible. Whewell's friend, the devout cleric Richard Jones, even suggested that 'if Newton himself were to come to life as a Master of Arts, the Heads would give it to French without hesitation, and perhaps with an additional relish from the mere weakness of the job'.

The younger wranglers, with a strong sense of their own virtues and those of their colleagues, shuffled themselves into polite competition. One was Joshua King, senior wrangler in the 1819 Tripos which Peacock had moderated, who was then embroiled in a disputed succession to the presidency of Milner's college. Eventually King would win both the college presidency and, from 1839, the Lucasian professorship. Meanwhile he would withdraw if either of the Trinity mathematics tutors, Peacock and Whewell, were to stand. Christopher Wordsworth, their Master, was as Vice-chancellor to preside over the election. Both Trinity men contemplated their candidacy. Jones thought Whewell would do better, but that it would be gracious if inconsequential to withdraw in Peacock's favour. The overly reformist 'Peacock is not popular in the University' and was thus expected to lose. Whewell hoped the chair would give him a chance for lectures on his metaphysics, but agreed that he would rather not compete in a 'desperate' contest and invite 'ridicule [by] asking for professorships indiscriminately'. At the top of this hierarchy was John Herschel, Johnian hero and exemplary wrangler. Peacock, Whewell and King would have retired in Herschel's favour. Peacock explained that the catastrophe of French's appointment would 'tend to lengthen the existence of the remains of the old mathematics of the University'. Herschel might embarrass French into withdrawal. St John's fellows lobbied Herschel to stand. Even James Wood, their veteran Tory Master, told Herschel that were French not a candidate he would have had his support. Whewell sent a list of the Heads to be canvassed but Herschel refused to play the game. He had a declared 'contempt against the university professorships' and told Whewell he did not 'wish to devote myself exclusively or par excellence to any one branch of science – perhaps, too, a consciousness that I prefer physical to mathematical science'. He saw himself pursuing science 'as an amateur rather than as a matter of duty and profession'. These were telling remarks by a gentlemanly natural philosopher now unconvinced, in contrast to his aims of 1820, that Cambridge professors could effectively pursue the vocation of science.[22]

By the end of October the field for the Lucasian chair was in turmoil. The Heads countered the rebels with a careful re-reading of the professorship's regulations. This implied that no tutor could stand, and thus excluded men such as Whewell and Peacock but neatly kept French in the frame. Herschel, out of the contest, started lobbying for a much more radical candidate, Charles Babbage. He tried to get support from Whewell and Peacock, and even more hopelessly from Wood. Babbage thought he would 'have a tolerable chance, if French can be got rid of, which seems not so impossible as it did'. Other Analytical Society veterans joined the Babbage campaign. 'Your name', Bromhead flattered Babbage, 'would add materially to the éclat of all your successors.'[23] From London, the loyal Thomas Colby reckoned Babbage's work on the calculating engine more important than a badly paid Cambridge job. In any case, the Trinity dons (some of whom Babbage had scorned when he moved from Trinity to Peterhouse after his freshman year) had their own man in mind, the precocious Airy. Peacock vanquished the young mathematician's cash worries by pointing out the possible security offered by the chair. Airy at once started canvassing the Heads without shifting their loyalty to French. But at the end of November the head of Airy's college, the

Vice-chancellor Wordsworth, intervened to remove even this obstacle, tactfully suggesting to French that the Lucasian chair and his Jesus mastership might be incompatible. Once battle was joined between Airy and Babbage, the issue was no longer in doubt.

Airy and Babbage were often to find themselves on opposing sides of the scientific disputes of early-Victorian Britain. In conflicts about the appropriate size of railway gauges, administration of scientific patronage or the design of telescopes, Babbage's energies were repeatedly countered by the assured competence of a man who made himself the nation's chief official scientist. In the 1840s, Airy reported to the government that Babbage's calculating engines were useless, unworthy of funding and unfit for calculating astronomical tables. There was undoubtedly a personal animosity here. Babbage loathed Airy's close friend Richard Sheepshanks, and raked his astronomical work and telescopic schemes over the coals. One London wit even supposed that Airy's criticisms of 'the Engine that Charles Built' was entirely due to Sheepshanks's venomous influence.[24] Having emulated Babbage's work in the early 1820s, in the election of 1826 Airy began defining differences.

Crucial to Airy's success was his commitment to reside in Cambridge, deliver lectures in mixed mathematics and thus to develop the programme of the *Mathematical Tracts*. Whewell deemed his pupil and colleague Airy 'the best they could have chosen'. Airy would 'reside and give lectures – practical and painstaking ones – who is par eminence a mathematician'. Babbage's metropolitan interests ruled out a Cambridge residence. During most of the campaign he was touring the industrial Midlands gathering data for the machine project. Furthermore, as his remarkable letters to the Master of Trinity revealed, he had a different model of Cambridge mathematics in view. Mathematical instruction was safe with the tutors. The professor should instead develop advanced science at the international research level. When the Vice-chancellor asked directly whether he would live and teach in Cambridge, Babbage stressed the importance of the calculating engine project and his plan to lecture in Cambridge from his *Essays on the Philosophy of Analysis*. Babbage proposed making the Lucasian chair the focus of an advanced seminar in the calculus of functions, mechanical analysis and innovative mathematical symbolism.

This was not a contrast between alien pure analysis and homespun applied mathematics – Airy had demonstrated his expertise in algebraic treatments of highly advanced topics, while Babbage was one of the principal philosophers of engineering. The contrast, rather, was between Babbage's dramatic change in disciplinary order, making a materialist philosophy of mind the master-science of all other mathematical and physical studies, and Airy's aims to reinvigorate professorial instruction in mixed mathematics, bringing the study of machinery and mechanics into the secure context of Cambridge training. Whewell knew Babbage had 'no chance whatever and it is mere extravagance his taking up the thing'. On 7 December, in French's diplomatic absence, the Heads under Wordsworth's direction gave eight votes to Airy and two to Babbage. That evening Airy began making notes on lecture apparatus, including Atwood's machine for demonstrations on mechanics, and devices from George Dollond for classes on wave optics.[25]

AIRY'S TREADMILLS, 1826–35

Critics of Cambridge mathematics, such as William Hamilton, compared its discipline to the awful effects of treadmills. The conscientious Airy made the analogy literal. These vast revolving cylinders, used to terrorize and discipline increasing prison populations in an era of economic dislocation and agrarian distress, were invented by Airy's close friend the engineer William Cubitt. Cubitt installed the first at Bury St Edmund's gaol in 1819, and another at Cambridge in autumn 1821, when he took the chance of contacting Airy at Trinity. In early January 1827, visiting his heavily indebted, ailing and deranged father at Bury St Edmund's, Airy obtained data there on the work-rate of Cubitt's prison treadmill to use in his lectures as Lucasian professor. The course he launched the following month, maintained when he

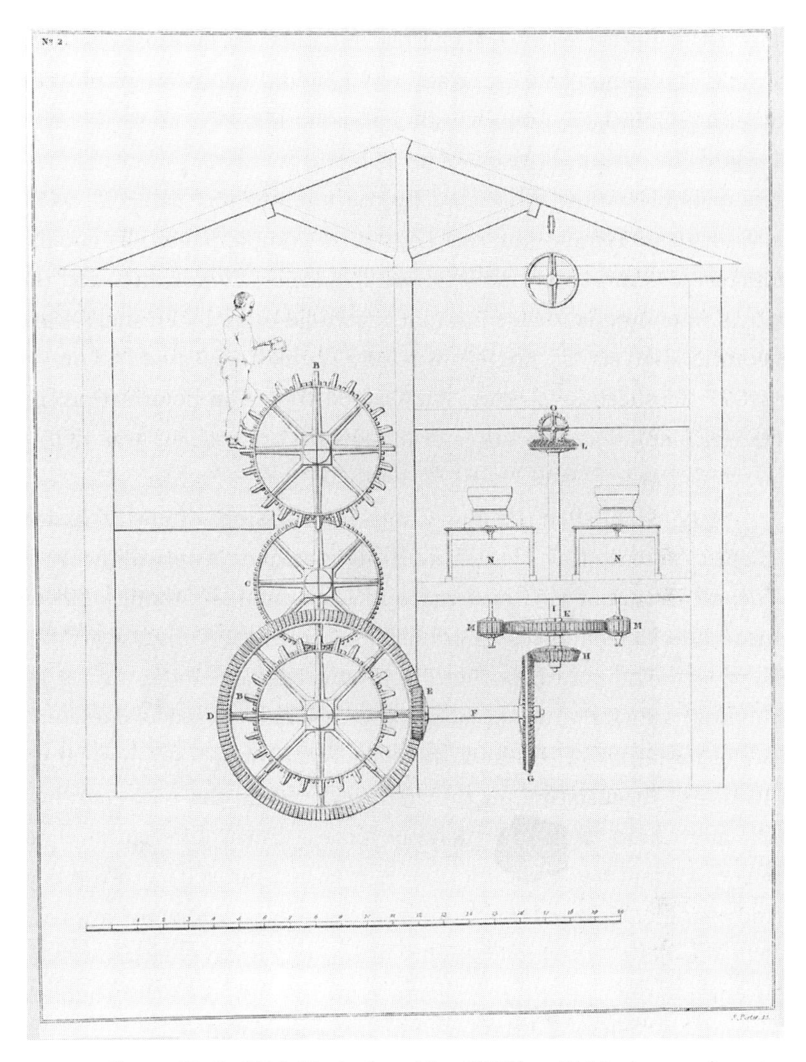

Figure 37 In 1819 Airy's close friend William Cubitt invented a prison treadmill and installed his first machine at Bury St Edmunds gaol. In January 1827, a month after becoming Lucasian professor, Airy visited the gaol to study the treadmill as an example for his new Cambridge lectures.

took over the management of the Observatory the next year, was a key element in the development of Cambridge mechanics. Airy restored and amplified the place which working machines would play both in the lecture-rooms and the Tripos. He bought his own apparatus at a cost of almost £200. His engineering tours and contacts paid off. 'Several tradesmen in Cambridge and London were well-employed', including Cubitt on stone-working and the London printer Thomas Hansard on steam-presses. Airy's lectures treated the classical principles of statics, illustrated by the roofs of the Cambridge Observatory and Trinity College. They were complemented by equally impressive demonstration classes in dynamics, hydrostatics and the optical phenomena reduced to mathematical order by Fresnel's wave theoretic analysis, an English version of which appeared in London in 1827. In successive years, Airy added topics on polarization, interference and wave theory, and made sure his analysis was tied to questions in the Senate House.[26]

Airy established the new Cambridge versions of optics, hydrodynamics and geodesy. He added a fresh treatment of optical theory to a second edition of his *Mathematical Tracts* in 1831, though Turton characteristically refused to publish this from the University Press. The work, Airy told Herschel, 'is intended quantum in me est [as much as is in me – a joke about the original version of Newton's laws of motion] to turn the Cambridge mathematics to physical applications'. By insisting on the analogy between the propagation of light and sound, both of which obeyed the linear differential equations of simple harmonic motion, Airy allied himself with the doctrine wranglers learned from their coaches. Airy claimed to have tried all of Fresnel's experiments. The authority of wave optics 'can be felt completely only by the person who has observed the phenomena and made the calculations.' But what he called 'the mechanical part of the theory' – speculations on the luminiferous æther – was excluded from his classes as 'far from certain', as were any themes in electricity or thermodynamics.[27] The topics excluded by Airy were what Whewell called 'progressive', topics not yet cast into the deductive form which would fit them for undergraduate consumption. Airy thus helped define the precedent for Lucasian lectures on hydrodynamics and optics, notably those delivered by George Stokes from 1851. Attendance at Airy's lecture demonstrations, which also generated income, was never vast but always impressive: sixty-four in his first year, rising to a peak of eighty-seven in Easter 1834. But even this disciplined energy did not commit Airy to the ill-endowed Lucasian chair. Within a few months of his election, he was already in search of a more secure and powerful position within the astronomical establishment.

The fathers of Babbage and Airy died within a few weeks of each other in spring 1827. Babbage's newfound riches prompted him to abandon a candidacy for a geometry chair at Oxford, but Airy was forcefully reminded by family catastrophe of the fragility of his situation. During the Easter vacation, barely three months after his inauguration as Lucasian professor, he went to Dublin in search of the vacant astronomy chair at Trinity College. When the fellows revealed that the brilliant young mathematician William Rowan Hamilton was the favoured candidate, Airy diplomatically withdrew. His frustration was brief. At the Board of Longitude, of which both he and Woodhouse were members ex officio, Airy headed a reform campaign against the regime of the conservative natural philosopher Thomas Young. The Board was the target of commercially minded astronomers who wanted to replace venal patronage with rational management. During 1827, Airy used solar tables from Greenwich and Paris to demonstrate systematic errors in the Board's Nautical Almanac. He explained these errors by a hitherto unknown perturbation with a period of 240 years due to the interaction of the motions of Venus and the Earth. This was seen as the first English improvement in solar tables since the time of Newton. Airy proudly slipped his calculations under Whewell's door early one morning, presented them to the Royal Society with applause and demanded the Board of Longitude change the Almanac. Young refused and Woodhouse exploded with fury. By spring 1828 the Board had been abolished for expensive inefficiency, and Airy lobbied hard to take over its Almanac. Babbage might once have aimed to automate these tables' production, but it was Airy who eventually manufactured them. Meanwhile, as Woodhouse's health finally collapsed after a long illness, Airy and Sheepshanks also began to arrange for his urgent replacement as Plumian professor. The Trinity cohort went into action. Sheepshanks planned to take charge of the university Observatory, but ruled out applying for the chair. When Peacock told Airy the news about Sheepshanks, he at once realized the advantages of the Observatory job – improved lectures, income, residence and thus the possibility of marriage. At the start of December 1827, the Observatory was put under the control of Airy, Sheepshanks, King and the St John's astronomer Thomas Catton. By 19 December Woodhouse was dead. Airy at once confidently moved to replace him as Plumian professor and Observatory director.

As Lucasian professor, Airy was one of the electors to the vacant Plumian chair. He used his power – and the considerable influence of Sheepshanks – to get his salary raised from £300 to £500, to hire an assistant recruited from the Navy and to overhaul the management of the Observatory. Airy's Observatory was a proving ground for an innovative administrative system. He obtained high-quality instruments from the London makers, rearranged the layout of observations and, using printed skeleton forms for observers, mechanized the efficient reduction and publication of stellar transit and planetary data. Airy toured German and Italian observatories, culminating with a visit to Arago in Paris. He got Troughton's firm to finish their expensive eight-feet mural circle, and another assistant, James Glaisher, was hired to work with it. Airy boasted that 'this introduction of an orderly system of exhibition not merely of observations but of the steps for bringing them to a practical result – quite a novelty in astronomical publications – had a markedly good effect on European astronomy in general'. Speeches in Trinity compared Airy with Newton. In 1835 Robert Peel, Whewell's political ally, offered Airy an annual civil list pension of £300. He had become, as it was said in Cambridge, 'the professor of stars'.[28]

The Trinity dons hoped that the appointment of a new Plumian professor would allow the further coup of electing Herschel as Lucasian professor. Whewell summarized the situation for Herschel at the end of November 1827 but conceded that 'so far as income is concerned, the Lucasian is rather a starving matter'. The St John's fellows, led by their observatory manager Catton, also courted Herschel, as did the Vice-chancellor, Martin Davy of Caius, all with the same disappointing result as the previous year. Once again, Herschel insisted on his own vocation as gentleman amateur in natural philosophy, denied he was a 'mathematician par excellence', and threw his considerable weight behind Babbage. Whewell, for one, doubted whether Babbage would offer himself or, having but an ordinary degree, would succeed. Away in Italy on his restorative and industrious tour, Babbage entrusted Herschel with directing the engineer Clement's demands for cash, time and materials. The difficulties of managing a recalcitrant workforce remotely via his ingenious machine symbolism were felt by its emotionally shattered designer. Babbage reported to his friend that whenever he thought about the difference engine 'it has been to wish I had never undertaken it; I shall finish the machine and be more than half ruined by it'. Babbage allowed his manifesto of autumn 1826, including plans for the public development of his philosophy of analysis, again to be circulated in Cambridge. Peacock reminded Babbage of his previous pledges to lecture, temporarily reside in Cambridge and work as an active professor in Airy's image. Once convinced of Babbage's aims, mainly by Herschel, the Trinity men mobilized behind him. Hints that Babbage would have to return from Italy for the election were quickly quashed. The night before the vote, Whewell and Peacock gave each Head a letter setting out Babbage's previous account of his aims and confirming his enthusiasm for the post. There were two rival candidates for the chair, but they could only count on the support of the heads of their own colleges.

On 6 March 1828 the Heads gave Babbage eight out of eleven votes. 'You sit without solicitation in the chair of Barrow and of Newton', one Trinity fellow congratulated him. Whewell, more callously, 'would be vexed if [Babbage] is not gratified, and now that he has no wife he may perhaps better like to live here part of the year'. Babbage bestirred himself in Rome quickly to reply enthusiastically, emphasizing his surprise and gratitude. But to Herschel he privately confided that he was still exhausted by the events of the previous year, preoccupied with the engine project, and so intended to delay return to Britain until the end of 1828. 'As far as Cambridge is concerned, at least, I shall forget the past and shall try to do credit to the recommendation of my friends.'[29]

BABBAGE'S DIFFERENCES, 1828–39

In the decade that he held the Lucasian chair Babbage neither forgot his past nor fulfilled the expectations of his backers. For many dons he became the antithesis of the ideal professor. Babbage had his own version of the academy, which he developed during the revolutionary conjuncture of the late 1820s and 1830s. Whatever his lassitude as Cambridge teacher, Babbage stood (unsuccessfully) for parliament twice, attacked the Royal Society and the Board of Longitude, helped set up the British Association for the Advancement of Science and the Statistical Society of London, and turned himself into an advisor on railway systems and telescope design. The aim was to define a new class of industrial experts who could direct economic policy and seize control from the hereditary nobility and the cloistered clergy. This project gathered momentum in 1830. An unsuccessful attempt was made to get Herschel made president of the Royal Society. In temporary alliance with the Edinburgh evangelical natural philosopher David Brewster, campaigns were launched for an overhaul of the rights of inventors and for the funding of specialist research at the moribund English universities by expropriating professorial endowments. The more cautious Herschel shared many of Babbage's views, but advised him to burn his attack on the Royal Society before it was published. Herschel then released his own rather irenic *Preliminary Discourse on the Study of Natural Philosophy*, a widely read handbook on the methods of dynamics and their implication for public welfare. In contrast, Babbage did not hesitate to publicize his more radical agenda. His philosophy of analysis, with its promise of unlocking the mechanisms of reason, was the justification of the new class's rights. In his professorial manifesto of autumn 1826, he had proposed the integration of his analytical philosophy of operational algebra into Cambridge culture. From 1828, he subjected the entire British system of manufacture to the same kind of operational analysis. The difference engine was the cause and case of this project. Whewell spotted this when Babbage came back to Cambridge to examine the Smith's prize in Lent 1829. 'I have really enjoyed his society much having seen him more closely than I have done before. His anxiety about the success and fame of his machine is quite devouring and unhappy.'[30]

Babbage's unhappiness focused on the rights of the intellectual aristocracy over those of labour and privilege. Clement and his team of young artisans designed unprecedentedly complex and precise components, demanded higher wages and asserted engineers' traditional ownership of the tools and gears they built. Babbage demanded the project be moved to a special workshop under his supervision and that the state meet its share of the costs. Babbage's colleagues at the Royal Society, including Herschel and Whewell, were temporarily prepared to back his claims. In 1830 the engine was nationalized. By summer 1834, after more than £17 000 had been spent, Babbage requisitioned plans and components of the engine for his own works in Marylebone, leaving the tools with Clement in Lambeth. His utopian vision of total management prompted more ambitious plans for analytical engines to compute higher algebraic functions. Though the difference engine was abandoned, this by no means halted Babbage's extraordinary programme of mechanical analysis and the rationalization of innovation. It provided a riposte to the conservatism of colleges, church and state.

Babbage launched a major public analysis of 'the Methods and Processes which are equally essential to the perfection of machinery' as soon as he became Lucasian professor. In winter 1827–8 he sent industrialists such as the younger James Watt copies of a preliminary essay on 'the application of machinery to manufacture', based on surveys of British and overseas industry conducted since 1822. Babbage's notebooks were full of remarks on 'causes of failure', which he believed could be solved by 'first' consulting a 'man of science on

Figure 38 Babbage's new calculus of operations could deal with both mathematical functions and machines. He wanted to introduce this technique into Cambridge mathematics teaching. When he asked August de Morgan for help with puzzles of mechanical computation, de Morgan made this image of Babbage fitting mathematical functions into his lathe.

the principle', and 'second, a practical engineer on mechanical difference'. The essay was quickly turned into a major article in the *Encyclopedia Metropolitana* (1829) and in summer 1832, a treatise *On the Economy of Machinery and Manufacture*. It was published by Charles Knight, the London agent for the reformers' useful knowledge campaign, and quickly sold almost 3000 copies. It ran to four editions by 1835, and was soon translated by Etienne Biot, son of Babbage's old friend. Knight, the publisher, reckoned that 'it is only by the extraordinary combinations of science with genius and capital that the working man is furnished with employment, and without a body of men whose labour consists simply of thinking, society would fall in pieces.'[31] These were the issues which Babbage, and his opponents, judged most at stake in their work on the philosophy of machinery.

Babbage's *Economy* was a patchy assembly of analytical essays on the economic and scientific facts of the factory system. It set the pace for a host of surveys of British industry. To his long encyclopedia entry of 1829, which analysed factory machines, work discipline, the division of labour and the social effects of intellectual work and trade unions, he added in his book of 1832 new chapters on the origin of manufactures and of price formation, cartels, engineering innovation and the methods of factory surveys. At the end of that year, in the book's second edition, he added to the work analyses of the role of money, the impact of technological unemployment and the utopian possibilities of profit sharing. Each topic was treated as an aspect of rational mechanics applied to economic behaviour. Mere makers became manufacturers by systematic cost minimization, and this demanded rational organization of the factory through economies of scale and perfected management. As in the calculating engines, each task must be allocated the right degree of skill in a minutely surveyed hierarchy. Babbage saw at least one enemy of rational economy – friction. Accordingly, he devoted space to engine lubrication and to uniform work-rates through routines of reliable, predictable performance. Strife with his own workforce and his industrial tours provided matter for his reflections on recalcitrant workmen and production mechanics. 'Workmen should be paid entirely in money; their work should be measured by some unbiassed, some unerring piece of mechanism; the time during which they are employed should be defined and punctually adhered to.' Transparency counted within the domestic economy of the factory. It also mattered to the surveys men like Babbage had to conduct, armed with statistical tools and lucid tables, and to the division of mental labour itself, of which mathematical computation provided the best example. This project put metrology, the production of precise standards and tabulated constants, at the centre of science. Babbage's French sources, such as Dupin, sought standard measures of work, and insisted that the value of work was set by its market value. Dynamic agents were standardized commodities. Babbage now argued that commodity production should be understood as dynamics. Capital's agency put skilled labour into action, money was a system of power transmission, and needless taxation was merely frictional. Since mental labour was the chief source of economic innovation, it must be understood through the analytical science of mechanics. According to Babbage, paper and brass could therefore master factories, colleges and pulpits. Surveillance mattered everywhere, in factories, colleges and in the clubrooms of scientific London too. 'Full publicity, printed statements of accounts, and general meetings are the only safeguards', he hectored the gentlemen of science, 'and a due degree of vigilance should be exercised on those who discourage these principles.'[32] All combinations against the interests of industry and commerce – from the clerical privilege which mystified the work of mind to the ritual of secretive artisans – would be abolished. In Babbage's political economy, the mechanization of reason went in hand with the rationalization of mechanism.

Babbage sought many ways of getting his programme to work. He presented copies to the London Mechanics' Institute, including his manifesto for profit sharing. The treatise was reviewed in mass-circulation radical magazines, and in November 1832 Babbage debated his creed on the hustings in a Canonbury pub with the leading socialist economist Thomas Hodgskin. His backers boasted that their parliamentary candidate was Lucasian professor. Babbage contemplated bringing the message to Cambridge where, during 1830, many were also concerned with the condition of public belief, and penned philosophical manifestos designed to correct it. That year, Whewell was commissioned by eminent churchmen and savants, acting on the will of the pious Earl of Bridgewater, to compose a natural theological treatise on astronomy and physics. He offered what amounted to a theological companion to Airy's new *Tracts*, including the claim that the universally distributed luminiferous æther was good evidence of divine wisdom. Because such a fluid would resist motion, Whewell reckoned that immaterial agency, ultimately that of God, was required to maintain the world order. Accordingly, Whewell tapped his friend Airy for summaries of up-to-date wave theories, and then wrote a chapter on the morality of friction, not as Babbage's obstacle to rational management but instead as the wondrous opportunity for free, nonmechanical endeavour. Both Whewell and Jones also soon completed important studies of political economy, lambasting the 'crasser dogmatizing' of utilitarians and materialists who reduced morality to mere economic interest. The apostles of mechanical deduction were 'downward mad people'. Whewell asked Jones for examples of the 'narrowmindedness and ineptitude of mathematicians'. The pious Jones' attack on utilitarianism soon gained him a chair in political economy at the Anglican King's College London, while the Bridgewater Treatise gained Whewell a national reputation as defender of the inductivist faith.

These were direct responses to imminent revolution. During autumn 1830, violent protest campaigns about rents, tithes and food prices spread throughout the south-eastern farmlands and campaigns for parliamentary reform gathered strength. In December 1830, Airy from the Observatory and Whewell from Trinity watched blazing farmland lit by rioters west of Cambridge and undergraduates armed to fight the rick-burners. 'This created the most extraordinary panic I ever saw', Airy recorded. 'I do not think it is possible without having witnessed it to conc[e]ive the state of men's minds.' In between their demands for troops to quash rural dissent and prevent the 'break up of society', Jones and Whewell swapped news of their work on economics and philosophy. 'We will live through the storm and teach the world wise things when the winds have lulled again', hoped Whewell. In 1831 there were even plans for establishing a permanent Cambridge chair of political economy. Then the university's Lucasian professor produced his own analysis of economics, machinery and manufacture.[33]

The dons saw the difference between Babbage's aims and their own projects. In Lent 1831, Whewell met Babbage on the Cambridge road, and learned that he planned to deliver his draft *Economy* as Lucasian lectures. Whewell soon published a fierce riposte to Babbage's and Brewster's attacks on English universities' failure to support original research. Babbage wanted to overhaul Cambridge teaching, make professors examine students in fields such as history and political economy, and teach the 'applications of science to manufactures'. 'We should be enabled effectually to keep pace with the wants of society', he argued, 'we should, at the same time, open a field of honourable ambition to multitudes'. These were certainly not Whewell's aims. In any case, the Trinity tutor reckoned professorial endowments were simply inadequate for the task at hand, while Airy moaned about the Lucasian professor's hypocrisy: 'I wish Babbage's non-lecturing could somehow be lugged into this controversy.' A year later, in early 1832, the situation was no better. Babbage invited his friends, such as Charles Lyell, Jones's opposite number as geology professor at King's, to visit the difference engine at the Southwark workshop, and there announced his plan to give lectures on the political economy of machinery at Cambridge.

Meanwhile, Jones scanned Babbage's draft treatise, thought it worse than the encyclopedia article, 'a strange collection of facts taken from his commonplace book', some 'trivial and uninteresting'. He sneered at Babbage's aim to offer an industrial version of Herschel's *Preliminary Discourse on the Study of Natural Philosophy*, though he conceded that 'the matter would have told well mixed up with lectures'. Whewell agreed and confirmed that Babbage still did not lecture on the economics of machinery, or anything else.[34] From spring 1833, the battle was publicly joined between the Trinity philosopher and the Lucasian professor, a controversy which would last until the end of Babbage's tenure and offer an apt summary of the Cambridge fights over mechanics, mathematics and morality in the Age of Reform.

In Lent 1833 Babbage again came up to Cambridge to examine the Smith's prize, and dined with Whewell and Airy at Trinity, where they chatted amiably about astronomical instruments for the Observatory. Then in early March Whewell's Bridgewater Treatise appeared, and found its way, too, to the library of the London Mechanics' Institute. The Treatise directly assailed Babbage's machine philosophy. Against the materialists, Whewell argued for a 'universal law of decay'. Without divinely spiritual sustenance things would fall apart, in nature and in society. The two men also differed, for example, in their use of Fourier's elegant new notions of thermal diffusion. In a major 1834 geological paper based on his tours at Pozzuoli and his engineering experiments, Babbage applied these thermal and mechanical principles to the entire history of the planet, which he saw as little more than a complex heat machine. In his *Economy* Babbage explained that steam engineering exploited but never undermined 'that equilibrium which we cannot doubt is constantly maintained throughout even the remotest limits of our system'. The entire world was a finely tuned economic system of balance and order. Whewell countered that all engineering in fact relied on disequilibrium, which could never be restored by merely material means. God's agency was indispensable. In particular, the historical emergence of intelligence from mere matter was a divine and thus miraculous act. 'We may thus deny to the mechanical philosophers and mathematicians of recent times any authority with regard to their views of the administration of the Universe.' Since deductive habits tended to atheism, the entire tradition from Laplace was damned. When the brilliant Paris savant Arago dined with Whewell, Airy and Peacock at Trinity, he regaled the receptive dons with anecdotes of Laplace's refusal to include the 'hypothesis' of a deity in his celestial mechanics. The task of the colleges was to extirpate such dangers and train wranglers in healthy philosophy.

By May 1833, Babbage had already worked out his reply to Whewell. To restore proper cosmological authority to mathematical mechanics, he would make his calculating machine mimic, thus rationalize, the miracles on which Whewell's theological posturing relied.

Brass machines could easily produce discontinuities which ignorant clerics might well mistake for signs of a sudden suspension of established laws. So Babbage designed an experiment in Marylebone using the difference engine to print the series of even integers up to 10 000 and then suddenly increase each term in steps of three. The switch in the numerical series was predictable to the analyst, yet surprising to the audience. In the same way, Babbage insinuated, apparent divine interventions, such as the origin of intelligence, were really law-like transformations in the rational order of the world. There was no need to invoke spiritual agents to explain the origin of complex structures such as machinery, mind or society. In answer to Whewell's boast that only induction, not machine analysis, might reveal the divine plan of the world, Babbage countered that the world-system was a macroscopic version of a factory. His house parties became showpieces of the mechanist faith. In June 1833 he was visited by Byron's widow, who had been coached in mathematics by William Frend, and her daughter, who would soon study mathematics with De Morgan and Babbage, and then become passionately involved in publicizing his engines. 'Both went to see the *thinking* machine (for such it seems)' and were treated to his miraculous show of sudden breaks in its output. 'There was a sublimity in the views thus opened of the ultimate results of intellectual power.'[35]

The mechanical metaphor for sudden innovation also impressed the sharper naturalists among Babbage's friends, including Lyell and his protégé Charles Darwin, who often attended Babbage's house parties after his return from the *Beagle* voyage. They learned that natural law could explain apparently discontinuous processes, such as the appearance of new species. Between 1833 and 1837, Babbage publicized his party trick with mechanical miracles in what he called *The Ninth Bridgewater Treatise*, and labelled it 'a fragment'. In this new treatise, cosmology and computation were carefully combined. 'The Devil is to have his due,' Lyell reported. To stir up the coals, Babbage repeated his Italian observations on vulcanology and printed two letters from Herschel on the uniformity of earth history and the law-like production of life. He sent copies of the book to figures of political eminence, including both the new Queen Victoria and the Piedmontese premier Cavour. An American mathematician told Babbage that 'when you carried me from the simple machine made by a man to the grand machine of the Universe I wish I could express to you one half of the enthusiasm I felt'. Lyell correctly predicted 'some people would not like any reasoning which made miracles more reconcileable with possibilities in the ordinary course of the Universe'. In May 1837 Whewell clarified his differences with Babbage, indicating the tendency of deductive mathematics to 'thrust some mechanistic cause into the place of God'. Whewell's conservative allies took the point. One Tory told him that 'the way in which you speak of the non-importance of mere mathematicians, coming from a person of your situation in science, is indeed very important'.[36]

So it was. In January 1839 Babbage resigned the Lucasian chair of mathematics and effectively sundered his tenuous relations with Cambridge. He broke, too, with the British Association, infuriated by the politicking of its genteel élite. He was absorbed by his new project to build more powerful calculating engines capable of executing the work of higher analysis. At the end of the 1830s, Babbage claimed that 'whenever engines of this kind exist in the capitals and universities of the world, it is obvious that all the enquirers who wish to put their theories to the test of number' would be subjected to the engines' sums.[37] By summer 1840 Babbage was again in Italy as a refugee from British storms, lionized at the Piedmont court and lecturing a receptive Turin congress of progressive scientists on these prospects of accurate metrology and analytical mechanization. The 'exultation' at his Cambridge election a decade earlier had gone.

CONCLUSION

With Babbage's departure from the Holy City of mathematics, the initiative lay with Whewell and Airy. The professoriate would be disciplined, lectures aligned with the Tripos and exacting training in wave optics and fluid dynamics reinforced. Whewell, soon wealthy through marriage into a manufacturing dynasty, was now powerfully placed under Tory patronage as Master of Trinity and professor of moral philosophy. He urged a permanent curriculum, with the works of Wood and his own mechanics texts as basis, and a progressive component, which graded wranglers' advance from the Analysts' *Examples* and Airy's *Tracts*. Airy agreed with the condemnation of the Analytical Society's youthful excesses, and played a decisive role from the Royal Greenwich Observatory in scotching any state plans to fund Babbage's engines. But in many ways the philosophy of machinery and manufacture mattered much in the reinforced Cambridge regime. In alliance with the prodigious engineer Robert Willis, Cambridge's successful professor of mechanism, Whewell issued his *Mechanics of Engineering* (1841) which turned Parisian theories of labouring force into academic doctrine. Unlike Babbage, he insisted that labouring force would always be lost without divine agency but, like Babbage, he made dynamic into economic values. The economic dynamics of Victorian submarine telegraphs and steam engines meant that training in electromagnetism and thermodynamics was a matter of considerable national importance, and thus eventually forced changes in Cambridge's mathematics curriculum. When, in the 1860s, the Cambridge mathematicians, led by Airy and Stokes, at last decided to incorporate electromagnetism and thermodynamics as permanent sciences, they helped establish a university teaching laboratory led by Clerk Maxwell, an institution explicitly committed to the virtues of metrology, the precision production of physical standards. Artisans and instrument-makers appeared alongside colleges and chapels.

In the end, Airy's treadmill stayed significant as object for analysis and as emblem of training. At his observatories, systems of mechanical discipline were imposed on large workforces to guarantee the good order of astronomy, meteorology and geomagnetism. Within Cambridge, these kinds of mental discipline were equally apparent. Despite some dons' best efforts, private coaches increasingly governed mathematical training. The leading coach William Hopkins was defeated by Joshua King in his attempt to succeed Babbage as Lucasian professor in early 1839. But though Hopkins lost the battle, he won the war. Coaches dominated the manufacture of wranglers. They taught that creation was an array of complex dynamical systems analysable by algebraic potential functions. Mathematical skill meant mastering rapid, correct manipulation of this algebra. By 1839, the dons had uncompromisingly ruled out any radical attempt to treat the properties of mind as outputs of a mechanical system. Yet they succeeded in reinforcing an academic regime which turned wranglers into something rather like calculating engines.

ACKNOWLEDGEMENTS

Thanks to the editors of this volume for their patient help and to the indispensable work of Harvey Becher and of William Ashworth on mathematics and politics in the Age of Reform. I am grateful also to the archivists and librarians of the Cambridge University Library, Royal Society Library, British Library and Wren Library, Trinity College Cambridge.

Notes

Abbreviations: AP: Airy papers, Royal Greenwich Observatory papers 6, University Library Cambridge; BP: Babbage correspondence, British Library, Additional Manuscripts; BPC: Babbage papers, University Library Cambridge MSS Add. 8705; HP: Herschel correspondence, Royal Society London MSS HS; WP: Whewell papers, Trinity College Cambridge, Additional Manuscripts.

1 G. N. Ray (ed.), *Letters and Private Papers of William Makepeace Thackeray*, (4 vols, Cambridge, MA, 1945–6), vol. 1, p. 104.

2 Herschel to Babbage, 9 March 1828, HP 2.222; Herschel to Babbage, 9 March 1828, in C. A. Chant, 'Two letters of a famous astronomer', *Journal of the Royal Astronomical Society of Canada*, 38 (1944), 225–7; Charles Babbage, *Passages from the Life of a Philosopher* (London, 1864), pp. 21–2.

3 Babbage, *Passages*, p. 29.

4 G. B. Airy, 'Report on the progress of astronomy during the present century', in *Report of the Second Meeting of the British Association for the Advancement of Science* (London, 1833), pp. 125–89, 184; Astronomer Royal's Report, 1853, AP.

5 Whewell to Jones, 24 March 1834, WP c.51/164; Turton to Whewell, 13 May 1837, WP a.213/167. Turton's 1826 hymn tune 'Bright was the guiding star that led' is at http://cyberhymnal.org/htm/b/r/brightwa.htm. Apart from the current incumbent, Turton is the only Lucasian professor whose work may be heard on the internet.

6 Charles Babbage, *On the Economy of Machinery and Manufactures*, 4th edn (London, 1835), p. 3; William Whewell, *On the Principles of English University Education* (London, 1837), pp. 138–9.

7 Wallace to Peacock, 1833, in Maria Panteki, 'William Wallace and the introduction of continental analysis into Britain', *Historia Mathematica*, 14 (1987), 119–32, 122.

8 James Clerk Maxwell, 'Introductory lecture on experimental physics' (1871), in W. D. Niven, (ed.), *The Scientific Papers of James Clerk Maxwell* (2 vols, (Cambridge, 1890), vol. 2, p. 31; Whewell to Jones, 13 October 1826, in Isaac Todhunter (ed.), *William Whewell: An Account of his Writings* (2 vols, London, 1876), vol. 2, p. 71; Babbage to Wordsworth, 14 November 1826, BP 37183 f. 366; Babbage, *Passages*, pp. 33–34.

9 Harvey Becher, 'Radicals, Whigs and conservatives: the middle and lower classes in the analytical revolution at Cambridge in the age of aristocracy', *British Journal for the History of Science*, 28 (1995), 405–26, 408–9 for Babbage's Acts; Whewell to Jones, 6 January 1828, in Todhunter, *Whewell*, vol. 2, p. 90; George Airy, *Autobiography* (Cambridge, 1896), pp. 22–45; Whewell to Morland, 3 October 1819, in Todhunter, *Whewell*, vol. 2, p. 35.

10 Whewell to Rose, 1 November 1818, WP, R.2.99/12; Whewell, *English University Education*, p. 137.

11 Whewell to Morland, 3 October 1819, in Todhunter, *Whewell*, vol. 2, p. 36; [William Hamilton], 'Study of mathematics: University of Cambridge', *Edinburgh Review*, 62 (1836), 409–55, 428; William Whewell, *Of a Liberal Education in General* (London, 1845), pp. 40–1.

12 Airy, *Autobiography*, p. 72; Babbage, *Passages*, pp. 32–3; Babbage's questions for Smith's prize, BPC no. 36; indignation about Babbage's examination papers, Whewell to Jones, 13 January 1831, WP c. 51/95.

13 Anthony Hyman, *Charles Babbage, Pioneer of the Computer* (Oxford, 1982), pp. 21, 33; Katherine Lyell, (ed.), *Life, Letters and Journals of Sir Charles Lyell,* (2 vols, London, 1881), vol. 1, p. 364; Airy, *Autobiography*, p. 71.

14 Babbage to Herschel, 3 April 1820 and Herschel to Babbage, 4 April 1820, HP 2.132-33; W. W. Rouse Ball, *History of the Study of Mathematics at Cambridge* (Cambridge, 1889), p. 120; Herschel to Whewell, 1 December 1819, WP a.207/5.

15 Charles Babbage, 'Observations on the notation employed in the calculus of functions', *Transactions of the Cambridge Philosophical Society,* 1 (1822), 63–76; Babbage, *Passages*, p. 429; Bromhead to Babbage, 4 March 1816, BP 37182, f. 52; Herschel to Whewell, 1 December 1822, WP a.207/6; Woodhouse's election recorded in Cambridge University Library MSS CUR 39.8 no.12.

16 Robert Woodhouse, 'On the necessary truth of certain conclusions obtained by means of imaginary quantities', *Philosophical Transactions of the Royal Society*, 91 (1801), 89–119, 106; Bromhead to Babbage, [?] 1813, BP 37182, f. 14; Peacock to Herschel, 3 December 1816, HP 13.247 and to Babbage, 10 December 1816, BP 37182, f. 79; Babbage, *Passages*, p. 196; Rouse Ball, *History*, pp. 118–19; Whewell to Herschel, 1 November 1818, in Todhunter, *Whewell*, vol. 2, p. 30 and to Rose, 1 November 1818, WP R.2.99/12; Bromhead to Babbage, March 1820, BP 37182, f. 201.

17 Charles Babbage, 'An essay towards the calculus of functions', *Philosophical Transactions*, 106 (1816), 179–256; Babbage's address to the Cambridge Philosophical Society, December 1821, in BP 37202, f. 5; Rose to Whewell, 5 October 1822, WP a.211/134.

18 Robert Woodhouse, *Treatise on Astronomy*, 2nd edn (2 vols, Cambridge, 1822), vol. 1, pp. xvii–xxii.

19 Airy, *Autobiography*, pp. 50–56, 60–69; Airy to Whewell, 26 February 1824, WP a.200/2; Airy to Babbage, 25 April 1826, BP 37183, f. 281.

20 Charles Babbage, 'On a method of expressing by signs the action of machinery', *Philosophical Transactions*, 116 (1826), 250–65 and draft in BPC no. 21; Babbage, *Passages*, pp. 142–6; Babbage, 'Of the influence of general signs in analytical reasoning', *Transactions of the Cambridge Philosophical Society*, 2 (1827), 325–77 (read 16 December 1821) and BP 37202 f.44.

21 Babbage to Dupin, 20 December 1833, BP 37188 f. 117; Babbage to Herschel, 24 November 1826, HP 2.205. Babbage italicizes the passage

from the 'Report from the select committee on the means of lessening the evils arising from the fluctuation of employment in manufacturing districts' in *Economy of Machinery*, p. 229. Dupin, *Géometrie et Mécanique des Arts et Métiers* (Paris, 1825–6) is cited in Lorraine Daston, 'The physicalist tradition in early nineteenth century French geometry', *Studies in History and Philosophy of Science*, 17 (1986), 269–95, p. 292.

22 Jones to Whewell, 16 October 1826, WP c.52/12; Whewell to Jones, 13 October 1826, in Todhunter, *Whewell*, vol. 2, p. 72; Peacock to Babbage, 14 October 1826, BP 37183 f. 343; Herschel to Hornbuckle, 13 October and to Whewell, 17 October 1826, HP 9.487 and 20.240.

23 Babbage to Herschel, 24 November 1826, HP 2.205; Bromhead to Babbage, 8 November 1826, BP 37183, f. 358.

24 Hyman, *Charles Babbage*, 146, 221; Augustus Bozzi Granville, 'This is the Engine that Charles built', 20 June 1851, AP 452, ff. 282–91.

25 Whewell to Jones, 10 December 1826, in Todhunter, *Whewell*, vol. 2, p. 80; Whewell to Rose, 12 December 1826, WP R.2.99/7; Whewell to Jones, 18 October 1826, WP c.51/29. Babbage's manifesto is two letters to Wordsworth, 14 and 21 November 1826, BP 37183, ff. 366–71. Votes in the election are in Cambridge University Library Registry MSS O.XIV.52, p. 5.

26 Airy, *Autobiography*, pp. 72–4, 90, 113–14.

27 Airy to Herschel, 16 July 1831, HP 1.39; George Airy, *Mathematical Tracts*, 2nd edn (Cambridge, 1831), pp. iii–iv.

28 Airy, *Autobiography*, pp.74–84, 105–7; George Airy, 'On the corrections in the elements of Delambre's solar tables required by the observations made at the Royal Obsevatory Greenwich', *Philosophical Transactions*, 118 (1828), 23–34; Sheepshanks to Whewell, 1 August, 3 and 15 October 1827, WP a.213/82-4; J. P. T. Bury, *Romilly's Cambridge Diary 1832–42* (Cambridge, 1967), pp. 33, 70.

29 Whewell to Herschel, 23 November 1827, in Todhunter, *Whewell*, vol. 2, p. 86; Thorp to Babbage, 6 March 1828, BP 37184, f. 116; Whewell to Herschel, 6 March 1828, in Todhunter, *Whewell*, vol. 2, pp. 90–1; Whewell to Jones, 17 March 1828, WP c.51/48; Babbage to Herschel, 2 April 1828, HP 2.224. Details of the vote are in Cambridge University Library MSS CUR O.XIV.52, p. 29 and CUR 39.8, no. 14.

30 Whewell to Jones, 4 February 1829, Trinity College Cambridge MSS Add. c.51/60.

31 Watt to Babbage, 25 February 1828, BP 37184, f. 110; Babbage's commonplace book, BPC no. 25, p. 10; Charles Knight, *The Rights of Industry, Capital and Labour* (London, 1831), cited in Maxine Berg, *Technology and Toil in Nineteenth Century Britain* (London, 1979), p. 91.

32 Babbage, *Economy of Machinery*, p. 310; Dupin, *Géometrie et Mécanique des Arts et Metiers* (1825–6) cited in Ivor Grattan-Guinness, 'Work for the workers: advances in engineering mechanics and instruction in France 1800–1830', *Annals of Science*, 41 (1984), 1–33, p. 10, 18; Babbage, *Reflections on the Decline of Science in England* (London, 1830), ch. 4.

33 Whewell to Jones, 2 April 1829, 23 July 1831, 31 October 1832, WP c.51/63, 110, 143; on the riots, see Whewell to Jones, 1 and 16 November, 7 December 1830, WP c.51/90-91, 93 and Airy, *Autobiography*, 92.

34 Charles Babbage, *Decline of Science*, ch. 1; [William Whewell], 'Cambridge transactions – science of the English universities', *British Critic*, 9 (1831), 71–90; Lyell, *Life*, vol. 1, p. 361; Airy to Whewell, 22 October 1831, WP a.200/11; Jones to Whewell, 17 February 1832, WP c.52/48.

35 William Whewell, *Astronomy and General Physics*, 4th edn (London, Pickering, 1834), 95, pp. 334; Babbage's experiment, 18 May 1833, BPC, no. 38, p. 38; Lady Byron to King, 21 June 1833, in Doris Langley Moore, *Ada Countess of Lovelace* (London, 1977), p. 44. For the material dynamics of Whewell and Babbage see M. Norton Wise with the collaboration of Crosbie Smith, 'Work and waste: political economy and natural philosophy in nineteenth century Britain, II', *History of Science*, 27 (1989), 391–449, 410–25.

36 Charles Babbage, *Ninth Bridgewater Treatise*, 2nd edn (London, 1838), pp. 32-43; Lyell to Herschel, 1 June 1836 and to Babbage, May 1837, in Lyell, *Life*, vol. 1, p. 467 and vol. 2, pp. 9–10; Bowditch to Babbage, 21 February 1835, BP 37189 f. 28; Whewell to Babbage, 30 May 1837, WP 266 c. 80.149/16; Rose to Whewell, 27 March 1833, WP a.211/143.

37 Babbage's resignation and King's election are recorded in Cambridge University Library MS CUR 39.8 no.15, O.XIV.53, p. 61. For analytical engines see H. W. Buxton, *Memoir of the Life and Labours of the Late Charles Babbage*, ed. A. Hyman (Cambridge, MA, 1988), ch. 8.

7 Arbiters of Victorian science: George Gabriel Stokes and Joshua King

David B. Wilson
History Department, Iowa State University, Ames, IA, USA

Sir George Gabriel Stokes (1819–1903) was known for his reticence. At a dinner party towards the end of his career, according to a story lingering in Cambridge long afterwards, Stokes was seated next to a young American woman. This quite different pair amazed the other guests by engaging in animated conversation – with *Stokes* doing most of the talking. When asked about the remarkable event later, the young woman said it was simple: she had just asked him whether he preferred algebra or geometry.[1]

It was mathematics and science that ignited Stokes, the thirteenth and longest-serving Lucasian professor. Another anecdote confirms the point. J. J. Thomson, discoverer of the electron and director of the Cavendish Laboratory in Cambridge, remembered a visit by Stokes and Lord Kelvin to the Cavendish. Long-time friends, Stokes and Kelvin were by this time the elder statesmen of British physics. As Thomson recalled:

> The temperaments of the two were, however, very different. When Kelvin was speaking, Stokes would remain silent until Kelvin seemed at any rate to pause. On the other hand, when Stokes was speaking, Kelvin would butt in after almost every sentence with some idea which had just occurred to him, and which he could not suppress. I once saw a curious reversal of this. They had come together to the Cavendish Laboratory, and I was showing

From Newton to Hawking: A History of Cambridge University's Lucasian Professors of Mathematics, ed. Kevin C. Knox and Richard Noakes. Published by Cambridge University Press.

> them some experiments I was engaged with at the time, on the electric discharge through gases. I happened to speak about atoms playing a part in one of the effects I was showing, when Kelvin said he did not believe in atoms but only in molecules. This was too much for Stokes. He began at once to give a charmingly clear account of the reasons why atoms as well as molecules must exist. He was so much in earnest that Kelvin for once could not get a word in edgeways: as soon as he started to speak, Stokes raised his hand in a solemn way and, as it were, pushed Kelvin back into his seat.[2]

Throughout his career, as this story illustrates, when Stokes did speak he spoke with authority. His stature in physics by mid-century had further consequences. First, it brought him the Lucasian professorship in 1849, and for the next few decades he was the authority for Cambridge undergraduates on what constituted physics. Second, in an age of contested ideas about religion and about its relation to science, Stokes gained a platform for the relevance of his own religious convictions. Third, Stokes's achievements in physics earned him a fellowship in the Royal Society of London in 1851 and led to his election as a Society secretary in 1854. Serving in that office for thirty years, he directly affected the content of the most prestigious scientific forum of the day, the *Philosophical Transactions of the Royal Society of London*. Thus, Stokes was an arbiter of proper scientific thinking and a force within the Victorian intellectual community. His reputation in physics extended his influence to such current subjects as evolutionary theory, spiritualism, materialism and the immortality of the soul. At times, his words involved disagreement and even conflict with other eminent intellectuals, especially Charles Darwin, T. H. Huxley and John Tyndall. Though *arbiter* hardly means *dictator*, Stokes's storied silence did not keep his voice from being heard.

This chapter analyses the life and career of George Gabriel Stokes. The first half of the chapter traces his rise from humble origins in Ireland to his appointment as Lucasian professor and his half-century in that chair. In doing so, it discusses the early-nineteenth-century reforms of the Cambridge curriculum, reforms which owed much to Stokes's predecessor as Lucasian professor, Joshua King (1798–1857), and which had a significant impact on Stokes's own mathematical training. While King was mainly concerned with the administration of Cambridge University, Stokes went much further and engaged the thought of his age. Accordingly, the second half of this chapter details and contextualizes Stokes's contributions to a wider range of scientific, religious and broader intellectual topics, including the nature of stellar aberration and the luminiferous æther, X-rays, radioactivity, the professionalization of science and the evolution of humans and the state of their souls. As Lucasian professor, Stokes commented extensively and passionately on all these momentous issues, and although his power in Cambridge dwindled over the years, his academic position reinforced his triumphs in physics to give him tremendous authority on a variety of subjects. Nevertheless, while Stokes's career had many dimensions, its centre was always Cambridge University.

STOKES AND KING: THE RESEARCHER AND THE ADMINISTRATOR

We can begin by noting Stokes's remarkable achievements on Cambridge's mathematical examinations.[3] In 1841, he achieved the university's highest undergraduate honours by becoming senior wrangler in the Mathematical Tripos and, in a subsequent competition, first Smith's prizeman. Those attaining such distinction could count on college fellowships and expect eventually to be in the running for college masterships and university professorships. If they took holy orders, they could also expect to be honoured with ecclesiastical preferment.

On his Tripos examination, Stokes faced questions concentrating on the Newtonian subjects of mathematics, astronomy, optics, mechanics and hydrostatics. In preparing for mathematical questions on Newton's *Principia* such as 'Enunciate and prove

Newton's eleventh lemma,' for example, students memorized large portions of Newton's treatise. In Stokes's student days, questions on astronomy still included questions on instruments, observational techniques and, above all, gravitational theory. Stokes could also expect to be examined on both the geometrical and physical components of optics. In addition to Newton's *Principia*, there was for the Tripos student a tradition of textbooks written by such Cambridge graduates as Samuel Vince (senior wrangler 1775), James Wood (senior wrangler 1782), William Whewell (second wrangler 1816) and George Biddell Airy (senior wrangler 1823).[4] Those previously successful in the Cambridge system guided the mathematically precocious to successes of their own.

In addition to these scientific and mathematical disciplines, Stokes was examined in classics, as well as moral and natural theology. Indeed, before 1824 there was only *one* Tripos, and during the four-day ordeal students had to demonstrate their competence at using Latin, recount a variety of arguments concerning the existence and nature of God and derive a number of moral arguments from first principles. By the time Stokes came up to Pembroke in 1837, classics and theology were part of the 'previous examination' that students took in their second year. A classical Tripos had also been introduced, but the classics student still had to earn an honours degree in mathematics before taking the classical examinations. Not until Stokes was Lucasian professor were further Triposes introduced, at mid-century.

Outside the formal undergraduate curriculum, Cambridge nourished non-Newtonian areas of the natural sciences. Adam Sedgwick (fifth wrangler 1808), professor of geology, and John Stevens Henslow (sixteenth wrangler 1818), later professor of mineralogy and then botany, were prime movers in 1819 in founding the Cambridge Philosophical Society for promoting the whole range of the natural sciences, including geology, biology and chemistry. The first two volumes of the society's *Transactions* contained articles by Sedgwick and Henslow on geology, by William Whewell on mineralogy, and by the professor of chemistry, James Cumming (tenth wrangler 1801), on electromagnetism produced by heat. Meanwhile, John Herschel (senior wrangler 1813), who had published an optical paper in the *Transactions*, helped to establish a new wrinkle in the Newtonian subject of physical optics. His long article on 'Optics' (1827) in the *Encyclopaedia Metropolitana* advocated the new wave theory of light, developed by Thomas Young and Augustin Fresnel, over Newton's corpuscular theory. In the years around 1830, George Biddell Airy bolstered the wave theory at Cambridge in professorial lectures, as well as in articles and a textbook. Airy introduced the wave theory into the Smith's prize examination, and it was also well established in the Mathematical Tripos by 1830 and 1831.

This was the university milieu that Stokes entered in 1837. His journey began, however, far from the Cambridge fens. Born and raised in Skreen, County Sligo, where his father was a rector in the Protestant Church of Ireland, Stokes had five older siblings. His sister Elizabeth, eight years his senior, cared for him in his youth and recalled that as a child George was

> [s]uch a pleasant-faced, plump, very fair, rosy baby . . . He grew into a dear good child, a little passionate, but with a strict sense of honour and truthfulness, I do not think he ever told a lie . . . He was exact even when a child; one day on being sent to see what o'clock it was he said, 'I can't tell you what o'clock it is now, but when I was at the clock it was such an hour.'[5]

Stokes's brothers all attended Trinity College, Dublin, and took holy orders. William Haughton Stokes (1802–84) also attended Cambridge, where he graduated sixteenth wrangler in 1828. Elected a fellow of Caius College, William remained in Cambridge for two decades before becoming a vicar in Norfolk.

George Gabriel was first taught by his mother, his father and the parish clerk until being sent in 1832 to the Reverend R. H. Wall's school in Dublin, where he stayed for three years. In 1835 he entered Bristol College. What no doubt took Stokes to Bristol was the fact that his brother's colleague from Caius was now principal of the college. As its governing council had decided a few years earlier, Bristol College's curriculum emphasized mathematics, classics, English literature and universal history. In mathematics, they intended to follow 'with no more alteration than can be avoided, the plan at present pursued in Trinity College, Cambridge'. Hence, it was Bristol College, modelled along the lines of the Cantabrigian curriculum and run by a graduate of the Mathematical Tripos, which began Stokes's immersion in Cambridge-style mathematics. When it was time for George to matriculate at Cambridge, William advised him to think carefully about his choice of college. The best college in preparing for a career in the church, William wrote, might be the worst in preparing for one in medicine.[6] For whatever reason, Stokes matriculated at Pembroke College, and he was to find Cambridge a comfortable setting.

As an evangelical, he was among others of like mind. He regularly took copious notes on sermons – the first four being those delivered by Whewell in 1837 and later published as *On the Foundations of Morals*. Like other undergraduates, Stokes also imbibed the theology of William Paley (fifth wrangler 1763). Paley's *A View to the Evidences of Christianity* (1794) was still a required text in Cambridge's examination system and remained so until 1920. The work sought to prove that Revelation was true, that it was indeed the word of the Creator, and that the 'probity and good sense' of the witnesses to Christ's miracles guaranteed their veracity. Stokes was thus ready to be examined on the means by which Paley demonstrated how those miracles constituted the best evidence for the truth of Christianity, and why the high moral character of Jesus was one of the 'auxiliary evidences for Christianity'. Stokes, like Charles Darwin, was also familiar with Paley's best-selling (but *not* required) *Natural Theology* (1802), which argued for the existence of a wise Creator from the purposeful design to be found in the natural (especially biological) world. Paley's *Moral Philosophy* (1785) was also required reading but, by the 1830s, Whewell and others were challenging its utilitarian morality, arguing that morality stemmed from a divinely implanted,

Figure 39 G. G. Stokes, probably at the time of his graduation in 1841. The watercolour was in the possession of Stokes's descendant, Mr Roger Wilding, who donated it to Pembroke College in 2000.

innate sense of right and wrong. Indeed, Stokes was so steeped in Whewellian moral philosophy that when he eventually published his own theory of morality, it closely resembled Whewell's conception of divinely implanted moral principles.

Along with his keen interest in religion, Stokes concentrated on the traditional mathematical and scientific topics that would win him university honours in 1841. He attended the lectures of James Challis,

Plumian professor of astronomy, and his notes on the lectures emphasized optics, including the wave theory. Stokes also enjoyed the private tuition of William Hopkins, the renowned mathematical coach whose students would eventually include such illustrious physicists as William Thomson (second wrangler 1845, later Lord Kelvin) and James Clerk Maxwell (second wrangler 1854). Hopkins tailored his coaching to the sanctioned content of the Mathematical Tripos, a system he strongly endorsed, not least because he regarded the Newtonian system of the world and the wave theory of light as the two crowning achievements of the mathematical investigation of nature.

Joshua King was one of those responsible for Cambridge examinations as they existed in 1841. Like his more famous contemporary Whewell, King hailed from Lancashire and matriculated as a sizar at Trinity College, although King soon migrated to Queens', evidently because of its evangelical character.[7] King's undergraduate career could hardly have been more successful. He won college prizes in both Latin and mathematics, and in 1819 became senior wrangler and first Smith's prizeman, taking his degree 'with higher distinction than perhaps any other man ever did'.[8] King naturally accepted the invitation to become a fellow of Queens', although he would have been unaware of how tumultuous the next decade would be at the college. With the death in 1820 of Isaac Milner, Lucasian professor *and* president of Queens', an extended power struggle ensued as fellows grappled for control of the college.[9] After twelve years of arguments and legal actions, King emerged in 1832 as the obvious man to succeed Milner's successor as president of Queens'.

At the end of the 1830s, King matched the example of the great Milner. When Charles Babbage vacated 'Newton's chair', eight heads of colleges met on 17 January 1839 to name his successor.[10] As president of Queens', King was one of the eight present. Two candidates were nominated – King and Stokes's coach, Hopkins – but King won the vote by a majority of seven to one. Even the Master of Hopkins's college voted for King, as, evidently, did King himself. As the new Lucasian professor, King was now one of the Smith's prize examiners,

Figure 40 Joshua King, posing with Newton's *Principia*, on the occasion of becoming President of Queens' College in 1832. The engraving is after an oil portrait, the latter having subsequently been trimmed to include King's head only.

while Hopkins retained the less prestigious, but more consequential, task of coaching.

King influenced Cambridge examinations in various ways. He was a member of syndicates in 1827 and 1832 whose recommendations made the Mathematical Tripos longer and its content more uniformly defined. From 1824 to 1831, he frequently examined in the Tripos and may have encouraged the wave theory of light.[11] He regarded his Lucasian professorship as neither a lectureship nor a research position, but primarily as an *examinership* for the Smith's prizes. In this sense he was following precedent. Milner, with whom King must have felt a special affinity, had not lectured during his two-decade tenure as Lucasian professor. During the 1820s, there had been a burst of lecturing by three of King's predecessors, but with the appointment of Charles Babbage in 1828, the status quo returned. Indeed, Babbage assumed that the chair carried few responsibilities and regarded his primary task as examining for the Smith's prizes. With such small remuneration from the Bedfordshire estates to pay the professor's salary, King could easily have shared Babbage's opinion.

In any case, 1841 was one of those years in which a future Lucasian professor was sitting examinations shaped by the current Lucasian professor. Stokes disappointed neither his coach nor his examiners. In the Mathematical Tripos in January, he faced 173 questions on the rather predictable topics of Euclidean geometry, the calculus, Newtonian mechanics, astronomy, hydrostatics and optics. Only a handful involved heat, electricity or magnetism. Well prepared by Hopkins, Stokes became Pembroke's first senior wrangler in six decades. Shortly afterwards, he also became first Smith's prizeman. In a suitably triumphant mood, Pembroke held a celebratory dinner in Stokes's honour and made him a fellow.

Stokes may have encountered predictable examination questions – for instance, none of his Smith's prize questions dealt with heat, electricity or magnetism, but three concerned the wave theory of light – but the 1840s witnessed serious debates about the Cambridge curriculum. Although King set only two questions on non-Newtonian subjects in his four years of examining for the Smith's prizes, other examiners were giving greater attention to subjects such as heat, electricity, magnetism and engineering. Whewell included these subjects on the Smith's prize examination the year after Stokes graduated and continued doing so in subsequent years. Like other Cambridge reformers, Whewell regarded the Smith's competition as a means to increase the number of subjects included in ordinary undergraduate studies.

Seeking to recreate Cambridge in his own image, Whewell also employed his power as Master of Trinity College (1841–66), as professor of mineralogy (1828–32) and later Knightbridge professor of moral philosophy (1838–55). Unlike King, he lectured incessantly and published voluminously on a wide range of subjects, including mineralogy, history, morality, philosophy, political economy, the tides and geometry. Unlike King, he transformed his professorships into profitable enterprises. By the time he stepped down from his Knightbridge professorship in 1855, he had changed it from a mediocre paying sinecure to an active position worth £700 per annum.

Moreover, these positions increased Whewell's influence in committees, where his savvy wrought even greater changes within the university. On 25 February 1843, for example, a syndicate headed by Whewell suggested ways 'to secure a correspondence between the Mathematical and Classical Examinations of the University, and the Mathematical and Classical Lectures of the University'. Whilst this committee included Hopkins and Challis, King was noticeably absent. Five years later, Whewell headed another committee that established not only a Board of Mathematical Studies, but also the Natural Sciences and the Moral Sciences Triposes. That syndicate included Hopkins and Challis, but, again, *not* King.[12]

King's declining health was probably one of the reasons for his absence from these syndicates. He suffered a stroke some time in late 1843 and by 1849 was ready to relinquish the Lucasian professorship, writing to the Vice-chancellor of the university that he would resign 'because of the continued infirm state of my health'. King resigned his professorship in a context much different from the one in which he

Figure 41 During the tenures of King and Stokes, 'The Backs' remained an important locale for gownsmen to exercise, relax and ruminate.

had assumed it. Now, more than ever, academic life focused on professorial lectures and new scientific subjects. Perhaps, in addition to his ill health, it was because King recognized his increasingly marginal status that he stepped down. Though remaining president of Queens', he resigned his professorship in September, hoping that the Bedfordshire estates would be more profitable to his successor 'than they have been to me'.[13]

Stokes, of course, was King's successor. The college Masters, including Whewell but not King, assembled at 10 a.m. on 23 October 1849 and elected Stokes unopposed.[14] Uncongenial for King, mid-century Cambridge suited Stokes to a tee. His interests fitted easily into plans to strengthen the university's curriculum. The *Cambridge University Calendar* for 1850 noted that 'The present professor intends to commence a course of lectures, theoretical and experimental, on Hydrostatics, Pneumatics, and Optics, with particular reference to the physical theory of Light.'[15] These were the main topics of both Stokes's lectures and his research. Though Stokes would be involved with many issues during the half-century of his Lucasian professorship, the foundation of his career was his research in physics.

Physics, and science more generally, changed dramatically during this half-century. Reintroduced to Britain in the 1820s, the wave theory of light was an established part of physics by mid-century. In the 1850s, Rudolf Clausius in Germany and Kelvin in Scotland formulated the two laws of thermodynamics, the first describing the conservation of energy and the second the 'dissipation' of energy. In 1867, Kelvin and P. G. Tait published a *Treatise on Natural Philosophy*, which embodied the new physics of energy. Charles Darwin's *Origin of Species* appeared in 1859 and his *Descent of Man* in 1871. Though largely persuasive, both of Darwin's expositions of biological evolution excited controversy among Victorians. In 1873, James Clerk Maxwell's *Treatise on Electricity and Magnetism* consolidated nearly two decades of his research that combined the wave theory of light and electromagnetic theory. By century's end, the ideas of Darwin, Kelvin and Maxwell defined new conceptions of man and nature. Most biologists contended that all forms of life, including humans, were subject to evolutionary change. Physicists understood that as energy changed from one form to another (sound to heat, for example) its total remained constant (or was conserved) although its availability for doing useful work declined (or dissipated). They understood also that the energy of light waves was electromagnetic in nature.[16]

Incorporating some of these changes, Stokes's own research in the 1840s and 1850s stemmed directly from his undergraduate studies. One of the questions that Stokes had encountered on the Smith's prize examination, for example, underscores the intimate tie between his education and his earliest research in hydrodynamics.[17] In this question, the examiner, James Challis, invited students to consider the connection between a particular differential equation and fluid motion. While Challis held that the equation would describe only rectilinear fluid motion, Stokes argued that the same equation would allow for curved motions within a fluid so long as a minute portion of the fluid – if one imagined it to be suddenly solidified – would possess no rotation. This meant that a fluid could move in curves but not exhibit whirlpools or eddies. Stokes's view eventually prevailed, this type of motion becoming known as 'irrotational' motion.

Stokes sought further to improve the science of hydrodynamics by investigating the internal friction, or viscosity, of fluids. What was so notable about Stokes's work was his conceptualization of the molecular cause of viscosity. Crucially, this conceptualization was closely tied to Stokes's epistemology of the unobservable. Briefly, Stokes concluded that, although difficult to obtain, knowledge of unobservable entities was possible. As long as a sufficient understanding of molecular interactions was known, then it was possible to make statements about the molecules themselves. Stokes was confident that he knew enough to do just this. Whatever their exact nature, molecules would give rise to a certain 'tangential' force as they moved with respect to each other. This force was the cause of viscosity and could be incorporated within a proper mathematical theory. The most well-known consequence of this work was 'Stokes's law' of a fluid's frictional resistance to a sphere moving through it. The law made analysis of a pendulum's swing more accurate and also explained the formation of clouds. In the former case, more accurate measurements meant a more precise determination of the acceleration due to gravity in different places, and these values were important to the science of geodesy, on which Stokes published important papers. In the latter case, falling drops of water would reach a 'terminal' velocity – the velocity at which further increase of velocity due to gravity was balanced by the friction of air. For small drops, the terminal velocity was so close to zero that they remained virtually suspended, forming clouds.

Probably Stokes's most fundamental research in this period dealt with the luminiferous æther. According to the wave theory of light, the æther was the putative physical medium that transmitted light. It appeared to be fluid-like because solid objects, like planets, moved through it. In fact, Cambridge scientists like Herschel, Whewell and, especially, Challis seemed to regard it *as* a fluid. Aether was therefore regarded as part of hydrodynamics as well as optics. Two related problems faced proponents of this basic physical concept. First, light waves seemed to be transverse, but transverse waves could not be transmitted by ordinary fluids. Second, and more serious, it seemed to Stokes that the well-known phenomenon known as stellar aberration was difficult to reconcile with the current theory of the æther.

Detected early in the eighteenth century, stellar aberration was first thought to result from the combined motions of the earth and the particles of light from stars. In other words, owing to the motion of the earth, a star appeared slightly offset from its actual position. With the acceptance of the wave theory of light, scientists explained stellar aberration by assuming the æther to be so subtle that the earth passed through it without disturbing the medium. This meant that the earlier combined-motions conception of stellar aberration was at least as compatible with the new wave theory as with the now discredited particle theory of light.

However, a problem remained. If the æther were *solid* enough to support transverse waves, how could it *not* be disturbed by the earth's motion? That was the question which Stokes asked and solved. His insight was twofold. He related the mathematics of fluid motion to that of elastic solids, and he postulated that the tangential force causing viscosity in fluids was the same as the 'restoring' force that enabled solids to support transverse waves. Appealing to a concept of continuity, he argued that all solids would have some degree of fluidity and all fluids would have some degree of solidity. He thus conceived the æther to be a fluid with some characteristics of a solid. Its solidity supported the very rapid, minuscule transverse light waves, while its fluidity allowed the relatively slow-moving earth to pass through it. But this movement did not leave the æther undisturbed. Rather, the æther 'flowed' around the earth. In fact, if this æthereal flow were irrotational, then, Stokes showed, stellar aberration was easily explained. Now it was unnecessary to posit that the æther and the earth failed to interact. Indeed, the æther's slight solidity would help maintain its irrotational motion because any tendency to form vortices would be immediately transmitted away by the æther.

Stokes's concept of a fluid-like, elastic–solid æther was widely adopted by Victorian physicists. Moreover, similarly to his theory of fluid viscosity, his model of the æther displayed what could be called a cautious realism towards the 'dynamical' theories that proliferated in mid-Victorian physics. Such dynamical theories generally explained observable phenomena in terms of underlying, unobservable motions. These theories were realistic in so far as British physicists thought that they could accurately represent reality. Stokes insisted that caution was necessary, however, so that one did not transgress the limits of reason. Stokes thought, for example, that contemporary evidence supported the existence of a fluid–solid, dynamical æther, but that this evidence was insufficient to determine whether the æther itself was continuous or particulate.

The same view was expressed in 1865 in one of Maxwell's seminal papers on electromagnetic theory, *A Dynamical Theory of the Electro-Magnetic Field.* Maxwell there presented electromagnetic phenomena as dynamical activity within a material medium. Yet, like Stokes, he insisted that due caution was necessary in correlating specific states or motions of this medium with specific observations of electromagnetic phenomena.

In 1849, Stokes published his own paper with 'dynamical' in the title, *A Dynamical Theory of Diffraction*. He offered a sophisticated experimental–mathematical study of the well-known phenomenon of diffraction – the optical effect in which light is bent by passing across the edge of an object. It solved a crucial question in the wave theory: what was the orientation of light's waves? Since the seventeenth century, physicists had been able to define experimentally the plane of polarization of polarized light. But physicists had been unable to determine whether the transverse waves of light were *in* that plane or perpendicular to it. In his 'dynamical' paper, Stokes concluded that they were perpendicular.

However, it was Stokes's research on fluorescence (a term he coined) which attracted the most attention. Investigating a familiar phenomenon, Stokes determined experimentally that fluorescent materials absorbed light of one wavelength and emitted light of a different wavelength. He demonstrated that the light emitted by fluorescent material always had a longer wavelength than the incident light, a relationship that is also known as 'Stokes's law'. In addition, Stokes cautiously speculated on the dynamical cause of such an unexpected result, suggesting that it probably lay in the interaction between the vibrating æther and oscillating particles of the fluorescent substance. These interactions, he emphasized, were consistent with the conservation of energy. Stokes published his experimental findings and provisional explanation in the *Philosophical Transactions* in 1852. The Royal Society acknowledged his research by awarding him its prestigious Rumford medal in the same year.

In the geography of Victorian physics, Stokes's Cambridge research of mid-century contrasted with that originating elsewhere. The non-Cambridge emphases flowed in large measure from the minds of the experimental physicists, James Prescott Joule in Manchester and Michael Faraday in London. Joule experimented on the conversion of heat and mechanical work, presenting a dynamical theory of heat which stated that heat could be reduced to the motions of material particles. Faraday discovered electromagnetic induction and the magneto-optical rotation of plane-polarized light (now known as the Faraday effect). With their Scottish (as well as Cambridge) university educations, the mathematical physicists Kelvin, Tait and Maxwell pursued these non-Cambridge subjects. Unlike Cambridge's mathematical studies, Scottish natural philosophy courses at mid-century did fully include heat, electricity and magnetism. In the early 1850s, Kelvin expanded Joule's ideas into the science of thermodynamics. Influenced by Kelvin's efforts to find a unified theory of Faraday's discoveries, Maxwell developed his electromagnetic theory of light in the 1850s and 1860s. As already noted, Kelvin and Tait's *Treatise* of 1867 and Maxwell's *Treatise* of 1873 were landmarks of this transformation of Victorian physics. Consequently, Kelvin, Tait and Maxwell were *Scottish*-educated Cambridge graduates extending the insights of the non-Cambridge experimental physicists Joule and Faraday.

We can thus place Stokes's detailed research in broader perspective. It was clearly conditioned and restricted by his studies at Bristol and Cambridge. His research may reflect a certain theoretical caution as well. One cannot easily foresee the most seminal areas of research in physics and simply adjust one's research agenda accordingly. Nevertheless, Stokes's *choice* to remain within the realm of Cambridge studies does contrast with the more aggressive and innovative theoretical insights of Kelvin and Maxwell. Indeed, contemporaries commented on Stokes's caution, especially in contrast to Kelvin's boldness.

But all this is merely to praise with a faint damn. Stokes's research carried him to the forefront of the highly competitive Cambridge milieu. He generally prevailed, for example, in his various disputes with his former professor, Challis, regarding hydrodynamical and optical theories. He took the Cambridge subjects to their highest theoretical level. His fluid–solid æther was a long-lasting theoretical concept. Not only did his research incorporate the conservation of energy, some of his ideas merged with the new physics of energy and electromagnetism. Physical concepts bear his name still. Stokes emerged from a specific Cambridge context to become a physicist of wide renown. He was the obvious choice in 1849 to be Lucasian professor.

The Lucasian professorship was an obvious way for Stokes to define proper science. There were limits, of course, because any teacher, however gifted, can always expect his best students eventually to surpass him. However, during the first half of his tenure, Stokes offered much of what Whewell had desired – lectures on physics directly relevant to the Mathematical Tripos. Indeed, in Cambridge during the third quarter of the century, Stokes best presented the desired blend of experimental–theoretical physics and, in doing so, attracted those students who would become the leaders of Victorian physics. He also supported university reforms which strengthened Cambridge physics, though they would help undermine his own position as Cambridge's premier physicist.

In the 1850s, the Lucasian chair's low income prompted Stokes to accept a second position, a lectureship in natural philosophy at the new Government School of Mines in London. There, his lectures included aspects of physics that were absent from the curriculum at Cambridge. For example, he covered the dynamical theory of heat and the conservation of energy, as well as Faraday's experimental researches on the interactions between magnetism and light. In 1861, however, by adding an annual stipend of £200 to Stokes's income as Lucasian professor, Cambridge University enabled him to relinquish his London lectureship.[18] Stokes's income was further augmented by the university in the 1880s to offset the reduction in his professorial income from the Bedfordshire estates by agricultural depression. But his financial concerns seem to have been ever-present. He had to weigh seriously, for instance, whether he could give up his £200 per year as Secretary of the Royal Society in order to take the presidency of the society, and in 1889 he wondered whether his modest finances would enable him to accept a baronetcy.

Stokes's London lectureship also underscores the differences between him and his most distinguished student, Maxwell. After attending Stokes's Cambridge lectures in 1854, Maxwell became a student transcending his teacher. In the same years that Stokes presented lectures on Faraday's investigations to his London classes, Maxwell used the same results as the springboard for his own revolutionary work leading to the electromagnetic theory of light. Maxwell thus enjoyed the benefits of Stokes's lectures on hydrodynamics and optics, but chose to investigate electricity and magnetism. Although Stokes could certainly teach these non-Cambridge subjects, as he did in London, his deeper thoughts reflected the Mathematical Tripos.

Cambridge's best undergraduates attended Stokes's lectures in large numbers. During the first quarter-century of his tenure, some eighty per cent of the top ten wranglers and all but two of the senior wranglers had attended Stokes's lectures. His lectures on optics were 'an epoch' in the life of Lord Rayleigh (senior wrangler 1865), and J. J. Thomson (second wrangler 1880) reminisced: 'The lectures I enjoyed the most were those by Sir George Stokes on Light. For clearness of exposition, beauty and aptness of the experiments, I have never

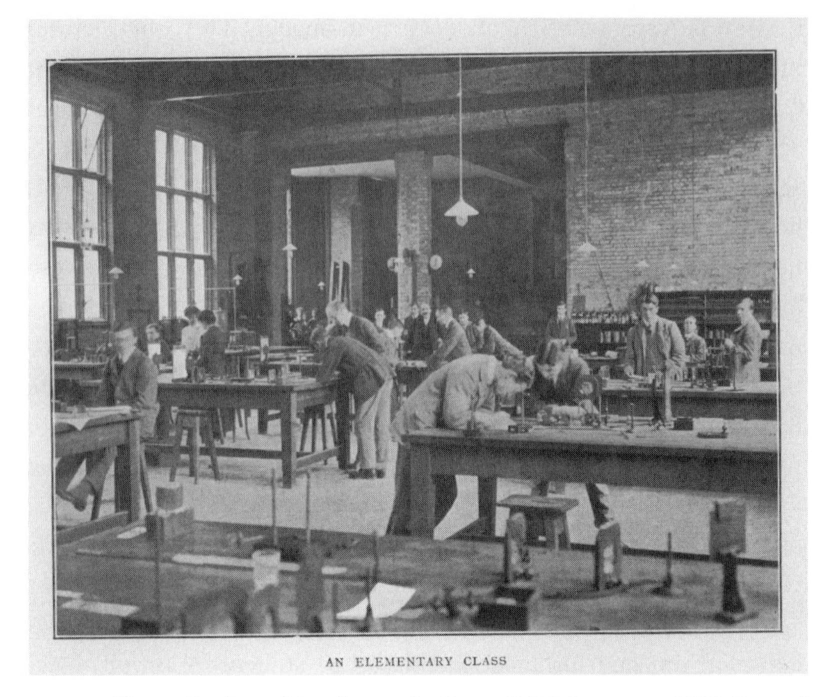
AN ELEMENTARY CLASS

Figure 42 A teaching class at the Cavendish Laboratory c. 1900. Opened in 1874, the Cavendish increasingly became the centre of Cambridge physics during the latter half of Stokes's tenure as Lucasian professor. From *A History of the Cavendish Laboratory 1871–1910* (1910).

heard their equal.'[19] During his fifty-four-year tenure as Lucasian professor, Stokes taught at least forty-five future fellows of the Royal Society, including the first three directors of the Cavendish Laboratory (Maxwell, Rayleigh and Thomson).

As part of a broader move for the reform of Cambridge physics, Stokes modified his papers for the Smith's prize examination over the years. For over three decades, Cambridge's very best mathematical students faced these questions posed by Stokes. He increased the number of questions on heat, electricity and magnetism from not quite two per cent to about ten per cent of the examination. He also increased the number of questions on optics, while decreasing those on astronomy and pure mathematics. Compared to the 1850s, therefore, Stokes's Smith's prize papers of the 1870s were decidedly more *physical* in content. Stokes's Smith prize papers thus supported that reform which his exceptional students – men like Maxwell, Rayleigh and Thomson – would implement, as they came to dominate Cambridge physics.

Despite Stokes's undeniable successes during the 1850s and 1860s as Lucasian professor, his course was not the most advanced in Britain. One looks instead to Scotland. By the 1850s, Kelvin was teaching the new physics at Glasgow University, and by the 1860s Tait was doing so at Edinburgh. As previously noted, the natural philosophy courses at those universities already housed the subjects of heat, electricity and magnetism, as well as light, hydrodynamics and gravitational theory. Thus, Kelvin and Tait were introducing new versions of the old subjects. In Scotland, the wave theory of light and the newer sciences of energy and electromagnetism existed together more comfortably than in the Cambridge curriculum or in Stokes's lectures. Indeed, the eventual reform of physics at Cambridge resulted essentially from the university's embracing this earlier Scottish point of view.[20]

THE COMPATIBILITY OF TRUE SCIENCE AND TRUE RELIGION: STOKES'S RELIGIOUS USES OF PHYSICS

Stokes's distinction in physics not only earned him the Lucasian professorship but also better enabled him to address religious issues of the time. Although he never followed in the footsteps of his father and brothers and took holy orders, his piety was never doubted. 'Though he was never narrow in his faith and religious sympathies, he always held fast by the simple evangelical truths that he learned from his father.'[21] Stokes adopted the design argument of William Paley and the moral philosophy of Whewell, and he retained his basic evangelical belief that eternal truths resided in Scripture. But while early-Victorian Britain had been 'the age of atonement', later in the century evangelical sentiments increasingly came under fire.

In addition to challenges to literal interpretations of the Bible from geologists and biologists, the 'higher criticism' of German biblical scholars entered British thought, primarily through the publication of the controversial theological work, *Essays and Reviews*, in 1860. Arguing for a metaphorical interpretation of the Bible, the *Essays'* Broad-church authors faced hostility in the 1860s but found their views more commonplace by the 1890s. Equally potent challenges to the established Church came from T. H. Huxley and his allies. They invented the term *agnosticism* to express their doubts about the existence of God. Britain was without question growing more secular, and yet Stokes remained relatively steadfast in his religious views.[22]

Of utmost importance to Stokes were the questions of human immortality and eternal punishment. In grappling with the latter thorny question, Stokes was not alone for the standard Anglican doctrine of everlasting damnation for unbelievers was much contemplated by Victorians. As a young man, for example, Darwin thought it such a 'damnable doctrine' that he could not see how anyone could wish Christianity to be true. As a boy, Stokes was greatly perturbed by the doctrine, speculating that his mathematical prowess had given him a better idea than others of the duration of eternity. If for Darwin the doctrine of eternal punishment was a good reason to abandon Christianity (though not belief in a God), for Stokes it was a good reason to re-examine the Bible more carefully.

Whereas Newton's biblical studies had led him to reject the concept of the Trinity in favour of Arianism, Stokes's ruminations led him to reject the doctrine of eternal punishment in favour of 'conditional immortality'. Immortality was not automatic, but contingent upon correct belief. Those who did not share these beliefs – that is, the unsaved – were not punished for eternity but, instead, were annihilated at the Final Judgement. Stokes looked to John 3:16 to support this belief: whoever believed in Christ 'should not perish, but have everlasting life'. The doctrine affirmed God's seriousness without making Him a relentless torturer.

Stokes's position concerning this doctrine also affirmed the seriousness with which he studied Scripture. As with Newton and other earlier Lucasian professors, such study led Stokes to unorthodox views. But whereas Newton was compelled to conceal his heterodoxy, and whereas William Whiston was dismissed for his anti-Trinitarianism, Stokes was free to express his views openly. Indeed, he was actually part of a movement within Victorian religion to promote this view of damnation and reward. Stokes was drawn to the ideas of the Congregational minister Edward White, whose *Life in Christ* (1846) argued for conditional immortality. He and White began a lengthy correspondence in the 1870s, and the 1878 edition of White's book included a supporting letter from Stokes, citing Scripture and insisting that 'to the wicked judgment will issue in condemnation, and they will receive their final doom'. A few years later, Stokes published a pamphlet which praised the efficacy of conditional immortality in missionary work, and he also attempted to persuade the Church Missionary Society to endorse the doctrine: 'Ye cannot indeed overthrow it; but how know you but that in the endeavour to stifle it by muzzling the evangelist sent out to the heathen you may be the occasion of preventing the salvation of many a soul for whom Christ died?'

Stokes also endorsed the radical idea of the tripartite nature of man. In addition to body and soul, Stokes believed that man possessed a spirit or an ego. After much reflection, he concluded that many biblical passages distinguished between spirit and soul. Spirit was the part of man that survived death and, for those who achieved salvation, was eventually provided with another body, as Scripture promised. 'What the nature of this body may be', Stokes wrote, 'we do not know, but we are pretty distinctly informed that it will be something very different from our present bodies.'

There was also the question of the state of this spirit during the period between death and the Final Judgement. Was it conscious or unconscious, evangelized or not? Though less certain on this point than the ones just mentioned, Stokes thought that spirit would be unconscious, arguing by analogy from his own experience of fainting. During a faint, one was totally unaware of the passage of time. This doctrine of the unconscious spirit was fair. Though quite unequal intervals of time from death to Judgement would be involved for different people, everybody would *appear* to proceed immediately from death to Judgement. Moreover, no dead spirit would have to endure years or centuries of worrying about its eventual fate.

Stokes thought that these doctrines rescued Christianity from the pernicious influence of neo-Platonism. For neo-Platonic Christians, such as the colleagues of Newton who had flourished in seventeenth-century Cambridge, man consisted of body and soul, with soul being *necessarily* immortal and able to think clearly once it was freed from the body. Since this dualistic belief led to the doctrine of eternal punishment, necessary immortality was the Platonists' worst mistake. But this error, Stokes maintained, had now been corrected by the doctrine of conditional immortality. Similarly, he was confident that the doctrines of man's tripartite nature and of his postdeath unconsciousness countered other Platonic blunders. For Stokes, the primacy of Revelation guaranteed the truth of these doctrines.

Meanwhile, Stokes had both religious and scientific grounds for doubting the most revolutionary theory of the day: Darwinian evolution. To a certain extent, many biologists themselves shared Stokes's concern about natural selection. But unlike Stokes, they generally accepted the occurrence of evolution, including human evolution, in the Darwinian 'branching' pattern with different species evolving from a 'common ancestor'. Moreover, within three decades of the publication of the *Origin* in 1859, Anglicans had begun to accommodate many evolutionary principles, viewing them as God's chosen method of creation.

Resisting this trend to assimilate Darwinian evolution within Anglicanism, Stokes pointed to the fossil record. Whereas Darwin imagined a gradual and continuous evolution, Stokes endorsed the scientists who claimed that fossils evinced discontinuity. Moreover, Stokes insisted that since there was scarcely any fossil record of man's development, accepting human evolution required an unwarranted and gratuitous extrapolation from other species to man.

In contrast to evolution, Stokes did accept many modern claims concerning man's antiquity – claims that might seem to endanger Scriptural accounts of creation. Long before the *Origin*, for example, geologists had established the earth's great age – estimated in the 1820s to be millions of years by the Oxford clergyman and geologist, William Buckland. Subsequent estimates had increased and decreased that number, but even the lowest values left the age at many millions of years. Moreover, increasingly precise palaeontological studies showed that the fossil record, although discontinuous, was generally progressive. That is, more advanced forms of life appeared in younger geological strata. In addition, by the 1860s geologists had established the astonishing antiquity of man: man had resided on earth considerably longer than the 6000 years derived from biblical hermeneutics. Stokes could accept all these claims without accepting biological evolution, for he endorsed the idea of God's successive creations. Over geological time, Stokes believed, God had created different organisms at different times and in different places, as disclosed by the discontinuous progress found in the fossil record.

Early Victorian theologians, Stokes declared, had mistakenly opposed geologists with a 'slavish' biblical literalism. Stokes thought that the Bible was true, even in these matters, but he did not slavishly adhere to literal interpretations of scripture. The story of the week-long creation in Genesis could even be reconciled with the nebular hypothesis – a conjecture describing the formation of the solar system from the gravitational contraction of a gaseous cloud – which Stokes himself endorsed. In the hot, misty conditions of the primeval earth formed thus, the sun would not be immediately visible, a circumstance that Stokes believed corresponded with Genesis's account of the sun's creation on the fourth 'day'. Hence, Stokes could readily accept recent spectroscopic evidence that some astronomical nebulae were gaseous in nature, a conclusion that lent plausibility to the nebular hypothesis.

However, relaxing the requirement of biblical literalism left in place two unequivocal religious truths which some Darwinians denied: God created the heaven and the earth, and He created man in His own image. Given these truths, Stokes argued that Darwinians had to provide more than the usual amount of empirical confirmation than was usually needed to establish scientific claims. Instead, militant agnostics such as John Tyndall and T. H. Huxley displayed hostility towards established religion by insisting on a materialist view of man and life. In his presidential address to the British Association in 1869, Stokes responded to a lecture that Tyndall had given to the Association the previous year. Disagreeing with Tyndall's conception of the origins of life, Stokes warned Britons that they should 'take heed that in thus studying second causes we forget not the First Cause'. Stokes also condemned the logical absurdities expressed by Tyndall in his infamous 1874 'Belfast Address' to the British Association. Stokes focused on Tyndall's illegitimate application of physical principles to life: 'In the attempt to deduce ourselves and our surroundings from that primeval condition of matter by mere evolution', Stokes stated, Tyndall was 'obliged to endow with emotion the ultimate molecules of matter in a fiery nebula, and to adopt a series of conjectures against which common sense rebels.' Stokes answered the determinism implicit in Tyndall's use of energy theory with what he called 'directionism'. With this notion, Stokes sought to show that the mind could influence the body by directing energy within it, requiring neither the mind to be material nor energy to be increased in violation of the principle of conservation of energy. Stokes believed, in addition, that even if the fossil evidence were ever sufficient to support the theory of human evolution, that evidence would not undermine God's role. The Creator could have intervened invisibly in the origin of man, in much the way He intervened in the Immaculate Conception.

Physics, as well as common sense, also rebelled against evolution. In fact, Stokes thought that an essential part of the controversy was a conflict *within* science. It was biology and physics that were in conflict, much more than science and religion. Physics, a discipline methodologically more rigorous than biology, actually supported religion. Stokes's nonmaterialistic concept of directionism, for example, was analogous to gravitational theory, according to which the *immaterial* force of gravity acted upon material objects. The content of physics, as this analogy demonstrated, was *not* on the side of extreme Darwinism. Stokes maintained that the cautious realism of physics was needed in biology, too. But to Stokes's chagrin, extreme Darwinians jumped to their conclusions, including unsupported 'consequences' of the law of conservation of energy. In their animosity towards religion and their enthusiasm for evolution, materialistic Darwinians misconstrued and misapplied the content of physics and, unlike physicists, disregarded the proper standards of scientific evidence.

True science and true religion, Stokes emphasized, could not contradict each other. Rightly understood, evidence from biology, geology, astronomy and physics threatened neither the creative power of God nor the unique quality of man. Most conflicts between scripture and science had been caused by dubious scientific reasoning or by extreme biblical literalism. Victorians, he explained, had committed mistakes of both sorts.

Because he had command of these important Victorian issues, Stokes found that he was in great demand. With regard to the apparent conflict between science and religion, Britons enjoyed hearing a prominent religious scientist with an irenic message. He was invited to address annual congresses of the Church of England, speaking, for example, on the 'Religious Benefits from Recent Science and Research' and 'The Harmony of Science and Faith'. From 1886 until his death, he was president of the Victoria Institute, an organization that was formed to combat *Essays and Reviews* and the *Origin of Species*. Nor was he simply a figurehead, for he delivered lectures at the Institute and engaged in lengthy correspondence on its affairs. Elsewhere, he accepted invitations to deliver the first series of the theologically oriented Burnett lectures in Aberdeen and the Gifford lectures in Edinburgh. Published in two volumes in 1891 and 1893, the Gifford lectures were especially difficult for Stokes as the terms of the lectures required him to discuss *natural* theology without reliance on the Bible. Unsurprisingly, given Stokes's Cambridge experience, Paley's natural theological perspective pervaded the lectures. For example, Stokes proclaimed 'that man's mental powers, as well as his bodily frame, were designed to be what they are', and, therefore, he saw 'no difficulty in supposing that man's innate sense of right and wrong was as much impressed upon him, as little the creation of his own will, as his bodily frame'. But, on the whole, Scripture, not nature, was Stokes's deeper source of religious truth.

In his views, Stokes was more or less at one with those revolutionaries in physics: Faraday, Joule, Kelvin, Tait and Maxwell. As a member of the Sandemanian sect, Faraday took a literalist view of the Bible. Joule thought that Genesis was an accurate portrayal of 'the dynamical theory of creation', and Maxwell evidently believed in the reality of Noah's flood as late as 1879.[23] Joule, Faraday and Kelvin thought that God conserved the 'forces' or 'powers' or 'energy' in nature. Hence, though they may have understood nature's *active* ingredient somewhat differently, they agreed that only God's action could destroy it. Tait turned Tyndall's ideas of continuity and energy conservation against him. He proposed a scheme in which continuous transfers of energy between the visible and 'unseen' universes involved the creation of the world, the resurrection of Christ and the immortality of humans. Maxwell declared that molecules bore the mark of 'manufactured' articles, thus demonstrating God's design in nature. Kelvin also endorsed Paley's argument, rejecting Darwinian natural selection as devoid of divine design. Though more amenable to the occurrence of biological evolution than Stokes was, Kelvin agreed that God must have created life directly. The enormous discontinuity between nonlife and life made a naturalistic connection between them inconceivable. Kelvin and Tait both declared God the Creator of human thought and, like Maxwell, they espoused the nondeterministic link between mind and body that Stokes would call 'directionism'. Consequently, although these men differed amongst themselves, collectively they endorsed a Christian theology of nature.

Proclaiming such a Christian theology of nature, Stokes, more than any other late Victorian, represented the ideal of a cultural balance between religion and science. In rebuking Tyndall, Stokes joined a struggle for cultural hegemony. Tyndall, Huxley and a few others, for example, had formed the X Club in London. They encouraged the professionalization of science and argued for a scientific naturalism that dispensed with religion. Professional scientists, they thought, should replace out-of-date clergymen in defining the national outlook. Moreover, urban masses had already been avoiding churches in large numbers. Stokes was thus part of the late-Victorian religious establishment which was defending its traditional place in society – evangelizing Englishmen in general and refuting Tyndall and Huxley in particular. Stokes contributed greatly to that defence. As a man of unquestioned scientific prominence, he spoke out in support of religion. Indeed, as president of the Royal Society of London from 1885 to 1890, Stokes was at the pinnacle of the newly significant profession.

SHAPING 'THE PLASTIC INTELLECT OF HIS TIME': STOKES AS ARBITER AND STATESMAN OF SCIENCE

The Royal Society was the main institutional basis for Stokes's influence on Victorian science. Around 1830, Charles Babbage and others had bemoaned the decline of British science. They targeted the Royal Society for its substandard membership and publications. In this climate, the British Association for the Advancement of Science was formed in 1831. Moreover, science was gaining importance both in Oxbridge and Scotland, as well as in the newer English universities. Acknowledging such changes, the Royal Society finally called for reform in the 1840s: new members had to meet higher standards, and, by century's end, only the most élite within the larger and more professional community of scientists could attain fellowship. Moreover, when the editorship of the *Philosophical Transactions of the Royal Society* was turned over to the society's two secretaries, a rigorous refereeing process for papers became routine. Early-Victorian gentlemanly science had become a late-Victorian, modern-looking profession of science.[24]

Stokes's ascent within the Royal Society was swift. Elected to fellowship in 1851 under the new regulations, he was appointed to the society's governing council in 1853. On becoming one of its two secretaries in 1854, he assumed editorship of the *Philosophical Transactions* for three decades until he became president in 1885, replacing Huxley. As secretary-editor, Stokes was exceedingly conscientious. He arranged for submissions to be refereed by other scientists, and he refereed some hundred papers himself. He often corresponded at length with the authors, whether biologists, geologists, chemists, mathematicians or physicists. In the process, he frequently became a kind of *de facto* collaborator, and occasionally this collaboration grew into an extensive partnership. Indeed, the editorship was so time-consuming that Kelvin urged Stokes to take the Cavendish chair partly to escape it, and Stokes's fellow secretary, Michael Foster, thought that his colleague had made the job much too burdensome. Stokes enjoyed the collegiality of his position, however, and continued his duties accordingly.

As a result, his role at the Royal Society, even more than his Lucasian professorship, made him a kind of gatekeeper of proper science. Research had to pass through him and his referees before entering the pages of the *Philosophical Transactions*. He could deal equally well with the genius of a Faraday as with men of more modest abilities. His thoroughness combined with his scientific expertise to influence and improve Victorian research. Stokes's pervasive presence appeared in his written reactions to papers submitted to the Royal Society by, for example, Henry Brougham, Faraday, and the analytical chemist, William Crookes, with whom he had his most far-reaching collaboration.

Stokes's referee's report on Brougham's paper in 1852 indicated his own criteria for experimental research acceptable for publication in the *Philosophical Transactions*. Brougham had opposed the wave theory of light for a half-century and submitted a paper to that effect in 1852. It was one of the papers Stokes was asked to referee before becoming secretary. An acceptable paper, he pronounced, should contain at least one of four things: significantly new methods of observation, striking experimental results, new means for comparing theory and experiment or experimental results at variance with received theory.

Brougham's results failed to meet Stokes's criteria. His paper might have been valuable three decades earlier, but no longer.[25]

Faraday fared better. Though now past his years of momentous discoveries, Faraday sent two papers to the Royal Society in the mid-1850s that Stokes warmly praised in his referee's reports. The first, Stokes reported, employed 'a very elegant method' of experiment to make various discoveries, including 'the remarkable result that unlike all other substances hitherto examined, cobalt *in*creases in magnetic capacity as the temperature increases'. Stokes did think the 'surmise' unlikely that the magnetic 'character of the body as a whole would become evanescent at the same time', and the published paper omitted that point. The second paper, Stokes said, dealt with an 'almost untrodden field of research', and, in fact, it would become an important contribution to what was eventually known as colloidal chemistry.

However, Stokes thought less of a third paper sent to him by Faraday in 1860. Following his conviction that the powers of nature were conserved, Faraday experimented on the conversion of gravity into other powers. His 1860 paper reported a null result. Stokes concluded that – even if Faraday's improbable theory were correct – a positive result with his particular experiment was itself improbable. Moreover, if the paper were published, Faraday would need to delete an incorrect sentence about gravity. Consistent with his earlier criteria for publication, Stokes wrote to Faraday that he judged the paper unsuitable for the *Philosophical Transactions*, suggesting that Faraday also consult someone else. Instead, Faraday withdrew the paper, writing to Stokes: 'I want no other opinion than yours and my own.'[26]

Before Stokes and Crookes formed their partnership, they had to overcome a rocky period in the early 1870s. They initially disagreed about what constituted proper science. Like many other Victorians, Crookes was enthusiastic about the scientific investigation of spiritualism. Experiments with the notorious medium, D. D. Home, in particular, had convinced him of the existence of a 'psychic force' emanating from the human body. In 1871, he submitted a paper on his research to the Royal Society. Stokes was instrumental in the Society's rejection of the paper, explaining to Crookes that he thought the evidence too weak to support his theory of this force. Crookes had some hard feelings about the matter, and he remained interested in spiritualism. In fact, he became a member of the Society for Psychical Research in 1882, serving as its president in 1897. His belief in the reality of spiritualistic phenomena directly conflicted with Stokes's religious conviction of the dead's unconscious sleep.

Nevertheless, by ignoring their spiritual differences and focusing on physics, they collaborated for over three decades. Stokes's more friendly reaction to Crookes's more conventional scientific research soon allayed Crookes's apprehensions. Crookes was the highly skilled experimentalist, Stokes the consummate theoretician. When other physicists explained the observed behaviour of Crookes's 'radiometer' with the new kinetic theory of gases in the 1870s, it was Stokes who helped Crookes grasp the argument. When Crookes thought that his experiments with cathode rays supported the old particle theory of light, it was Stokes who explained to him that the results agreed with the wave theory of light. Stokes devised an experiment to decide whether the cathode rays themselves were particles or waves. Crookes performed the experiment, and concluded that it supported the particle theory. While German physicists contended that cathode rays were waves in the æther, Stokes and Crookes persuaded British physicists that the rays were negatively charged gaseous molecules which were repelled by the negative terminal in a cathode-ray tube.

Stokes's conscientious assistance to others was an acknowledged feature of mid- and late-Victorian science. Noting Stokes's work with others, Rayleigh wrote that the Royal Society's publications 'abound with grateful recognitions of help thus rendered, and in many cases his suggestions or comments form not the least valuable part of memoirs which appear under the names of others'. An obituarist in the *Times* agreed: 'He was a vivifying influence ever operating upon the plastic intellect of his time, and his work is to be looked for in the minds he stimulated, helped, and directed.' Similarly,

Figure 43 Engraving of Stokes accompanying P. G. Tait, 'Scientific Worthies', in *Nature* from 1875. Tait complained that the 'mole-eyed State' had 'virtually forced' administrative burdens upon Stokes.

Arthur Smithells, who corresponded extensively with Stokes during the 1890s on the chemistry of flames, proclaimed: 'What Stokes did for his generation can hardly be estimated.'[27]

Stokes had famous rivals and none more so than T. H. Huxley, with whom he served as fellow Secretary of the Royal Society from 1872 to 1881. Darwin's 'bulldog' and Tyndall's fellow X Club member, Huxley was Stokes's antithesis and they had several areas of disagreement. In 1864, for example, Huxley complained about the manner in which the Royal Society awarded Darwin its Copley medal. It had been awarded for Darwin's research other than that contained in the *Origin of Species*, and the statement read by Stokes at the ceremony said so. Huxley wrote to Stokes that this explicit statement insulted Darwin and, moreover, that Stokes had made it worse by reading that the *Origin* had been 'expressly excluded' instead of 'expressly omitted'. Stokes replied by defending the Royal Society, saying that the *Origin* was too important to pass over in silence. A more dramatic clash took place in 1887 when Huxley published an anonymous editorial criticizing Stokes for agreeing to represent Cambridge University in parliament while he was still President of the Royal Society. Huxley argued that a President of the Royal Society should be neither an MP nor President of the Victoria Institute, because such combinations of offices could imply that science was sanctioning particular political or religious views. In fact, Huxley agreed with Stokes – the Irish Tory – in opposing Home Rule for Ireland, the controversial policy associated with the Liberal party and its leader, W. E. Gladstone. When Huxley told Stokes that he had written the editorial, Stokes was genuinely surprised, for he had thought that it 'must have been written by some hot Gladstonian'.[28]

Seeing no conflict among his offices, Stokes remained in parliament until 1891. Conscientious in his attendance, he kept his eye on bills affecting the university, but was said usually to be silent. As correspondence between them indicates, Huxley and Stokes remained on good terms before, during and after their joint secretaryships. Huxley and others' intense opposition to Stokes became neither personal animosity nor denial of what all regarded as his positive contributions. Stokes also, of course, remained President of the Royal Society.

By the time Stokes's presidency ended in 1890, his Royal Society offices had thrust him into several controversies. He had tried to define and defend established religion. He had opposed the evolutionary agnosticism of Tyndall and Huxley and had himself represented a continued cultural role for religion, alongside science. He had also tried to define and defend established science. He was especially influential regarding the content of the physical sciences. But he also sought to keep spiritualism and evolution at bay, outside the realm of accepted science. He was more successful in combating Crookes's spiritualism than Darwin's evolution, but he vanquished neither. These various disputes all reflected the increasingly prominent professional role of scientists within Victorian society and, consequently, the rising cultural visibility of the President of the Royal Society.

AN AGEING LUCASIAN PROFESSOR IN A CHANGING UNIVERSITY

The final decade of Stokes's life began with the high honour of the Royal Society's Copley medal awarded to him in 1893. The medal recognized a distinguished career, but his career was not finished. Though he was no longer in the vanguard of Cambridge studies, he continued to deliver professorial lectures. He also continued to deliberate and speak on theological matters and, most significant, to provide scientific insights, especially regarding those spectacular new phenomena of the 1890s: X-rays, radioactivity and electrons.

Thanks to the research by Crookes and Stokes, British physicists could take for granted that cathode rays consisted of electrically charged particles. Stokes and Kelvin, for example, were incredulous when they learned that German physicists believed cathode rays were æthereal waves. Kelvin even wrote to Stokes in 1896 expressing the 'perversity' of the German view. Working on cathode rays from the British perspective, J. J. Thomson discovered the electron in 1897 and subsequently formulated his electron model of the atom. That model soon led to the 'nuclear' atomic model proposed in 1911 by Ernest Rutherford, who had studied at the Cavendish under Thomson, and then to the 'quantum' atomic model proposed in 1913 by the Dane Niels Bohr, who had divided a postgraduate year between Thomson's Cavendish and Rutherford's laboratory in Manchester.

Linked theoretically to Stokes's earlier thoughts on cathode rays, his theory of X-rays was also influential. X-rays, first observed in 1895 by Wilhelm Roentgen, were understandably the subject of much interest and debate. Kelvin, for example, first thought they were a particular kind of wave that would help confirm his favourite theory of the æther. Before long, however, Kelvin and Stokes agreed that

X-rays were like light waves, only with vastly shorter wavelengths. Stokes then went on to develop his 'pulse' theory of X-rays. Because these rays were produced when cathode-ray particles were suddenly brought to a halt, Stokes reasoned, they were more probably irregular pulses in the æther rather than regular trains of waves. 'Contrasted with the other idea', Stokes wrote to Kelvin in 1896, 'this would be like the sound arising from the hedge fire of a regiment as contrasted with the ring of rapidly . . . struck bells.' Stokes presented a more fully formed version of his 'hedge row' analogy in public lectures of 1896 and 1897, and it became a leading theory of X-rays.

Stokes also proposed, in conjunction with Crookes, what he called his 'wagtail' theory of radioactivity. He postulated that irregular vibrations *within* radioactive atoms produced erratic vibrations in the æther that constituted the rays of radioactivity. His wagtail theory was related to his other theoretical work in that irregular vibrations of radioactivity would be located midway, as it were, between regular vibrations of light and pulses of X-rays. In other words, given the existence of pulses and regular vibrations, Stokes thought that it was logical to posit the existence of a middle state of irregular vibrations. When experiments demonstrated that radioactive rays consisted of particles as well as waves, Stokes concluded that the atom's wagtail vibrations produced the waves, while molecules of air, somehow repelled by the radioactive atom, probably constituted the particles. Though experimental–theoretical advances in radioactivity soon eclipsed their ideas, Stokes and Crookes had contributed significantly to the study of the fundamental problems of radioactive energy, the nature of radioactive emanations and the link between them.

Unlike his sophisticated contributions to the physical sciences, Stokes's Lucasian lectures in the 1890s attracted less attention than earlier for at least four reasons. First, in the early 1880s, the Smith's prizes ceased being awarded for performance on examinations. Strong students thus no longer had Stokes's Smith's prize paper as an incentive for attending his lectures. Second, Stokes's lectures on hydrodynamics and optics did not represent current physics as well as they had done thirty years before. It was not that Stokes was mistaken, but that other areas of physics had become more advanced and relevant than they had been before. Though Stokes generally supported Maxwellian electromagnetic theory, for example, he did not add electricity and magnetism to his lectures. Even his Smith's prize questions in the 1870s on electricity and magnetism had not touched Maxwell's new ideas. Third, Cambridge now boasted several more instructors in mathematics and physics. Four university lecturers had been appointed in these subjects in the 1880s (independent of the Cavendish), and J. J. Thomson's Cavendish staff numbered seven by the 1890s. Fourth, Cambridge Triposes had changed considerably. From the 1870s, the Mathematical Tripos had included the non-Stokes subjects of heat, electricity and magnetism. In the early 1890s, Thomson succeeded in reforming the Natural Sciences Tripos so that it included a fair measure of mathematical physics. Physics students thereafter preferred the experimental–mathematical physics in the Natural Sciences Tripos and naturally gravitated to the Cavendish and its large staff rather than to Stokes's lectures. Undergraduates who were prospective mathematicians and astronomers now dominated the Mathematical Tripos, but Stokes's lectures emphasized neither pure mathematics nor astronomy. Although Stokes's lectures were certainly still sound, their Cambridge context had changed.

Another development in late-Victorian Cambridge was the education of women. Girton and Newnham Colleges were founded around 1870.[29] Though women would not be granted degrees until the next century, they soon gained admission to university lectures and permission to take Tripos examinations. From the early 1880s, their results for both the Mathematical Tripos and the Natural Sciences Tripos were published. The list of women's names was separate from that of the men, but their rankings with respect to the men were indicated. Famously, Newnham's Philippa Fawcett was placed 'above the Senior Wrangler' in 1890. These events generally opposed the usual Victorian ideas that women were less intellectual than men, that they were suited primarily for the domestic roles of wives and mothers and that they were not suited for higher education.

Though hardly on the cutting edge of this academic change, Stokes did not reject it either. His daughter, Isabella, recorded that her father was inclined to deny the request to open his lectures to women. But he changed his mind when she pointed out that he might be overlooking a future Mrs Somerville, the well-known early-Victorian popularizer of science. He became interested in the women attending his lectures, Isabella said, and he always knew how much they understood. He was especially amused by one increasingly puzzled woman who eventually dropped out of his lectures. 'Some of the ladies got on splendidly', however, 'and he was much pleased when a Newnham lady who had attended his lectures brought him some original work which he approved.'[30]

Based on Victorians' customary view of women, Stokes's initial reluctance is historically understandable. Unlike Emily Davies, the founder of Girton, Stokes had not rejected his childhood evangelicalism. For him, as we have seen, Scripture was true. He quoted Old Testament verses to his daughter, for example, in explaining his views of the impropriety of a man's marrying his deceased wife's sister. On the subject of men, women and education, Paul had written to Timothy: 'Let the woman learn in silence with all subjection. But I suffer not a woman to teach, nor to usurp authority over the man, but to be in silence. For Adam was first formed, then Eve. And Adam was not deceived, but the woman being deceived was in the transgression.' To the Corinthians Paul had written: 'Let your women keep silence in the churches: for it is not permitted unto them to speak; but *they are commanded* to be under obedience, as also saith the law. And if they will learn any thing, let them ask their husbands at home . . . ' If regarded as God's truth, such statements would, of course, have rendered Victorian 'custom' not *merely* custom but something of a divine dictum. Furthermore, Paul's scriptural truth would have been reinforced by everyday considerations, as experienced by Stokes – as well as Darwin. Indeed, Darwin's experience with women, especially his wife Emma, supported his conclusion in the *Descent of Man* that women were superior to men in moral qualities such as tenderness and unselfishness but inferior to them in intellect.[31] Stokes's experience would have been similar. He regarded himself as retiring and even 'cold', but he was a superior mathematical physicist. The women in his life were the reverse. Hence, though Stokes and Darwin would have explained them quite differently, they would no doubt have agreed that there were *observable* differences between men and women. For Stokes, as with his theory of successive creations to explain the fossil record, so also, with his view of women, experience and Scripture would have agreed with each other.

But Stokes was less dogmatic on these subjects than the biblical Paul, and one can surmise his assessment of women's Tripos results. Of the women obtaining honours in the Mathematical Tripos during the 1880s and 1890s, about ten per cent reached the status of wrangler, compared to about thirty-one per cent of the men. Of the women taking the advanced part II of the Natural Sciences Tripos during this period, about twenty-six per cent earned 'firsts', compared to about forty-six per cent of the men.[32] Such results may have confirmed Stokes's conviction that women generally were intellectually inferior to men, at least in mathematics and the natural sciences. On the other hand, many Cambridge women did perform quite well. One can imagine that, just as Stokes had already accommodated modern ideas in the sciences, he would have concluded that the Bible was true, but not 'slavishly' so, in regard to the questions of women's intelligence and education. Exactly how such a conclusion would have translated into university policy is, obviously, another matter. Perhaps one might best regard Stokes as a sceptical convert to some form of higher education for women, possibly to that examinations-but-not-degrees version that actually existed during his final decade.

At any rate, by century's end Stokes had actively fostered at Cambridge what was the unquestioned centre of British physics. He had supported reform of both the Mathematical Tripos and the Natural Sciences Tripos, and the foundation of the Cavendish

Laboratory. He had encouraged Maxwell to take the Cavendish chair. Maxwell, Rayleigh and Thomson, as well as Thomson's Cavendish staff of the 1890s, had attended Stokes's lectures. Cambridge had assimilated the previously 'Scottish' university subjects of heat, electricity and magnetism. Moreover, it was the Cambridge physicists, not the Scottish, who pursued and developed Maxwellian electromagnetic theory. At Glasgow, Kelvin had resisted Maxwell's specific version of electromagnetic theory and Tait, himself something of a 'bulldog' for Kelvin's ideas, had been slow to include Maxwell's theory in his course at Edinburgh. In addition to electromagnetism, Cambridge physicists – including Stokes – also probed the new subjects of X-rays, radioactivity and electrons. Indeed, Cambridge's dominance continued well beyond Stokes's lifetime. Of the thirty-three physicists who presided over Section A (mathematics and physics) of the British Association between 1890 and 1935, nearly two-thirds had studied at Cambridge. Only two were from a Scottish university. Though it reduced his own importance at Cambridge, Stokes had helped effect the shift from Lord Kelvin's mid-nineteenth-century Scotland to J. J. Thomson's early-twentieth-century Cambridge.

Stokes's final decade also saw his most substantial work regarding his chief theological concern, *Conditional Immortality: A Help to Sceptics* (1897). That same year he lectured to the Church Congress on the inductive method in theology. At about the same time, Stokes was disclosing some of his deepest biblical insights in letters to the Cambridge clergyman, Henry Paine Stokes (no relation), seeking to persuade him, for example, that the principle of conditional immortality was valid. Stokes also explained to him that he thought 'the most probable view to be that the state of death is one of sleep'.[33] Stokes remained convinced that conversations with the dead were impossible as they were unconscious.

On 12 January 1903, Stokes, his health failing, wrote to Reverend Stokes: 'I do not like the idea that the thoughts which have been so satisfactory to me should be buried with me in the grave, and am desirous of imparting them to some one else. I should be rather glad if some day when you happen to have leisure you would drop in that we might have a further talk on the subject.' Having received this gentle expression of anxiety, Reverend Stokes evidently did drop in, later asking Stokes for notes on their conversation. He must have been especially gratified that he had made the visit. Only a week and a half later G. G. Stokes died, perhaps entering that state of sleep which he had so long anticipated.

CONCLUSION

Comparing and contrasting the minor figure of Joshua King with his much more illustrious successor, G. G. Stokes, illuminates how the Lucasian professorship changed dramatically during the nineteenth century. While both King and Stokes achieved the highest undergraduate distinction, it was Stokes whose research advanced Cambridge mathematical physics. His work in hydrodynamics and optics was a cornerstone of Victorian physics. In London, he was a crucial element of the increasingly professional Royal Society for four decades. His unselfish collaboration with others greatly enriched Victorian science and helped to define the new physics of electrons and X-rays. While King was primarily head of a Cambridge college with local influence, Stokes became a university professor of international and enduring reputation. Both King and Stokes were evangelicals. But whereas King's religious views were the norm in late-Georgian and early-Victorian Cambridge, Stokes's views appeared increasingly obsolete towards the end of his life. Nevertheless, Stokes's position secured audiences for his religious ideas that King could not have gained. The religious establishment appreciated hearing an eminent scientist and deeply religious man declaring that the conflict between science and religion was merely an aberration.

King's and Stokes's respective tenures as Lucasian professor were quite disparate. While King did not lecture, Stokes did, becoming the mainstay of Cambridge physics education for a quarter of a century and teaching the leaders of later Victorian physics. In King's day science was primarily the domain of gentlemen. By the end of Stokes's career, science was almost exclusively a professional enterprise. The Mathematical Tripos was for King a liberal training for the educated man, but during Stokes's time it increasingly became (along with the Natural Sciences Tripos) a specialized preparation for a scientific career. Natural philosophy gradually yielded to mathematical physics. Turn-of-the-century Cambridge dominated physics much more than a half-century earlier. In myriad ways, Stokes had promoted these various changes and they, in turn, augmented and eventually diminished his own role at Cambridge. His was a legacy profound and complex and lasting.

ACKNOWLEDGEMENTS

I am especially grateful to Jayne Ringrose of the Cambridge University Library and also to Elisabeth Leedham-Green, former Cambridge University archivist, for their help in dealing with archival materials. I had very helpful discussions about this chapter with Peter Harman and Simon Brook. The editors have generously provided me with information that is now part of the chapter. The editors, the anonymous referees and Julia Wilson have made several valuable suggestions. Murray Milgate, senior tutor and archivist for Queens' College, and Jonathan Smith at Trinity College Library have also provided help. I am grateful to the Master and fellows of Trinity College, to the Syndics of Cambridge University Library, and to the Royal Society of London for permission to quote from manuscript materials they hold.

Notes

1 I was told this story in recent years by the late Sidney Kenderdine, occupant of Stokes's rooms at Pembroke College. In her admiring memoir written shortly after his death, Stokes's daughter emphasized his 'very quiet and silent disposition'. (Mrs Laurence Humphry, 'Notes and recollections', in Joseph Larmor (ed.), *Memoir and Scientific Correspondence of the Late Sir George Gabriel Stokes* (2 vols, Cambridge, 1907), vol. I, p. 41.)

2 J. J. Thomson, *Recollections and Reflections* (London, 1936), pp. 50–1.

3 For histories of Cambridge University and of King's and Stokes's respective colleges, see Peter Searby, *A History of the University of Cambridge*, vol. III: *1750–1870* (Cambridge, 1997); Elisabeth Leedham-Green, *A Concise History of the University of Cambridge* (Cambridge, 1996); A. Rupert Hall, *The Cambridge Philosophical Society: A History, 1819–1969* (Cambridge, Cambridge Philosophical Society, 1969); John Twigg, *A History of Queens' College, Cambridge, 1448–1986* (Woodbridge, 1987); A. V. Grimstone (ed.), *Pembroke College, Cambridge: A Celebration* (Cambridge, 1997).

4 For discussions of Cambridge studies and examinations, see June Barrow-Green, ' "A corrective to the spirit of too exclusively pure mathematics": Robert Smith (1689–1768) and his prizes at Cambridge University', *Annals of Science*, 56 (1999), 271–316; Harvey Becher, 'William Whewell and Cambridge mathematics', *Historical Studies in the Physical Sciences*, 11 (1981), 1–48; Menachem Fisch, 'The making of Peacock's treatise on algebra: a case of creative indecision', *Archive for History of Exact Sciences*, 54 (1999), 137–79; David B. Wilson, 'Experimentalists among the mathematicians: physics in the Cambridge Natural Sciences Tripos, 1851–1900', *Historical Studies in the Physical Sciences*, 12 (1982), 325–71; and *idem*, 'The educational matrix: physics education at early-Victorian Cambridge, Edinburgh and Glasgow Universities', in P. M. Harman (ed.), *Wranglers and Physicists: Studies on Cambridge Physics in the Nineteenth Century* (Manchester, 1985), pp. 12–48.

5 Humphry, 'Notes' vol. I, pp. 3–4.

6 W. H. Stokes's letters to his brother are in the Stokes Collection, Cambridge University Library, Add. MS 7656. The Stokes Collection is the main repository of Stokes's manuscripts. For a description, see David B. Wilson (ed.), *Catalogue of the Manuscript Collections of Sir George Gabriel Stokes and Sir William Thomson, Baron Kelvin of Largs in Cambridge University Library* (Cambridge, 1976). For published editions of Stokes's correspondence, see Larmor, *Memoir*, and David B. Wilson (ed.), *The Correspondence between Sir George Gabriel Stokes and Sir William Thomson, Baron Kelvin of Largs* (2 vols, Cambridge, 1990).

7 On King, see Twigg, *A History*; Frederic Boase, *Modern English Biography* (Truro: Netherton and Worth for the author, 1879) vol. II, p. 226; *The Gentleman's Magazine and Historical Review* (July to

December 1857), p. 469; and J. P. T. Bury (ed.), *Romilly's Cambridge Diary, 1832–42* (Cambridge, 1967). On Whewell, see Menachem Fisch and Simon Schaffer (eds), *William Whewell: A Composite Portrait* (Oxford, 1991).

8 Senior Wrangler [Solomon Atkinson], 'Struggles of a poor student through Cambridge', *The London Magazine and Review*, 1 April 1825, p. 504. Atkinson severely criticized the Cambridge system.

9 Reports of Cases Argued & Determined in the High Court of Chancery, Commencing in the Sittings before Easter Term, 1. Geo. IV. 1821. The Case of Queens' College Cambridge.

10 Minutes of the meeting are in Cambridge University Archives (CUA), O.XIV.52, pp. 60–1. The Vice-chancellor of the university and the Masters of the colleges were the electors for the professorship.

11 Reports of the two syndicates are in CUA, Min. VI.I, vol. 4, pp. 69, 99. There were four examiners per year for the Tripos, and King actually examined more frequently than any other person in the first half of the nineteenth century. Though King was one of the examiners for the Tripos in 1830 and 1831, it is not clear which examiner set the questions involving the wave theory. On Babbage, see Charles Babbage, *Passages from the Life of a Philosopher* (London, 1968, reprint of 1864 edn), pp. 29–34.

12 The syndicates' reports are in CUA, CUR 28.6.1/7 and CUR 28.6.1/13.

13 King to Vice-chancellor, 2 February 1849, 29 September 1849, CUA, O.XIV.278. King's poor health was undoubtedly also the reason why he examined for the Smith's prizes only four times (early in his professorship), having substitutes the other years. In October 1843 he had to write to Whewell as Vice-chancellor to excuse himself 'on account of my health' from attending events that were part of Queen Victoria's visit to the university (King to Whewell, 21 October 1843, Whewell Papers, Trinity College Library). For the date of King's stroke, see Twigg, *A History*, p. 274.

14 Minutes of the meeting are in CUA, O.XIV.53, pp. 76–7.

15 Published each year, *The Cambridge University Calendar* contains information about college fellows and university professors, requirements for undergraduates, descriptions of examinations, lists of graduates and examination questions.

16 On these developments in science, see Peter Bowler, *Charles Darwin: The Man and His Influence* (Oxford, 1990); W. H. Coleman, *Biology in*

the Nineteenth Century (Cambridge, 1977); P. M. Harman, *Energy, Force, and Matter: The Conceptual Development of Nineteenth-Century Physics* (Cambridge, 1982); *idem*, *The Natural Philosophy of James Clerk Maxwell* (Cambridge, 1998); Daniel M. Siegel, *Innovation in Maxwell's Electromagnetic Theory* (Cambridge, 1991); Crosbie Smith, *The Science of Energy: A Cultural History of Energy Physics in Victorian Britain* (Chicago, 1998); Crosbie Smith and M. Norton Wise, *Energy and Empire: A Biographical Study of Lord Kelvin* (Cambridge, 1989).

17 On Stokes's research, see Jed Buchwald, 'Why Stokes never wrote a treatise on optics', in P. M. Harman and Alan Shapiro (eds), *The Investigation of Difficult Things* (Cambridge, 1992), pp. 451–76; F. A. J. L. James, 'The conservation of energy, theories of absorption and resonating molecules, 1851–1854: G. G. Stokes, A. J. Ångström, and W. Thomson', *Notes and Records of the Royal Society of London*, 38 (1983), pp. 79–107; R. K. DeKosky, 'George Gabriel Stokes, Arthur Smithells and the origin of spectra in flames', *Ambix*, 27 (1980), 103–23; Marjorie Malley, 'A heated controversy on cold light', *Archive for History of Exact Sciences*, 42 (1991), 173–86; B. R. Wheaton, *The Tiger and the Shark: Empirical Roots of Wave-Particle Dualism* (Cambridge, 1983); David B. Wilson, *Kelvin and Stokes: A Comparative Study in Victorian Physics* (Bristol, 1987); Maria Yamalidou, 'Molecular ideas in hydrodynamics', *Annals of Science*, 55 (1998), 369–400. Stokes's published research papers are collected in his *Mathematical and Physical Papers* (5 vols, Cambridge, 1880–1905), vols IV and V edited by Joseph Larmor.

18 The report of the relevant syndicate and Stokes's letter agreeing to its terms are in CUA, CUR 39.8, ff. 18–19. For the 1880s, see 'Report of the council on the Lucasian and Sadlerian professorships' (17 May 1886) in CUA, CUR 39.8, f. 20.

19 Robert John Strutt, Fourth Baron Rayleigh, *John William Strutt, Third Baron Rayleigh* (London, 1924), p. 32, and Thomson, *Recollections*, p. 48.

20 David B. Wilson, 'P. G. Tait and Edinburgh Natural Philosophy, 1860–1901', *Annals of Science*, 48 (1991), 267–87, and *idem*, 'Scottish influences in British natural philosophy: rise and decline, 1830–1910', in Jennifer J. Carter and Donald J. Withrington (eds), *Scottish Universities: Distinctiveness and Diversity* (Edinburgh, 1992), pp. 114–26.

21 H. P. Stokes, 'Reminiscences of Sir George Stokes', *The Cambridge Chronicle* (13 February 1903).

22 On Victorian religion, see Owen Chadwick, *The Victorian Church* (2 vols, New York, 1966, 1970) and Bernard M. G. Reardon, *From Coleridge to Gore: A Century of Religious Thought in Britain* (London, 1971).

23 J. P. Joule, *On Some Facts in the Science of Heat Developed since the Time of Watt* (no place or publisher [1865]), p. 14, as quoted (at greater length) in David B. Wilson, 'Victorian science and religion', *History of Science*, 15 (1977), 55–6. In addition to works cited in note 16, see P. M. Heimann, 'The *Unseen Universe*: physics and the philosophy of nature in Victorian Britain', *British Journal for the History of Science*, 6 (1972), 73–9, and *idem*, 'Conservation of forces and the conservation of energy', *Centaurus*, 18 (1974), 147–61. As a Sandemanian, Faraday probably differed most from the other religious physicists. For example, he evidently did not think that God's existence was proven by design in nature, independently of the Bible. See Geoffrey Cantor, *Michael Faraday, Sandemanian and Scientist: A Study of Science and Religion in the Nineteenth Century* (New York, St Martin's Press, 1991). On this group's cultural opposition, see Joe D. Burchfield, 'John Tyndall – a biographical sketch', in W. H. Brock *et al.* (eds), *John Tyndall: Essays on a Natural Philosopher* (Dublin, 1981), pp. 1–13; Frank M. Turner, 'John Tyndall and Victorian scientific naturalism', in *ibid.*, pp. 169–80; Ruth Barton, '"An influential set of chaps": the X-Club and Royal Society politics 1864–1885', *British Journal for the History of Science*, 23 (1990), 53–81; *idem*, 'Huxley, Lubbock, and half a dozen others': professionals and gentlemen in the formation of the X Club, 1851–1864', *Isis*, 89 (1998), 410–44.

24 See Marie Boas Hall, *All Scientists Now: The Royal Society in the Nineteenth Century* (Cambridge, 1984) and Jack Morrell and Arnold Thackray, *Gentlemen of Science: Early Years of the British Association for the Advancement of Science* (Oxford, 1981).

25 Stokes's referee's report is in the Royal Society of London (RSL), RR.2.35, 8 June 1852. For a discussion of the episode and for comments by the other referee, G. B. Airy, see G. N. Cantor, *Optics after Newton: Theories of Light in Britain and Ireland, 1704–1840* (Manchester, 1983), p. 177.

26 Faraday to Stokes, 11 June 1860, in Larmor, *Memoir*, vol. I, pp. 151–2. Stokes did say that Faraday's paper might appear in the Society's *Proceedings*, a journal of lesser prestige. (Stokes to Faraday, 8 June 1860,

in *ibid.* vol. I, pp. 150–1.) Stokes's referee's reports on Faraday's earlier papers are in the RSL, RR.3.104, 20 December 1855, and RR.3.106, 28 May 1857. On Faraday's research, see L. Pearce Williams, *Michael Faraday: A Biography* (New York, 1965), pp. 465–74.

27 These tributes are in Rayleigh, 'Obituary notice', in Stokes, *Papers*, vol. V, pp. xix–xx, and Larmor, *Memoir*, vol. I, pp. 266, 306. See also the remarks by the astronomer, William Huggins (*ibid.*, vol. I, pp. 103–4); G. D. Liveing, Cambridge's professor of chemistry (*ibid.*, vol. I, p. 96); and Kelvin ('The scientific work of Sir George Stokes', in Kelvin, *Mathematical and Physical Papers* (6 vols, Cambridge, 1882–1911, vols IV–VI edited by Joseph Larmor), vol. VI, p. 344).

28 See correspondence between Stokes and Huxley in Larmor, *Memoir*, vol. I, pp. 168–80. See Adrian Desmond, *Huxley: From Devil's Disciple to Evolution's High Priest* (Reading, MA, 1999), which discusses a disagreement in 1873 between Huxley and Stokes regarding a paper on evolution for the *Philosophical Transactions*.

29 See Christopher N. L. Brooke, *A History of the University of Cambridge*, vol. IV: *1870–1990* (Cambridge, 1993), ch. 9; Gillian Sutherland, 'The movement for the higher education of women: its social and intellectual context in England, *c.* 1840–80', in P. J. Waller (ed.) *Politics and Social Change in Modern Britain* (Sussex, 1987), pp. 91–116; and Daphne Bennett, *Emily Davies and the Liberation of Women, 1830–1921* (London, 1990).

30 Humphry, 'Notes', vol. I, pp. 26–8.

31 For discussions of the historiographical and other complexities involved in this subject, see Evelleen Richards, 'Darwin and the descent of women', in D. Oldroyd and I. Langham (eds) *The Wider Domain of Evolutionary Thought* (Dordrecht, 1983), pp. 57–111 (which makes the point about Emma Darwin) and, more recently, *idem*, 'Redrawing the boundaries: Darwinian science and Victorian women intellectuals', in Bernard Lightman (ed.), *Victorian Science in Context* (Chicago, 1997), pp. 119–42.

32 The Tripos lists are in J. R. Tanner (ed.), *The Historical Register of the University of Cambridge* (Cambridge, 1917). For a discussion of the women who took the Natural Sciences Tripos, see Roy MacLeod and Russell Moseley, 'Fathers and daughters: reflections on women, science and Victorian Cambridge', *History of Education*, 8 (1979), 321–33.

33 Cambridge University Library, Add. MS 8699. Acquired in 1986 by Cambridge University Library, the collection includes G. G. Stokes's letters to H. P. Stokes, as well as H. P.'s extensive biographical writings (in a difficult hand) on G. G. Letters from H. P. Stokes to G. G. Stokes are in the Stokes Collection (see note 6). In 1891, H. P. Stokes became vicar of St Paul's Church, Cambridge, where G. G. Stokes had been church warden for two decades.

8 'That universal æthereal plenum': Joseph Larmor's natural history of physics

Andrew Warwick

Centre for the History of Science, Technology and Medicine, Imperial College, London, UK

> [Larmor was] a man whose heart was in the nineteenth century, with the names of Faraday, Maxwell, Kelvin, Hamilton, Stokes ever on his lips – as though he mentally consulted their judgement on all the modern problems that arose. He would often say that all scientific progress ceased about 1900 – or even earlier, for his own fin de siècle effusion was only dubiously qualified.
>
> A. S. Edddington, 1944[1]

Joseph Larmor was by far the most important contributor to the development of mathematical electromagnetic theory in Britain during the 1890s. He refounded James Clerk Maxwell's field theory of electromagnetism on a new fundamental unit of electric charge (the electron) and used this foundation to develop a powerful theory of the electrodynamics of moving bodies. Despite these considerable achievements, however, his work has been marginalized in the folk memory of modern physics. One reason for this is that Larmor's work had much in common with that of the Dutch physicist, H. A. Lorentz, who was exploring similar issues at roughly the same time. Where Lorentz's work came to be seen as an important precursor to Albert Einstein's special theory of relativity of 1905, Larmor's continued interest in the notion of an electromagnetic æther led to him being dismissed by many as a reactionary.[2]

From Newton to Hawking: A History of Cambridge University's Lucasian Professors of Mathematics, ed. Kevin C. Knox and Richard Noakes. Published by Cambridge University Press.

In this chapter I shall trace the development of Larmor's major work in electrodynamics through the 1890s, and then discuss his view of the nature and purpose of physical theory. In doing so I hope to cast light on why Larmor not only held the concept of an all-pervading, dynamical æther in such high regard, but continued to do so even after the majority of continental physicists had followed Einstein in abandoning the concept as superfluous. This is an important issue because Larmor's seemingly stubborn unwillingness to abandon the æther concept has tainted his reputation and even, on occasion, led to him becoming a figure of fun in the physics community.[3] Yet Larmor's commitment to the concept sprang neither from nostalgia for nineteenth-century mechanical theories nor from a lack of appreciation of early-twentieth-century theories such as relativity and quantum mechanics. They sprang rather from his deeply held views of the nature and purpose of physical theories, views that he felt it important to air in public following his appointment to the Lucasian chair. In Larmor's opinion, the rise of relativity and quantum mechanics marked a sharp break in what had previously been the gradual evolution of physics as a unified field of intellectual endeavours. Despite the empirical success of the new theories in accounting for experimental data, Larmor believed them to lack the dynamical foundations and intuitive conceptual clarity that he thought a physical theory should display. Although he did his best to keep abreast of the new work, and strongly encouraged younger physicists to participate directly in its development, Larmor himself made little attempt to contribute. As we shall see, his reasons for doing so were closely related to his understanding of the historical process by which physics as a discipline had developed.

Born in Country Antrim (Ireland) in 1857, Larmor grew up in a family which, although relatively humble, was strongly committed to self-improvement through education. His father ran a grocer's shop in Belfast while his mother was 'ardently ambitious for the intellectual advancement of her family'. An academically able child, Larmor was educated at the Royal Belfast Academical Institution and, like

Figure 44 Larmor photographed c. 1912. From the *Mathematical Gazette* (1911–1912).

his four brothers (of which he was the eldest), attended Queen's College (Belfast).[4] At the Academical Institution he was taught by a Cambridge-trained mathematics master, Randal Nixon, before going on to obtain the 'highest honours' in mathematics and natural philosophy at Queen's College.[5] Larmor's precocious ability in mathematics took him next to St John's College (Cambridge) where, supported by a series of mathematical scholarships, he was trained by the greatest of the Cambridge mathematical coaches, Edward Routh. Routh subsequently wrote in a testimonial that, having known Larmor 'intimately' and overseen his work 'day by day', he was prepared to 'certify that [Larmor's] mathematical talent [was] of no ordinary kind'. As several other of his Cambridge teachers attested, even by Cambridge standards Larmor 'possessed in an eminent degree the faculty of rapidly and correctly thinking out the answers to the most difficult questions'.[6] His acquisition of the senior wranglership and Smith's prize in 1880 won him a fellowship at St John's College and the professorship of natural philosophy at Queen's College (Galway). In 1885 Larmor returned to St John's College as a lecturer in mathematics, a position he retained until 1903 when he succeeded George Stokes in the Lucasian chair.

Larmor's scientific reputation was built almost entirely on his research in electrodynamics carried out between 1893 and 1900. In this work he sought to derive all electromagnetic and luminous phenomena from the properties of an all-pervading, dynamical æther. The project ultimately proved unsuccessful but in the process of working it out Larmor made a number of very important contributions to the mathematical theory of electrodynamics. He introduced the concept of discrete electric charges, or 'electrons', to electromagnetic theory, developed a new theory of the electrodynamics of moving bodies, derived an expression for the energy radiated by an accelerating charge (now known as 'Larmor's formula'), derived an important theorem concerning the motion of charged particles in a magnetic field (now known as 'Larmor's theorem') and derived a now widely used expression for the frequency with which atoms precess in a magnetic field (now known as the 'Larmor frequency'). Larmor was also the first to derive what would later be known as the 'Lorentz transformations'. These transformations, linking the space and time coordinates of systems in states of relative uniform motion, were later rederived by Lorentz, and subsequently became central to modern physics through their role in Albert Einstein's special theory of relativity.

The most productive period of Larmor's research career was also accompanied by a stream of dramatic developments in experimental physics that would soon challenge received ideas concerning the nature of atoms and radiation. In 1896, Wilhelm Roentgen discovered a new kind of radiation (which he dubbed 'X-rays') capable of passing through many otherwise opaque materials, and Henri Becquerel announced that crystals of uranium salts spontaneously emitted a number of different kinds of ray. Investigation of these phenomena over the next two decades played a major role in the foundation of experimental atomic and nuclear physics, and in the formation of quantum mechanics. Of more immediate relevance to Larmor's work was the announcement the same year by Pieter Zeeman that a powerful magnetic field could broaden the spectral lines of sodium. This result lent powerful experimental support to the theory that such lines were generated by the motions of tiny charged particles, or 'electrons'. The following year, 1897, Larmor's Cambridge contemporary and professor of experimental physics at the Cavendish laboratory J. J. Thomson added further weight to this claim. He argued persuasively that cathode rays were composed of subatomic 'corpuscles' that were constituent parts of all atoms. Larmor showed how these latter developments lent powerful support to his electronic theory of matter, and provided a persuasive explanation of Zeeman's results in terms of rapidly orbiting electrons.

It was in recognition of the above contributions that Larmor was elected to the Lucasian chair in 1903 and knighted in 1909. Larmor was also a Unionist member of parliament for Cambridge University from 1911 until 1922. He made only about a half-dozen speeches in the House of Commons, most of which were concerned with academic

Figure 45 By the Edwardian era Joseph Larmor was recognised as a conservative in politics and physics. In this photograph from the 1910s, he is seen (right) conversing with the eminent Tory statesman Arthur Balfour at a garden party. From the *Proceedings and Report of the Belfast Natural History and Philosophical Society* (1942–43).

matters and the issue of Home Rule for Ireland. Larmor served as secretary of the Royal Society from 1901 to 1912 and as president of the London Mathematical Society in 1914 and 1915. The Royal

Society awarded him its Royal and Copley medals in 1915 and 1921 respectively. Following his appointment to the Lucasian chair, Larmor made no sustained contribution to the development of any branch of theoretical physics. His scientific research thereafter was confined mainly to a handful of papers clarifying specific aspects of his earlier work and occasional, though sometimes very significant contributions to the theory of dynamics, geodesy and the propagation of wireless waves. As we shall see, much of his effort during the second half of his career was devoted to clarifying his view of the nature and purpose of physical theory.

THE DEVELOPMENT OF THE ELECTRONIC THEORY OF MATTER (ETM)

In order to understand Larmor's general approach to physics we must begin with his research in the 1880s. During this early part of his career the vast majority of his original investigations was concerned with the mathematical methods and physical applications of analytical dynamics. Foremost among the mathematical methods was the principle of least action, a technique that Larmor believed to represent the clearest, most compact and most general means of expressing any physical problem. Larmor was convinced that this form of expression best revealed the formal mathematical connections that existed between the 'different departments' of mathematical physics, and so facilitated the analogical solution of a wide range of problems.[7] He also had a sound knowledge of James Clerk Maxwell's new field theory of electromagnetism which he had learned as an undergraduate in the Trinity College classroom of W. D. Niven. According to Maxwell, electromagnetic phenomena derived from the physical state of an all-pervading dynamical æther, a view which led him to propose that light was an electromagnetic wave in the æther. Larmor was in fact one of a small group of British physicists – recently dubbed the 'Maxwellians' – who sought during the 1880s and 1890s to explore the insights contained in Maxwell's 1873 exposition of his mature theory, a *Treatise on Electricity and Magnetism*.[8] This loosely affiliated group included

Oliver Heaviside, Oliver Lodge, J. J. Thomson and, most significantly for Larmor's work, the Irish physicist George FitzGerald.

It was Larmor's combined interests in dynamics, electrodynamics and the principle of least action that led him in the early 1890s to begin searching for a dynamical foundation for electromagnetic phenomena. This line of research was initially inspired by a paper of FitzGerald's that Larmor studied while preparing a report on magneto-optics for the British Association. FitzGerald had noticed that a remarkable formal analogy existed between the expression given by another Irish mathematician, James MacCullagh, for the mechanical energy stored in his rotationally elastic æther, and those given by Maxwell in his *Treatise on Electricity and Magnetism* for the energy stored in the electromagnetic field. By replacing the mechanical symbols in MacCullagh's theory with appropriate electromagnetic symbols, and applying the principle of least action to the resulting Lagrangian, FitzGerald was able to follow MacCullagh's analysis and obtain an electromagnetic theory of the propagation, reflection and refraction of light. Through his work on the analytical dynamics of magneto-optic rotation, and through reading FitzGerald's paper, Larmor became convinced that MacCullagh's æther could provide a common dynamical foundation for Maxwell's synthesis of electromagnetic and luminiferous phenomena.

The fruits of Larmor's research were published by the Royal Society as *A Dynamical Theory of the Electric and Luminiferous Medium* (hereafter *Dynamical Theory*) in three instalments (with various appendices) between 1894 and 1897, but during this period his theory changed considerably. The first instalment came in for some powerful criticism from FitzGerald, who acted as a referee for the Royal Society. During a period of intense correspondence between the two men in the spring and summer of 1894, FitzGerald persuaded Larmor to introduce the concept of 'discrete electric nuclei' or 'electrons' into his theory. But the introduction of the electron did far more than solve the immediate problems that troubled Larmor's theory; over the following three years it also had a profound effect upon his understanding of the relation between the electromagnetic æther and gross matter. In Maxwell's scheme an electric current was not thought of as a material flow of electrical substance, but as a breaking-down of electric displacement which, by some unknown mechanism, converted the electromagnetic energy stored in the æther into heat energy in the conductor. Prior to the introduction of the electron, the electromagnetic æther and gross matter were largely distinct concepts, and the mechanism of their interaction was seldom discussed. Likewise the effects produced by the motion of a charged body was something of a side issue, not perceived as having any real relevance to the much more central problem of the nature of the conduction current. But with the introduction of the electron as the fundamental carrier of a natural unit of electric charge the situation changed considerably.

The nature of the magnetic field generated by a moving charge had first been investigated by J. J. Thomson. In 1881 Thomson had drawn attention to the 'extra mass' that such a body would appear to have due to the energy of self-induction stored magnetically in the field. With the introduction of the electron this work assumed much greater significance. According to Larmor's fully developed theory, the universe consisted of a sea of æther populated solely by positive and negative electrons. These electrons could be thought of mechanically as point centres of radial strain in the æther and the sole constituents of ponderable matter. This view of the universe partially solved the problem of the relationship between æther and matter by reducing all matter to movable discontinuities in the æther. Furthermore, if all matter was constructed solely out of electrons, then perhaps its inertial mass was nothing more than the electromagnetic mass of the electrons as they are accelerated with respect to the æther. Thomson's 'extra mass' now became the *only* mass, and the mechanics of the universe was reducible to electrodynamics. I shall refer to this theory as the electronic theory of matter (hereafter ETM).[9]

The ideal of reducing mechanics to electrodynamics gained credence with a number of European physicists towards the end of the nineteenth century, and has been characterized by Russell McCormmach as the *Electromagnetic View of Nature*. Like Larmor, who was actually the first to suggest that the entire mass of the electron might be electromagnetic in origin, these physicists imagined a universe made only of æther and electrons, and whose dynamics was governed solely by the equations of electromagnetism. But Larmor's ETM differed in one very important aspect from the continental electromagnetic view of nature. Larmor, unlike some of his European contemporaries, continued to believe that the equations of electromagnetism were not themselves fundamental, but were ultimately to be derived from the dynamical properties of the underlying æthereal medium by the application of the principle of least action.

A further consequence of Larmor's introduction of the electron was that the previously minor problem of the electrodynamics of moving bodies was brought to centre stage. If electrical conduction and associated electromagnetic effects were due solely to the motion of electrons, and if matter was itself composed exclusively of electrons, then virtually every problem, both in electrodynamics and matter theory, became a problem in the electrodynamics of moving bodies. Indeed, these two previously distinct realms of physical theory became virtually inseparable. Such well-known effects as the electric polarization and magnetization of matter – which previously had been ascribed to changes in the dynamical properties of the æther somehow brought about by the presence of matter – could now be explained in terms of the electronic microstructure of matter. Moreover, by 1895 Larmor had shown that if the universe consisted of nothing but positive and negative electrons whose motions were governed solely by the equations of electrodynamics (that is, if the ETM was correct), then moving matter would contract minutely in its direction of motion through the æther. The importance of this contraction was that it offered a natural means of explaining the puzzling null result obtained in an important experiment undertaken by Albert Michelson and Edward Morley. Since this experiment was crucial to Larmor's theory I shall briefly explain its origin and interpretation.

Two important claims made by Maxwell's field theory of electricity and magnetism were that light was an electromagnetic phenomenon and that all electromagnetic effects were ultimately attributable to physical states in an all-pervading æther. It was Maxwell himself who suggested in 1878 that it might be possible to detect the earth's orbital motion through the æther by comparing the times taken by a light beam to travel with and against the earth's motion.[10] In the early 1880s, the American experimenter, Albert Michelson, proposed to try a modified form of this experiment using a device that would come to be known as an interferometer. Michelson's plan was to split a light beam into two perpendicular rays which would then be made to traverse equal distances (along perpendicular arms of the interferometer) before being reflected back to recombine and form an interference pattern. If the rays happened to travel respectively parallel and perpendicular to the earth's direction of motion through the æther they ought to take different times to make the round trip, a difference that would reveal itself through shifts in the interference pattern when the interferometer was rotated.

In a series of experimental trials undertaken between 1881 and 1887, those from 1885 being carried out in collaboration with Edward Morley, Michelson invariably obtained a null result; the earth's motion through the æther did not affect the journey time of light rays in the expected manner. One way of explaining this result was to assume that the æther was dragged along, or 'convected', at the earth's surface, but this assumption did not appeal to the leading theoreticians of electromagnetism of the time. Instead, the Irish physicist, George FitzGerald, and the Dutch physicist, H. A. Lorentz, suggested independently in 1889 and 1892 respectively that matter (and hence one arm of the interferometer) contracted minutely in its direction of motion through the æther, thereby concealing an effect that would otherwise be detectable.[11] The suggestion came to be known as the Lorentz–FitzGerald contraction hypothesis. FitzGerald's suggestion was initially derided in Britain as theoretically untenable, but following the advent of Larmor's ETM it quickly acquired a new status.[12]

The accommodation of the contraction hypothesis within Larmor's theory is of such importance that I shall briefly explain the two lines of investigation by which it occurred. The first concerns the phenomenon of light convection. When a beam of light passes into a moving, transparent medium, part of the medium's velocity is imparted to the beam. Larmor constructed a theoretical model of this effect and used it to investigate the behaviour of the rays in Michelson and Morley's interferometer. He found he could explain the null result if he employed new space–time transformations to relate the electromagnetic variables of the stationary-æther frame of reference to those measured in the rest-frame of the moving interferometer. Larmor also found that Maxwell's electromagnetic field equations were invariant to second order under these new space–time transformations provided that new electromagnetic vectors were introduced in the moving system.[13] This important finding was eventually explained by Larmor in the following way. He claimed that the electromagnetic fields measured by observers in moving systems (on the earth's surface, for example) were not the physically real fields that would be measured by observers who were stationary in the æther. Rather, they were subsidiary fields generated partly by the earth's motion through the æther. It just happened that these subsidiary fields also satisfied Maxwell's equations. Larmor noted that the new space–time transformations had a strange property; they seemed to imply that moving electrical systems contracted in precisely the manner suggested by FitzGerald.

The second line of research that led Larmor to his explanation of Michelson and Morley's null result concerned the potential surrounding a moving electron. By 1897, Larmor was convinced that positive and negative electrons were the sole constituents of gross matter. In that year, he derived a general expression for the total electromagnetic potential surrounding a moving electron. This expression indicated that the potential surfaces surrounding the electron would contract by exactly the amount predicted by FitzGerald (and by the new space–time transformations). In order to remain in electromagnetic equilibrium, a stable group of electrons moving through the æther would therefore have to move closer together by the same amount. Since Larmor believed gross matter to be solely electronic in construction, it seemed reasonable to him to suppose that moving matter must also contract by this amount at the macroscopic level. Larmor now had a physical explanation for the contraction implied by the new space–time transformations. A very important point to note in this context is that, during 1897, the relationship between the ETM and the Michelson–Morley experiment changed dramatically. From this point on, Larmor claimed that the null result was actually *predicted* by his theory and so provided important experimental evidence that matter was electrically constructed and that his theory was correct.

Towards the end of 1897 Larmor collected and clarified the most important results of the new electrodynamics for his *Theory of the Aberration of Light*, the winning entry for Cambridge University's prestigious Adams prize essay competition. In this essay he gave the first systematic account of the ETM including the following new space–time transformations. If (x, y, z, t) $(\mathbf{E}, \mathbf{B})$ and (x_1, y_1, z_1, t_1) $(\mathbf{E}_1, \mathbf{B}_1)$ represent respectively the space–time and electromagnetic-field variables measured in the stationary æther frame of reference and a frame of reference moving through the æther with velocity v, then these variables are related as follows:

$$x_1 = \beta(x - vt), y_1 = y, z_1 = z, t_1 = \beta(t - vx/c^2) \quad (1)$$

$$\mathbf{E}_1 = (E_{x1}, E_{y1}, E_{z1}) = \beta[\beta^{-1} E_x, (E_y - vB_z)', (E_z + vB_y)] \quad (2)$$

$$\mathbf{B}_1 = (B_{x1}, B_{y1}, B_{z1}) = \beta[\beta^{-1} B_x, (B_y + v/c^2 E_z)', (B_z - v/c^2 E_y)] \quad (3)$$

where $\beta = (1 - v^2/c^2)^{-1/2}$ and c is the velocity of light. These were the transformations later given independently by Lorentz and by Einstein, equation (1) now being known as the 'Lorentz transformations'. It was a revised and expanded version of Larmor's Adams prize essay that appeared in 1900 as Larmor's best-known and most widely read publication, *Aether and Matter*.[15]

Larmor was the first mathematical physicist to employ the above transformations but his interpretation of their physical meaning was very different from that subsequently given by Einstein. The transformations were used by Larmor to 'compare a system shrunk in the FitzGerald–Lorentz manner and convected through the æther, with the same system unshrunk and *at rest*'.[16] This understanding is difficult for the reader trained in modern physics to grasp for two reasons. First, following Einstein, we are used to thinking of motion as a purely relative quantity, so that *any* inertial frame can legitimately be considered to be *at rest* by an observer moving with it. Second, the modern reader likewise recognizes no privileged description of the electromagnetic fields surrounding a charged system – only different accounts given by observers in different states of uniform relative motion. For Larmor, however, electromagnetic fields were the observable manifestations of real physical states of a dynamical æther. He accordingly used the above transformations in the following way: he first imagined an electromagnetic system *at rest* in the æther and surrounded by the electric and magnetic fields ($\mathbf{E}$, $\mathbf{B}$); he then employed the transformations (1), (2) and (3) to calculate the new fields ($\mathbf{E}_1$, $\mathbf{B}_1$) that would surround the *same system* (as measured in the stationary æther frame) if it were moving through the æther with the uniform velocity v (with the moving earth, for example). Larmor referred to systems related in this way as 'correlated systems'.

The fields ($\mathbf{E}_1$, $\mathbf{B}_1$) were not of course those that would be measured by an observer in the rest frame of the moving system ($\mathbf{E}_1$, $\mathbf{B}_1$). The moving observer might assume ($\mathbf{E}_1$, $\mathbf{B}_1$) to be real fields but they actually owed their existence to the following three effects upon ($\mathbf{E}_1$, $\mathbf{B}_1$): first, the motion of the observer through the æther, which had the effect of 'mixing up' the real electric and magnetic fields so that part of the electric field appeared as if it were a magnetic field and vice versa; second, the physical contraction of the moving matter which led the moving observer to adopt a new scale of measurement; and third, the new standard of time measurement unknowingly adopted by the moving observer. It was for Larmor a mere peculiarity of the fields ($\mathbf{E}_1$, $\mathbf{B}_1$) that they too satisfied Maxwell's equations (at least to second order) and so concealed the earth's proper motion through the æther. Only by referring back to calculations made according to coordinates that were stationary in the æther could the real fields be calculated.

From a mathematical perspective, Larmor's application of the new transformations was similar to a range of analytical and geometrical mapping techniques routinely employed by Cambridge-trained mathematicians to solve problems in various branches of mathematical physics. By transforming the known solution of a problem in electrostatics, for example, it was often possible to find the solution to another previously unsolved problem of a similar nature. Maxwell had discussed numerous examples of this kind in the first volume of the *Treatise*, and Routh's lecture notes reveal that he taught such methods to his students. Routh also employed these methods to solve problems in mechanics and even referred to dynamically related bodies as 'correlated' bodies. Larmor made the method applicable to moving electrical systems, thereby simplifying calculations that had previously been extremely complicated.

From a physical perspective, the stationary æther frame of reference remained for Larmor a truly privileged frame, compared to which moving bodies actually became shorter and moving clocks ran more slowly. Conversely, a system that was stationary in the æther would have to appear *longer* to an observer moving through the æther. That Larmor really did think of his system in this way is highlighted by the fact that he sometimes referred not to a 'contraction' of moving matter, but an 'elongation' of the system when brought to rest; and, on at least one occasion, he even referred to the FitzGerald–Lorentz 'elongation hypothesis'. Larmor made little explicit comment on the physical meaning of his new nonstandard space–time transformations, presumably because he did not regard them as problematic. The transformation $x_1 = \beta(x - vt)$ expressed the contraction of moving matter, but this was to be expected in the ETM. The transformation $t_1 = \beta(t - vx/c^2)$ received no specific physical interpretation but would, according to Larmor, be 'unrecognizable' because time in any frame was 'isotropic'. By this Larmor appears to have meant that an observer in a moving frame (as on the earth's surface) would be bound to employ clocks made of electrons and that these clocks would naturally record t_1. But, even if clocks really did run more slowly in a moving system, this local time could in theory be related back to the 'real' time measured in the stationary æther frame. We should therefore think of Larmor's transformations as being *in* space and time rather than *of* space and time.

Finally, we must briefly consider why Larmor thought it not only possible, but desirable, that the earth's motion through the æther would eventually be detected experimentally. Larmor's mathematical formulation of the ETM did not rule out the possibility that the earth's motion would remain undetectable to all orders of v/c. However, from a physical perspective this possibility required one also to assume that positive and negative electrons could legitimately be treated as point sources in the æther. The null result of the Michelson–Morley and other experiments designed to detect the earth's motion indicated that this assumption was empirically admissible up to second order. But if the electron constituted a finite structure in the æther, as Larmor believed it might, then it was possible that matter would not contract in the predicted way when electrons were brought into close proximity with one another. It would then be possible not only to detect the earth's proper motion through the æther at higher orders – and hence to calculate the real electromagnetic fields in the stationary æther – but also to learn something about the electron's size and structure. These considerations led Larmor always to approximate his own calculations to second order, since he believed them to be empirically unfounded at higher orders. It was this habit of approximating calculations that prevented him from noticing until after the publication of *Aether and Matter* that Maxwell's equations actually transformed exactly under the Lorentz transformations.

Larmor's new electrodynamics, based on the electron and incorporating new space–time and electromagnetic transformations, marked an important break, both conceptually and geographically, with the British Maxwellian tradition of the 1880s and early 1890s. From a theoretical perspective, the ETM transcended the work of other British scholars working on Maxwell's theory by positing the electron as the natural origin and unit of positive and negative electric charge and by offering a new unification of optics, electrodynamics and matter theory. The ETM embodied what would shortly become known as the 'Lorentz transformations' and relied heavily upon the null result of the Michelson–Morley experiment to provide empirical support for its central claims that matter was electrically constituted, that all mass was electromagnetic in origin and that matter contracted minutely in its direction of motion through the æther. Larmor's theory also provided the theoretical tools necessary to tackle a wide range of problems in the electrodynamics of moving bodies.

From a geographical perspective, the emergence of the ETM was accompanied by an effective localization of British research in mathematical electromagnetic theory in Cambridge. During the 1880s several non-Cambridge men, most notably George FitzGerald in Dublin and Oliver Heaviside in London, mastered Maxwell's *Treatise* and made extremely important contributions to its theoretical interpretation. However, the most important theoretical results achieved by these men were not passed by them to a new generation of students. Only in Cambridge, and partly through Larmor's own teaching, was a new generation of students trained during the 1890s to undertake further research on electromagnetic theory. Larmor's extremely technical papers on the ETM subject found few, if any, competent British readers beyond Cambridge, but the theory was effectively transmitted to a new generation of Cambridge mathematicians during the first decade of the twentieth century.

THE ROLE OF THE ÆTHER IN LARMOR'S PHYSICS AFTER 1900

I noted above that the publication of Larmor's book *Aether and Matter* in 1900 marked the end of his major contribution to the development of electrodynamics. This was in part because the book made a natural conclusion to his earlier research, but his lack of productivity thereafter can also be attributed to his increasing workload as an administrator both within and beyond Cambridge University – especially in his capacity as a secretary of the Royal Society. By the mid-1900s, moreover, a younger generation of Cambridge-trained mathematicians was developing the ETM, sometimes by combining Larmor's work with aspects of Einstein's special theory of relativity. Larmor himself took little part in these developments, not least because, even in Cambridge, the new generation of theoreticians tended to play down the role of a dynamical æther in the ETM. However, Larmor's steadfast commitment to the æther concept was no mere obsession with finding a micromechanical explanation of macroscopic electromagnetic effects; it went rather to the very heart of his epistemology of physics. In the remainder of this chapter I shall discuss why Larmor believed the æther to be completely indispensable to physics and show that his defence of the concept preceded the advent of Einstein's theory of relativity.

A good place to begin exploring Larmor's notion of a dynamical æther is with one of his best known claims concerning the æther's physical status:

> All that is known (or perhaps need be known) of the æther itself may be formulated as a scheme of differential equations defining the properties of a continuum in space, which it would be gratuitous to further explain by any complication of structure; though we can with great advantage employ our stock of ordinary dynamical concepts in describing the succession of different states thereby defined.[17]

Larmor seems to be suggesting in this passage that the mathematical equations describing the æther's state are virtually coextensive with the concept of the æther itself. On this reading the æther would be little more than a didactic aid to be shrugged off as soon as the equations were mastered. Yet this reading is inconsistent with the bulk of

Larmor's writing on the æther during the period. In the above passage, as elsewhere, Larmor is seeking to attack the old concept of a mechanical æther – as espoused by William Thomson, for example – to which could be ascribed a definite mechanical structure capable of producing dynamical effects. Larmor's vision of the æther as the fundamental dynamical essence of the universe rendered such projects obsolete, since the mechanical properties of gross matter were but one manifestation of the dynamical properties of the underlying medium. It was in this sense that his work had altered the ontological status of the æther. There was, of course, a long tradition in Britain of theorizing over the possible nature and structure of the luminiferous æther. Indeed, it was in the work of men such as George Stokes, George Green and George MacCullagh, who investigated the properties of the æther as a mechanical continuum, that much of the mathematical theory of nineteenth-century Cambridge physics had been developed. But once Maxwell had shown that light could be treated as an electromagnetic phenomenon, those who were committed to the æther concept began to seek a common electromagnetic *and* luminiferous medium, capable of underpinning Maxwell's electromagnetic theory. Furthermore, at least one of these men, George FitzGerald, had considered the possibility that the universe might, ultimately, be electromagnetic in origin, so that this new æther might, one day, provide the basis for a unified theory of the physical world. But as Bruce Hunt has shown, FitzGerald gradually became more concerned with purely electromagnetic aspects of Maxwell's work than with its dynamical foundations.[18]

Larmor, on the other hand, was more typical of the Cambridge mathematical tradition. It is true that many important elements of his ETM originated in contact with other scholars of Maxwell's work beyond Cambridge, but the mathematical methods he employed were those routinely used in Cambridge for investigating the dynamical properties of mechanical continua, and his search for a single, dynamical, luminiferous and electromagnetic medium remained central to his research throughout his life.[19] It was Larmor alone who first developed the ETM, and only in his work did a unified dynamical theory of the whole physical world appear to become a real possibility. Only Larmor's theory predicted in full mathematical detail that the FitzGerald contraction would actually occur, and only Larmor's theory made the æther responsible for *all* the phenomena of the physical world.

The nature of this ultimate dynamical essence was not wholly determined by the mechanical properties of gross matter because they, like electromagnetic phenomena, were but partial manifestations of its dynamical nature. In Larmor's own words it was the differential equations which were 'defining the properties of a continuum in space', and these required only that we accept the principle of the conservation of energy, and the principle of least action. One could not therefore expect to construct any kind of purely mechanical model of the æther. In his presidential address before the British Association in 1900, Larmor clarified this point as follows:

> we should not be tempted towards explaining the simple group of relations which have been found to define the activity of the æther, by treating them as mechanical consequences of concealed structure in that medium; we should rather rest satisfied with having attained to their exact dynamical correlation.[20]

For him, it was the *assumption* that the æther was the common physical foundation of all of the manifest forces of nature which rendered their dynamical correlation possible. The æther represented the physical embodiment of the whole mathematical structure that described electromagnetic processes, and these mathematical structures exhausted our direct knowledge of its properties. But until the fundamental equations correlated *all* of the known forces of nature in every detail, the exact dynamical nature of the æthereal medium had not been completely ascertained. The further development of a unified theory of the physical world thus depended crucially upon our continued faith in the existence of the æther and, as we shall shortly see,

Larmor went so far as to state that it was the concept of the dynamical æther which made physics possible.

But in order to appreciate fully the extraordinary importance which Larmor attached to the concept of the dynamical æther it is necessary to trace its unifying role beyond the embodiment of the equations of dynamical physics. In addition to rendering seemingly disparate branches of physical theory accessible to unification within theoretical mechanics, there were two further themes in Larmor's vision of a unified physics which drew upon the æther concept: first, it stood as a recurrent theme in the evolution of theoretical mechanics; and, second, through its mechanical origins it provided the physicist with a natural means of comprehending the hidden unity of the physical world. In the balance of this section I will develop the meaning of these subtle themes and describe how they interacted to make the æther practically indispensable for Larmor.

In *Aether and Matter*, Larmor underlined the reliance of physics on the æther concept in the following words: 'the reason why progress [in physics] is possible at all is that an individual molecule is not an isolated thing ... jostling among its neighbours, but a nucleus in that universal æthereal plenum'.[21] And he expressed very similar sentiments in his 1900 address to the British Association for the Advancement of Science:

> [T]he complication of the material world is referable to the vast range of structures and states of aggregation in the material atoms; while the possibility of a science of physics is largely due to the simplicity of constitution of the universal medium through which the individual atoms interact on each other.[22]

One crucial role of the æther, then, was to represent the simple, analysable substratum, which rendered the chaotic motions and diverse structures of matter accessible to theoretical unification. But this definition seems to contain an inherent ambiguity; does Larmor mean that it is the *physical reality* of the æther which renders the whole natural world accessible to theoretical mechanics, or is he

28 A. Einstein

Einführung eines „Lichtäthers" wird sich insofern als überflüssig erweisen, als nach der zu entwickelnden Auffassung weder ein mit besonderen Eigenschaften ausgestatteter „absolut ruhender Raum" eingeführt, noch einem Punkte des leeren Raumes, in welchem elektromagnetische Prozesse stattfinden, ein Geschwindigkeitsvektor zugeordnet wird.

Die zu entwickelnde Theorie stützt sich — wie jede andere Elektrodynamik — auf die Kinematik des starren Körpers, da die Aussagen einer jeden Theorie Beziehungen zwischen starren Körpern (Koordinatensystemen), Uhren und elektromagnetischen Prozessen betreffen. Die nicht genügende Berücksichtigung dieses Umstandes ist die Wurzel der Schwierigkeiten, mit denen die Elektrodynamik bewegter Körper gegenwärtig zu kämpfen hat.

I. Kinematischer Teil.

§ 1.

Definition der Gleichzeitigkeit.

Es liege ein Koordinatensystem vor, in welchem die Newtonschen mechanischen Gleichungen gelten.*) Wir nennen dies Koordinatensystem zur sprachlichen Unterscheidung von später einzuführenden Koordinaten systemen und zur Präzisierung der Vorstellung das „ruhende System".

Ruht ein materieller Punkt relativ zu diesem Koordinatensystem, so kann seine Lage relativ zu letzterem durch starre Maßstäbe unter Benutzung der Methoden der euklidischen Geometrie bestimmt und in kartesischen Koordinaten ausgedrückt werden.

Wollen wir die *Bewegung* eines materiellen Punktes beschreiben, so geben wir die Werte seiner Koordinaten in Funktion der Zeit. Es ist nun wohl im Auge zu behalten, daß eine derartige mathematische Beschreibung erst dann einen physikalischen Sinn hat, wenn man sich vorher darüber klar geworden ist, was hier unter „Zeit" verstanden wird. Wir haben zu berücksichtigen, daß alle unsere Urteile, in welchen die Zeit eine Rolle spielt, immer Urteile über *gleichzeitige Ereignisse* sind. Wenn ich z. B. sage: „Jener Zug kommt hier um 7 Uhr an," so heißt dies etwa: „Das Zeigen des kleinen Zeigers meiner Uhr auf 7 und das Ankommen des Zuges sind gleichzeitige Ereignisse.**)

Es könnte scheinen, daß alle die Definition der „Zeit" betreffenden Schwierigkeiten dadurch überwunden werden könnten, daß ich an Stelle der

*) Gemeint ist: „in erster Annäherung gelten".

**) Die Ungenauigkeit, welche in dem Begriffe der Gleichzeitigkeit zweier Ereignisse an (annähernd) demselben Orte steckt und gleichfalls durch eine Abstraktion überbrückt werden muß, soll hier nicht erörtert werden.

Figure 46 Larmor defends the æther from Einsteinian physics. A page from Larmor's annotated copy of H. A. Lorentz, A. Einstein and H. Minkowski, *Das Relativitätsprinzip: eine Sammlung von Abhandlungen* (Leipzig: B. G. Teubner, 1913).

suggesting that it is the *idealized concept* of a dynamical æther which is a necessary prerequisite to the construction of a unified theory of nature? Certainly, much of the criticism directed at æther theory, both contemporary and more recent, attributes to Larmor something like the former of these points of view, but I suggest that it is the latter position that more consistently characterizes his position.

Larmor divided physical science into three areas: experimental observations; empirical laws (such as the second law of thermodynamics); and theoretical mechanics. These areas were, of course, intimately related since experiments might be suggested by a theoretical prediction and empirical laws were largely based upon observation. But at the very top of Larmor's hierarchical structure stood theoretical mechanics. It was theoretical mechanics that provided the concepts within which a unified theory of the physical world could be constructed, and upon which any physical explanation of a phenomenon had ultimately to be based. Morever, theoretical mechanics had itself taken almost 200 years to reach maturity and Larmor believed that understanding its modern dynamical form required knowledge of its mechanical origins and the course of its evolution. He argued that the laws of mechanics had received their first clear statement in Newton's *Principia*. They had then evolved in the form of analytical mechanics to become even more widely applicable and ever less dependent on any detailed knowledge of the specific forces at work in the system to be explained. Larmor acknowledged the great contributions made by Lagrange and Hamilton in founding mechanics on variational or minimum action principles, but he believed theoretical mechanics had reached its most all-encompassing formulation in the 'dynamical theory' set down by Thomson and Tait in their *Treatise on Natural Philosophy* of 1867. In this form the fundamental principles of mechanics – based upon the conservation of energy and the principle of least action – were expressible without reference to any detailed mechanical structure or mechanism. Larmor believed that the realm of theoretical mechanics would continue to expand until it encompassed every known phenomenon of nature.

This view of the gradual 'evolution' of physical theory in the form of mechanics is in some ways analogous to Thomas Kuhn's concept of 'normal science'. A central core of theoretical assumptions (theoretical mechanics) is retained and used to predict and to accommodate new physical phenomena. But in Larmor's scheme there was no counterpart to the Kuhnian 'revolution' that would follow the accumulation of apparently inexplicable results. Larmor believed, rather, that theoretical dynamics was itself in a state of gradual evolution and that it was sufficiently flexible to accommodate all physical phenomena given sufficient ingenuity. Consider, for example, the following passage from *Aether and Matter*:

> Theoretical mechanics is thus an abstract science engaged in the application to natural phenomena of principles which are themselves in a state of development mainly as regards detail, arising from the gradual reclamation from an empirical fringe surrounding the settled domain of science. This fringe now extends a long way into molecular phenomena as distinct from mechanical, chiefly in the regions of the theory of gases and of radiant heat and electrical actions.[23]

Thus theoretical mechanics had already transcended what were regarded as obviously mechanical phenomena, and might itself be further modified as it was used to explain new phenomena in the theory of gases, thermodynamics and electromagnetic theory. In the case of gas theory, for example, it was currently possible to explain the distribution of molecular speeds using certain mechanical models and statistical considerations. But, as with the second law of thermodynamics, this represented for Larmor a statement of an observed law of nature rather than its explanation. Statistics was, he remarked:

> a method of arrangement rather than of demonstration. Every statistical argument requires to be verified by comparison with the facts, because it is of the essence of this method to take things as fortuitously distributed except in so far as we know to the contrary.[24]

Larmor thought that statistics was the mathematics of fortuitous distribution, but that this was not an appropriate model through which to understand and explain the order of the physical world. For the science of gases to advance, the equilibrium distribution of molecules needed to be explained in terms of the ætherially mediated interactions between the molecules. This explanation would be complete when it was based on the principle of least action and the concept of the conservation of energy, and made predictions that accorded with experiment.

Furthermore, Larmor would not entertain the possibility that the principles of theoretical mechanics might be inadequate to describe some aspects of the world. This was, in part, because he believed theoretical mechanics to be *definitive* of theoretical physics, but also because of the way in which he believed physics to have evolved. The gradual expansion of the domain of theoretical mechanics was an evolutionary process whereby both dynamical theory and the concept of the æther were gradually adapted so as to bring unity to disparate realms of physical theory. The goal of physics was therefore the production of an ultimate theoretical mechanics capable of accommodating every known physical phenomena through the agency of the dynamical æther.

According to Larmor, theoretical mechanics and the concept of an æther had evolved coextensively, the latter fulfilling two important functions along the way. First, it had inspired some of the most fruitful concepts in physics. The æther concept had, for example, led Young and Fresnel to develop the seemingly unlikely idea that light was a wave motion, one of the greatest innovations of the nineteenth century. Second, it had acted as the conceptual medium that had gradually made all of the other disparate regions of physical theory accessible to unification under theoretical mechanics. Through the concept of the æther, the powerful mathematical physics developed to describe the passage of waves through matter and the motion of fluids had been extended to encompass heat theory, gas theory and the theories of electricity and magnetism. Most remarkably of all, Larmor thought, the concept had enabled Maxwell to identify light as an electromagnetic phenomenon. Larmor's own excursion into electromagnetic theory was inspired by the promise of a single electric and luminiferous medium which could physically underpin Maxwell's brilliant insight. By 1897 this æther had become the lone physical reality which embodied the principles of theoretical mechanics and acted as the medium through which the diverse forces of nature would be correlated. The æther was 'real' for Larmor in that it had been the mental image guiding the unification of mathematical physics for more than a century. He believed that the concept had to be preserved if similar progress were to be made in the future. Only a dynamical æther could render the mathematical correlations which had been found to exist between disparate branches of physics comprehensible to the human mind.

In order to appreciate fully the depth of Larmor's commitment to the æther it is important to realize that, despite his frequent use of evolutionary metaphors, his 'evolving physics' was not advancing by a series of chance events. We have already seen that the æther was the medium that had enabled the unification of physics through dynamics, but for Larmor this was not merely one of many possible ways of unifying physical phenomena. His notion of an æther-based theoretical mechanics was *naturalistic* in the sense that he assumed its evolution to represent the inevitable development of man's faculty for understanding the physical world. In 1900 he wrote:

> It might be held that this conception of discrete atoms and continuous Aether really stands, like those of space and time, in intimate relation with our modes of mental apprehension, into which any consistent picture of the external world must of necessity be fitted.[25]

For Larmor the concept of discrete atoms in a continuous æther was as natural and necessary to human thought as were those of space and time; without them physics could neither exist nor progress. In his British Association address he specifically warned against what he regarded as the folly committed by those who sought to retain the equations describing a particular natural phenomenon while claiming to abjure any further metaphysical speculation. Quite apart from the necessity of such speculation for the continued progress of physics there was, he argued, no means of drawing a reliable distinction between the hard experimental facts, and the 'intellectual scaffolding' from which they originated and drew meaning.[26] He emphasized this point by stating explicitly that it was the natural concepts of space, time, æther, discrete structures in the æther, and theoretical mechanics which alone *imposed* our perceived order upon nature. And he continued:

> Is the mental idea or image, which suggests, and alone can suggest, the experiment that adds to our concrete knowledge, less real than the bare phenomenal uniformity which it has revealed? Is it not, perhaps, more real in that the uniformities might not have been there in the absence of the mind to perceive them?[27]

Likewise in *Aether and Matter* he explained that the æther had to be retained as a working concept because the formal mathematical analogies which existed between different branches of physics 'perhaps [originated] as much in the necessary processes of thought as in the nature of external things'.[28] Only through the concept of the dynamical æther could the mind grasp the origin and meaning of these formal analogies and relate them back to the simple mechanical concepts of motion and force.

It would therefore have been practically impossible for Larmor to have drawn any sharp distinction between the *real* nature of the physical world, and the mode by which it was *apprehended*. The æther and its discrete structures were as fundamental to his physics as were the concepts of space and time. He could hardly have posed the question 'does the æther actually exist?', since for him it would have been virtually synonymous with the question 'does theoretical physics exist?'

THE NATURAL HISTORY OF PHYSICS AND THE CONCEPT OF FORCE

Having investigated the sense in which the æther rendered physics *possible* for Larmor, we are now in a position to understand two additional strands of his epistemology that further supported his belief in the necessity of the æther concept. The first of these concerns the status that he ascribed to the history of physics. Since the co-development of theoretical mechanics and the æther concept was an inevitable and natural process in the intellectual mastery of the physical world, the history of physics became for Larmor a kind of *natural history* and, as such, worthy of careful study in its own right. Just as mechanical illustrations of some of the æther's dynamical properties were of considerable use in the teaching of æthereal physics, so the 'study of the historical evolution of physical theories is . . . essential to the complete understanding of their import'.[29] Furthermore, Larmor regularly adopted arguments from the history of physics in defending his ETM. In 1907, for example, he addressed the problem of why the positive electron had not been produced (and the implied attack upon the concept of the electromagnetic inertia of the atom) by arguing that:

> all who have appreciated the course of evolution of the principles of modern physical explanation through Newton, Lagrange, Young, Fresnel, Faraday, Maxwell, Helmholtz and Hertz, to mention only a few names of the past, will still hold that even if the atom itself is intrinsically unfathomed, yet the interaction between atoms separated in space is in its larger features understood and has its seat in their sub-electric connections with the æthereal medium.[30]

For Larmor, the natural history of physics revealed that the unification of the physical sciences during the eighteenth and nineteenth centuries had been carried out under the aegis of the dynamical æther, a unification that had culminated in the ETM. I shall shortly show how Larmor drew upon this argument to defend his theory.

The second strand in Larmor's epistemology that reinforced his adherence to the concept of an æthereal medium was his belief that the physical experience of 'force' was fundamental to our intuitive understanding of what was meant by a 'dynamical theory'. He believed that the clarity of such mechanical concepts as momentum, force and energy, through which students were introduced to the more subtle concept of an ultimate dynamical medium, derived from the *physical experience* of force and motion that forces could produce. Larmor took this physical experience to have been the starting point for mechanics, arguing in *Aether and Matter* that:

> As mechanics took its origin in the equilibration of tendencies to motion of the various types that can be recognized, its chief concept lay to hand in muscular effort, which suggested a common standard of measurement. To make use of this concept for scientific purposes, precision in the method of measurement is, as usual, all that was required.[31]

He accordingly rejected any claim that force could meaningfully be regarded as a mere number in an equation defining the relationship between mass and acceleration. The formal mathematical relationships and analogies which had been found to exist between the various branches of physics could not, by themselves, impart any real understanding or grasp upon the underlying unity of nature. Only by tracing these relationships back to their common æthereal origins, and then to the physical sensation of force, could this seemingly mysterious coincidence be comprehended. Since the æther was not accessible to direct human experience, comprehending its dynamical essence required one to return by analogy to the mechanics of gross bodies from which dynamics had originated. Only mechanics could be intuitively grasped by the mind, via the physical sensation of force, and in this sense it formed a vital link between experience and the universal dynamical æther. In *Aether and Matter* Larmor discussed the fundamental nature of force and rejected the claim that it was somehow a conventional concept in the following way:

> To say, as is sometimes done, that force is a mere figment of the imagination which is useful to describe the motional changes that are going on around us in nature, is to assume a scientific attitude that is appropriate for an intelligence that surveys the totality of things ... when a person measures the steady pull of his arm by

> the extension of a spring, where or what, for example, are the motions of which the pull is only a presentation? The only way of gradually acquiring knowledge as to what they are, is to develop and make use of all the exact concepts that examination of the phenomena suggests to the mind.[32]

As we saw above, force was the scientifically quantified equivalent of muscular effort, and it was ultimately to this concept that one had to return when trying to grasp the nature of the æther. This also highlights the value of mechanical models capable of imitating some of the properties of the æther; they acted both as heuristic guides in the application of theoretical mechanics to some new area of physics, and as models by which we might dimly glimpse some of the diverse mechanical properties of the ultimate dynamical reality.

Larmor's vision of what I have called the *natural history of physics* led him to attach great significance to historical precedent in science. Moreover, recognition of this fact enables us to understand better his apparent obsession with the past (see epigraph to this chapter). Larmor did not, as Eddington suggested, simply mourn the passing of the old physics; for him Kelvin, Maxwell and their like were the great men of science who had been privileged to help bring the true physics to maturity, and the wisdom of their judgement could not be lightly set aside. Larmor had learned this dynamical physics from his Cambridge coach, Edward Routh, and from the great British treatises of the late nineteenth century – Thomson and Tait's *Treatise on Natural Philosophy* (1867), Maxwell's *Treatise on Electricity and Magnetism* (1873) and Rayleigh's *Theory of Sound* (1877). It is within this tradition that both the mathematical methods and epistemology expressed in *Aether and Matter* are most readily understood.[33] Several of Larmor's contemporaries at Cambridge produced now classical investigations of different aspects of physical theory using the same dynamical approach, but Larmor's own contribution was much more daring and potentially far-reaching. He had transformed the æther to make possible a unified theory of the whole physical world within which all dynamical theories might find a common physical foundation. But, as we have seen, the continued development of physics required that the practitioners remained sensitive to the dynamics of physical theory itself. The essence of Larmor's æther could not be stated simply in words or in bare mathematical form, but had to be divined through the practice of its mathematical methods, mechanical illustration and a knowledge of its evolutionary history. The reader of *Aether and Matter* was therefore primed with a long 'Historical Survey' which recounted the history of physics in such a way that it made the concept of an electrodynamic æther populated with tiny centres of intrinsic strain seem the inevitable outcome of the development of physical theory. The abandonment of this cumulative evolutionary process could only mean the end of real progress in physics.

We can now begin to see why Larmor never entertained the idea of abandoning the æther, even if it could not be explicitly identified by experiment. For him the very success and unity of nineteenth-century electromagnetic theory attested to its necessary existence. Larmor nevertheless continued to believe that motion relative to the æther would one day be detected and that this would eventually reveal the structure of electrons.

CONVENTIONALIST EPISTEMOLOGY VS THE NATURAL HISTORY OF PHYSICS

Larmor's commitment to the view outlined above is very nicely illustrated in his response to the conventionalist epistemology espoused by the French mathematician Henri Poincaré. Poincaré outlined his position in his book *La Science et l'Hypothèse*, and when it was translated into English in 1905, the fledgling fourteenth Lucasian professor was invited to write an introduction. Larmor took the opportunity to disagree with Poincaré on several important matters and, in doing so, provided a useful summary of his own reasons for defending the æther concept. One of the fundamental problems identified and addressed by Poincaré in *Science and Hypothesis* concerned the relationship between the status of scientific knowledge, and the history of its acquisition. Like Larmor, he considered physics to be in a state of evolution, but wondered how, if physics was continually changing, its truth content could be determined at any point in time? In other words, if the most recent theories were not necessarily the correct ones, in what sense could they be said to be better, or more correct, than those they had replaced? Furthermore, the history of science, as a record of scientific evolution to date, seemed to demonstrate that even the most fruitful theories of the past had later been discredited, so that one might well infer that all scientific theories were ultimately worthless. Poincaré summed up this natural scepticism of the 'superficial' observer as follows:

> The ephemeral nature of scientific theories takes by surprise the man of the world. Their brief period of prosperity ended, he sees them abandoned one after another; he sees ruins piled upon ruins; he predicts that the theories in fashion to-day will in a short time succumb in their turn, and he concludes that they are absolutely in vain. This is what he calls the *bankruptcy of science*.[34]

Naturally, he did not accept this conclusion, but having claimed that the actual history of physics undermined the common-sense concept of scientific knowledge, he needed to ensure that his own epistemology was supported by the historical record. Poincaré therefore tied his understanding of the way in which physics progressed to a particular reading of the historical development of physics, and this had an important consequence. If access to the true nature of scientific knowledge was only to be had via the proper interpretation of the historical record, then the history of science was likely to become as contentious an issue as that of the status of contemporary science itself. And since, as we have already seen, Larmor's concept of scientific progress was also closely linked to a particular reading of the history of science, it was almost inevitable that they would clash on this matter. In the following two examples I shall show how Larmor drew upon his concept of the evolutionary nature of scientific progress in attempting to refute Poincaré's arguments.

Poincaré's response to the apparently ephemeral nature of scientific theories was to shift the focus of attention away from theories themselves towards the interrelationships they had revealed *between* the phenomena of the natural world. The *value* of a theory could then be judged according to its ability to correlate observations and to predict new and experimentally verifiable relationships between the 'objects' of the natural world. The ultimate nature of the objects themselves would remain beyond our ken. Poincaré thus placed considerable emphasis upon the mathematical formalism by which these relationships were expressed, but stopped short of a fully instrumentalist vision of science by insisting that the mathematical relationships between the 'images' did represent real relationships between the objects of nature.

For Poincaré, the proper goal of physics was to reduce all the known relationships between the phenomena of nature to the smallest number of equations required to predict the known experimental results. Thus *experiment* was the final arbiter of scientific truth, while scientific theories, beyond their bare mathematical formalism, were merely heuristic devices which, however appealing to the mind, were of no intrinsic value. Poincaré's scepticism extended beyond the highest-level theories however, and he also believed that many of the most fundamental concepts adopted in physics were of a *conventional* rather than an *empirically testable* nature. Physicists might find it convenient to adopt such concepts as absolute simultaneity, Euclidean geometry, the laws of mechanics and the conservation of energy, but these were merely useful conventions which, though not quite as dispensable as high-level physical theories, could be replaced by other conventions if a simpler mathematical description of the experimental facts would thereby be achieved.

The tenth chapter of *Science and Hypothesis* was entitled 'The theories of modern physics', and, during a discussion of the possible utility of the concept of the æther, Poincaré suddenly made a digression in order to express his personal belief that the earth's motion through the æther would never be detected and that electromagnetic theories ought to incorporate this fact. He then attacked æther-based electromagnetic theory by presenting its development in the following way: when it had been shown that the earth's motion could not be measured to first order, Lorentz had demonstrated that this was to be expected, but when it was further shown that the motion could not be measured to the second order, a new and 'opportune' hypothesis had had to be invented (the contraction hypothesis) in order to explain the result. Poincaré thought that this left too important a role to chance and insisted that 'the same explanation must be found for the two cases, and everything tends to show that this explanation would serve equally well for the terms of the higher order, and that the mutual destruction of these terms will be rigorous and absolute'.[35] He believed that the concept of the æther, particularly in its role as the frame against which absolute motion might be measured, had outlived its usefulness. The experimental evidence, he argued, indicated that motion through the æther could not be measured to *any* order, and he therefore suggested that a 'principle of relativity' would be more fruitful in the long term.[36] Although these remarks were directed primarily at the work of Lorentz, Larmor assumed that they would be construed in Britain, and especially in Cambridge, as an attack upon his own electromagnetic theory.

In 1904, a year after he was appointed Lucasian professor, Larmor responded to these remarks in the original French edition of Poincaré's book. He defended the basic principles of the ETM and went on to observe:

> Such a train of remarks indicates that the nature of the hypotheses has been overlooked. And if indeed it could be proved that the optical effect is null up to the third order, that circumstance would not demolish the theory, but would rather point to some finer adjustments that it provides for; needless to say the attempt would indefinitely transcend existing experimental possibilities.
>
> As, then, the theory contains no further power of immediate adaptation, what are the hypotheses on which it rests, and how

> far are they gratuitous hypotheses introduced for this purpose alone?[37]

Larmor thought that experiments accurate to third order of v/c might well give positive results, but he was pointing out that, even if the results were null, one would still not be entitled to infer that motion through the æther was not measurable *in principle*. This clearly contrasts Larmor's concept of the gradual reclamation of the 'empirical fringe' surrounding established physics with Poincaré's concept of the purely 'relational' nature of scientific knowledge. It was precise experimental quantification which provided the parameters by which new physical phenomena were brought within theoretical mechanics, and to jump to unwarranted conclusions would simply undermine the whole evolutionary progress of physics. Larmor acknowledged Poincaré's claim that there were discrepancies between the predictions of current dynamical theory and measured experimental values but, he argued, this did not mean that dynamical theory was to be abandoned. It was fully to be expected that in an evolving theory 'want of absolute, exact adaptation [could] be detected'; and he added the warning, 'let us beware of counting limitation as imperfection, and drifting into an inadequate conception of the wonderful fabric of human knowledge'.[38]

Here Larmor was making two important points. First, that experiments represented the only way of telling whether a particular theory was correct, and since in the case of the earth's motion it was not possible to carry out experiments to third order, his theory of the electrodynamics of moving bodies could not for the moment be further adapted. Second, that we should not consider this a fatal imperfection of the method and begin introducing empirically unfounded assumptions that were inconsistent with the concept of the æther. By so doing we would lose the mechanical intelligibility of the whole theory and be parting company with the collective evolutionary wisdom which has brought physics to its present state of maturity.

In examining the hypotheses upon which his theory of the electrodynamics of moving bodies stood, Larmor argued that in order to explain first-order null results one had to assume only that electricity was atomic in nature. But, he argued, even the 'facts' known to Ampère and Faraday were sufficient to demonstrate that 'no other conception of electricity [was] logically self consistent'. And when it came to explaining second-order null results he naturally did not agree that one had to *assume* the contraction hypothesis, but rather that one had to *assume* that matter was 'constituted electrically', an assumption from which the contraction hypothesis followed automatically. For this assumption (the electrical construction of matter) Larmor admitted there was no independent evidence 'except perhaps the general simplicity of the correlations of physical law', and continued: 'This principle does not yet, so far as one can see, stand in the way of any other branch of physical science, while it accounts for the very remarkable absence of influence of the Earth's motion through space on the most sensitive phenomena, and is almost led up to thereby'.[39] Once again we notice that Larmor was using the null results to *underpin* his claim that the world was, ultimately, electromagnetic in nature; but now he had added the further important argument that the ETM could also be justified, and perhaps most effectively so, on the grounds of its simplicity.

My second example of Larmor's reaction to Poincaré's work is drawn specifically from his introduction to the English edition of *Science and Hypothesis*, the former probably written in January or February of 1905. This short essay took the form of a review of Poincaré's book, and while Larmor curtly described the translation as 'doubtless required', he stoutly defended the 'British school' throughout. Indeed, we can be virtually certain that he only agreed to write the introduction in order to preface the work with a statement of his own objections to Poincaré's epistemology and to discourage British readers from taking the book too seriously.[40] In his preface Larmor was prepared to grant that the work was interesting and ingenious, if in a somewhat mischievous way, but he continually urges the reader to treat the book as highly speculative and of little relevance to the practice of physics. I have, in fact, already used several quotations from this text above, but I shall now consider what is perhaps the most important theme in Larmor's argument.

In seeking to undermine or at least to counterbalance the main thrust of conventionalist epistemology, Larmor clearly thought that challenging Poincaré's reading of the history of physics would prove the most effective strategy. In his introduction he therefore reviewed a number of episodes from the recent history of science, repeatedly emphasizing the essential role played by 'mechanical and such-like theories – analogies if you will'. Finally, he summed up his argument as follows:

> The aspect of the subject which has here been dwelt on is that scientific progress, considered historically, is not a strictly logical process, and does not proceed by syllogisms. New ideas emerge dimly into intuition, come into consciousness from nobody knows where, and become the material on which the mind operates, forging them gradually into consistent doctrine, which can be welded on to existing domains of knowledge. But this process is never complete: a crude connection can always be pointed out by a logician as an indication of the imperfection of human constructions.[41]

Here again Larmor is appealing to his concept of the evolution of physics. Progress had not been attained by the application of logic or any strict scientific methodology to the natural world, but was, rather, a mysterious, transcendental process. The 'forging' of new ideas into 'consistent doctrine' – the reclamation of the empirical fringe – was an endless process, but one which relied upon theoretical mechanics and the 'mental image' of the æther for its possibility. I have argued, however, that Larmor went further than this; he believed that the evolution of physics was not a chance or random process but the unique path by which human beings had developed the faculty for comprehending the physical world as a dynamical unity – and that the history of science was an invaluable record of this natural history. Poincaré's conventionalism represented a powerful attack on this argument since it implied that there were many ways by which man's relational knowledge of the physical world *might* have been obtained and organized, and that the mechanical or dynamical world-view had simply happened to appear first.

Larmor could have countered this claim in a number of ways. We have already seen, for example, that he denied the distinction between 'experimental fact' and 'intellectual scaffolding'. He might therefore have argued that there was no such thing as a theory-independent empirical record which could be relationally represented within different conventional frameworks. In *Science and Hypothesis*, however, Larmor chose instead to argue that the natural history of physics was a unique record of the *actual development of physics*, and that as such it should be privileged above all ingeniously concocted stories of how science *might* have developed. Thus he wrote:

> In M. Poincaré's earlier chapters the reader can gain very pleasantly a vivid idea of the various and highly complicated ways of docketing our perceptions of the relations of external things, all equally valid, that were open for the human race to develop. Strange to say, they never tried any of them; and, satisfied with the very remarkable practical fitness of the scheme of geometry and dynamics that came naturally to hand, did not consciously trouble themselves about the possible existence of others until recently.[42]

Our very grasp upon the meaning of a physics unified by the concepts of the æther and theoretical mechanics was largely dependent upon our retaining a sense of the historical development by which it had matured. To deny the unique status of this natural history was potentially as dangerous as abandoning experiment as the unique guide to the fine adjustment of dynamical theory itself – Poincaré was guilty on both of these counts.

CONCLUSION

Larmor's debate with Poincaré casts new light on the former's commitment to the concept of the æther. Larmor's defence of the concept is generally associated with Einstein's famous claim in 1905 that the æther was 'superfluous' to physics; yet, as we have seen, Larmor mounted a general defence of the æther before the publication of the special theory of relativity. Unlike Einstein, Poincaré did not claim in *Science and Hypothesis* that the æther had to be *abandoned* as a concept, merely that motion through the æther should be assumed undetectable *in principle to all orders of magnitude*. Larmor thus felt it necessary to defend the place of the æther in physics even when its *existence* was not at issue. What Larmor understood Poincaré to be attacking was the ETM and the notion that a dynamical æther formed the *necessary* foundation of physics. For Larmor, the æther represented the ontological reality of the principles of theoretical mechanics, principles that were definitive of the unity of mathematical physics. For more than a century the notion of a dynamical æther had informed the evolution of theoretical mechanics and brought about the gradual unification of the diverse forces of nature. Furthermore, it was through the primitive sensations of force and motion, in which mechanics found its origin, that physicists could dimly divine the nature of the ultimate dynamical essence. In this sense the æther was the guarantor of the continued good conduct of the physicist. Abandoning the æther would for Larmor have been an immoral act since it would have deprived physics of physical intelligibility, and because it would have ended the evolutionary unification of the forces of nature.

In the light of these beliefs we can see that Larmor's rejection of the special theory of relativity was predictable, and that he was unlikely to have become reconciled to Einstein's approach with the passage of time. On the one occasion when Larmor responded directly to Einstein's claim that the introduction of a light-æther 'with special properties' was superfluous to the approach he was developing, Larmor underlined the phrase just quoted and remarked: 'Resting æther [is] the ideal coordinate system and necessary to thought' (Figure 46). He likewise believed that Einstein's general theory of relativity of 1915 – which attributed gravity to curvature in the four-dimensional space-time continuum – was overly mathematical and lacked the intuitive, dynamical foundation that was proper to a comprehensible theory of gravitation. Indeed, in 1939, seven years after retiring from his chair, he wrote that it seemed to him 'more and more strange that the Einstein scheme has survived so long – largely by ignorance of values'. By this time, however, Larmor was well aware that the scientific 'values' he had imbibed in Ireland and Cambridge in the 1870s had been displaced by those of a new and international modern physics. In the final years of his life he felt increasingly isolated, the last survivor of a race of mathematical physicists that was now almost extinct. In 1941, the year before his death, he confided despairingly to a friend that he had 'outlived [his] generation' and 'appeare[d] to be lingering in a world to which [he] did not belong'.[43]

Larmor's influence on the development of physics in Cambridge during the first two decades of the twentieth century went beyond his immediate contribution to electromagnetic theory. Although a somewhat disorganized lecturer who taught, and wrote, in a 'curiously involved style', as Lucasian professor he nevertheless generated a small group of research students in Cambridge who built upon his work. Most notable among this group were Ebenezer Cunningham and Harry Bateman, who combined Larmor's electrodynamics with some aspects of Einstein's early papers to produce a peculiarly Cambridge hybrid of the ETM and the special theory of relativity. Larmor took no active role in these developments but through them he helped indirectly to interest a younger generation of Cambridge-trained mathematicians in Einstein's work. Among these was the young Arthur Eddington, the man who would later be instrumental in bringing Einstein's general theory of relativity to Cambridge. Perhaps more surprisingly, Larmor also used his position as Lucasian professor to support experimental physics in Cambridge. Although no experimenter himself, he appreciated that the 'school' associated with the Cavendish laboratory was one of the 'essential assets' of the university. At the end of World War I, Cambridge sought to bring Ernest Rutherford to the Cavendish as director to build a new school in radioactivity and atomic physics. Larmor played a key role in these delicate negotiations by convincing the incumbent professor of experimental physics, J. J. Thomson, that there was room for two professors in the laboratory.[44]

With the benefit of hindsight it is clear that Larmor stands as an important transitional figure between the different worlds of physics represented by his predecessor and successor in the Lucasian chair; respectively, George Stokes and Paul Dirac. Stokes played a major role in developing the mathematical theories of hydrodynamics and the wave theory of light, work that helped to establish continuum mechanics in Cambridge as the foundation of future physical theory. It was firmly in this tradition that Maxwell made a dynamical æther the conceptual centerpiece of a new, unified theory of optics and electromagnetism. Larmor sought to explore the dynamical foundations of Maxwell's vision, and, in the process, took an electron-based, electromagnetic theory to its classical limits. Indeed, it was his work, like that of his better-known Dutch counterpart, Lorentz, that prepared the ground for Einstein's special theory of relativity. Dirac, by contrast, was one of the first of a new generation in Cambridge that was trained from the start in an Einsteinian approach to electrodynamics and gravitational theory. According to Dirac, the new physics was built on the insight that the fundamental laws of nature did not govern the world 'as it appears in our mental picture in any very direct way'. He believed that further progress lay 'in the direction of making our equations invariant under wider and still wider transformations'.[45] It was this essentially mathematical approach, lacking what Larmor regarded as an intuitive, mechanical foundation, that he so disliked about quantum mechanics and relativity theory. Yet in deriving and applying the so-called Lorentz transformations in the late 1890s, it was he who had helped to reveal the extraordinary power of mathematical transformations for tackling new problems in physics.

Notes

1 A. S. Eddington, 'Obituary of Joseph Larmor', *Obituary Notices of Fellows of the Royal Society*, 4 (1944), 197–207, 205.

2 *Ibid.* p. 198. Eddington's remarks are interesting in this respect as he was a younger contemporary of Larmor's and responsible for popularizing relativity theory in Britain.

3 *Ibid.* p. 206. Eddington recounts an occasion early in the twentieth century on which a paper of Larmor's, read to the British Association in the author's absence, reduced the audience to laughter.

4 W. B. Morton, 'Sir Joseph Larmor', *Proceedings of the Belfast Natural History and Philosophical Society*, 1942/43, pp. 82–90, 83. Larmor also had two sisters. His brother, Alexander, graduated eleventh wrangler in the Mathematical Tripos of 1889, and became a fellow of Clare College (Cambridge) and professor of natural philosophy at Magee College. Another brother, John, became an engineer and assisted Joseph with mathematical calculations.

5 *The Cambridge Review*, 4 February 1880, p. 5.

6 Almost all of the testimonials from Larmor's Cambridge teachers comment on his extraordinary problem-solving ability. The citations are from letters by Routh and Ernest Temperly. Larmor Box, St John's College Library, Cambridge.

7 Joseph Larmor, *Mathematical and Physical Papers* (2 vols, Cambridge, 1929), vol. 1, p. 31.

8 See, for example, Bruce J. Hunt, *The Maxwellians* (Ithaca, 1991).

9 For Larmor's electronic theory of matter see Andrew Warwick, 'On the role of the FitzGerald–Lorentz contraction hypothesis in the development of Joseph Larmor's electronic theory of matter', *Archive for History of Exact Sciences*, 43 (1991), 29–91.

10 The history of the Michelson–Morley experiment is discussed in L. S. Swenson, *The Ethereal Aether: A History of the Michelson–Morley Aether-Drift Experiment, 1880–1930* (Austin, 1972).

11 The contraction factor was $(1 - v^2/c^2)^{1/2}$, where v is the matter's velocity through the æther and c the velocity of light.

12 Warwick, 'On the role of the FitzGerald–Lorentz contraction', pp. 48–56.

13 *Ibid.* pp. 83–91. The term 'order' refers to the power of the ratio v/c, where v is the velocity of the electrical system through the æther and c is the velocity of light. Thus v/c is first order, $(v/c)^2$ second order,

and so on. Maxwell's equations are actually invariant to *all* orders under these transformations but Larmor did not notice this until around 1904.

14 Lorentz and Einstein introduced these transformations in 1904 and 1905 respectively. The transformations (1) were first referred to as the 'Lorentz transformations' by the French mathematician, Henri Poincaré, in 1905. A. I. Miller, *Albert Einstein's Special Theory of Relativity: Emergence (1905) and Early Interpretation (1905–1911)* (Massachusetts, 1981), p. 80.

15 Joseph Larmor, *Aether and Matter* (Cambridge, 1900). The transformations (1), (2) and (3) actually preserve the form of Maxwell's equations to all orders of v/c but Larmor did not notice this property until after the publication of *Aether and Matter*.

16 Quoted in Warwick, 'On the role of the FitzGerald–Lorentz contraction', pp. 29–91, 62.

17 *Ibid.* p. 78.

18 Bruce J. Hunt, 'How my model was right': G. F. FitzGerald and the reform of Maxwell's theory' in Robert Kargon and Peter Achinstein (eds) *Kelvin's Baltimore Lectures and Modern Theoretical Physics: Historical and Philosophical Perspectives* (Cambridge, MA, 1987), pp. 299–321.

19 For a discussion of FitzGerald's work on MacCullagh's æther and its subsequent development by Larmor see Bruce J. Hunt, *The Maxwellians* (Ithaca, 1991), chs 1 and 9.

20 J. Larmor, Presidential address before the Mathematics and Physical Sciences Section of the British Association for the Advancement of Science (Bradford, 1900), reprinted as 'The methods of mathematical physics' in Joseph Larmor, *Mathematical and Physical Papers* (Cambridge, 1929), vol. II, pp. 192–216, 200.

21 Larmor, *Aether and Matter*, p. 272.

22 Larmor, 'The methods of mathematical physics', p. 199.

23 Larmor, *Aether and Matter*, p. 272.

24 Larmor, 'The methods of mathematical physics', p. 215.

25 *Ibid.* p. 202.

26 *Ibid.* p. 213.

27 *Ibid.* p. 214.

28 Larmor, *Aether and Matter*, p. 70.

29 See Larmor, 'Introduction', in H. Poincaré, *Science and Hypothesis* (London, 1905), pp. xi–xx, xvii.

30 J. Larmor, 'The physical aspects of atomic theory', *Mathematical and Physical Papers* (Cambridge, 1929), vol. II, pp. 344–372, 362.

31 Larmor, *Aether and Matter*, p. 271.

32 *Ibid.* p. 272.

33 W. Thomson and P. G. Tait, *Treatise on Natural Philosophy* (Oxford, 1867); J. C. Maxwell, *Treatise on Electricity and Magnetism* (Oxford, 1873); Lord Rayleigh, *The Theory of Sound* (London, 1877).

34 H. Poincaré, *Science and Hypothesis* (London, 1905), p. 160.

35 *Ibid.* p. 172.

36 Poincaré was not advocating a 'relativity principle' in Einstein's sense. Poincaré entertained the principle because the earth's motion had not been measured. If it were measured the principle would have to be abandoned. For discussion of Poincaré's relativity principle, see J. Giedymin, *Science and Convention: Essays on Henri Poincaré's Philosophy of Science and the Conventionalist Tradition* (Oxford, 1982), ch. 5.

37 J. Larmor, 'On the ascertained absence of effects of motion through the æther', *Mathematical and Physical Papers*, vol. II, (1904), pp. 274–280, 277.

38 J. Larmor, 'Introduction', p. xx.

39 J. Larmor, 'On the ascertained absence of effects', pp. 277–8.

40 On 3 March 1905, Larmor wrote to Oliver Lodge stating that he had initially declined to comment on Poincaré's book but had then changed his mind because it had been 'a sort of Gospel for the people here who know no better for sometime'. See University College (London) Archive, Lodge Collection, Ms Add. 89/65 (v).

41 J. Larmor, 'Introduction', p. xvii.

42 *Ibid.* p. xix.

43 Morton, 'Sir Joseph Larmor', p. 89.

44 Lord Rayleigh [Robert John Strutt], *The Life of Sir J. J. Thomson* (Cambridge, 1942), pp. 216–17.

45 P. A. M. Dirac, *Principles of Quantum Mechanics* (Oxford, 1930), p. v.

9 Paul Dirac: the purest soul in an atomic age

Helge Kragh
History of Science Department, University of Aarhus, Aarhus, Denmark

Paul Dirac, Nobel laureate of 1933, is widely recognized not only as Britain's greatest theoretical scientist in the twentieth century but also as one of the world's most influential physicists. It can even be argued that modern physics is essentially rooted in Dirac's great discoveries of the 1920s and 1930s: 'Paul Dirac had a much bigger impact on modern science in the 20th century than Albert Einstein', wrote Antonino Zichichi, a leading contemporary particle physicist.[1] If this claim seems somewhat exaggerated, one should consider his remarkable contributions to science, from quantum mechanics to cosmology.

Throughout his distinguished career, Dirac occupied himself with fundamental problems of theoretical physics. His scientific work was always marked by a high degree of independence and originality. Although his conception of which problems were fundamental and hence worthwhile to address was not always shared by his colleagues, this mattered little to him. Dirac was self-contained and the inspiration he received was mainly his own. He never compromised on what he thought was right, even if this caused him to become isolated from mainstream developments, which was what happened in the postwar period of physics. Any idea of joining popular trends, or otherwise bending his ideas in order to adapt to majority views, was foreign to him.

Niels Bohr once remarked that 'of all physicists, Dirac has the purest soul'.[2] Dirac's intellectual purity was a great scientific

From Newton to Hawking: A History of Cambridge University's Lucasian Professors of Mathematics, ed. Kevin C. Knox and Richard Noakes. Published by Cambridge University Press.

and moral strength, but to some critics it was a social weakness. In 1978, the German physicist Walther Elsasser characterized Dirac's single-minded, almost fanatical occupation with fundamental physics. Dirac, Elsasser wrote,

> had succeeded in throwing everything he had into one dominant interest. He was a man, then, of towering magnitude in one field, but with little interest or competence left for other human activities ... In other words, he was the prototype of the superior mathematical mind; but while in others this had coexisted with a multitude of interests, in Dirac's case everything went into the performance of his great historical mission, the establishment of the new science, quantum mechanics, to which he probably contributed as much as any other man.

The characteristic feature of 'purity' in Dirac's mental constitution manifested itself in many ways: in his science, his lifestyle and his aloofness from students and Cambridge research. Indeed, it was the major reason why his influence on Cambridge physics was relatively modest. But 'purity' is not a well-defined term, and there were areas where Dirac's approach and ideals cannot be described as either pure or purist. For example, his use of mathematics in physics did not always satisfy mathematical purists. But, of its many connotations, 'rationality' is probably the one that best encapsulates the kind of purity associated with Dirac. Unlike Cambridge philosophers such as Ludwig Wittgenstein and G. E. Moore, Dirac believed it was possible to stand outside society, to think independently and to remain uncorrupted by social conventions. And though he probably agreed with his contemporary Cantabrigian, C. P. Snow, who famously proposed in the 1950s that the intellectual world had been divided irrevocably into 'two cultures', one literary and the other scientific, it is doubtful if he would have conceived himself to belong to either 'culture': for Dirac imagined that *any* culture would necessarily be tainted with irrational behaviour. Dirac's obsession with thinking and acting rationally was a feature noticed by his colleagues. While the eminent

Figure 47 Dirac *c.* 1933, a shy and solitary postgraduate whose emerging theories of quantum mechanics owed more to his meetings with Bohr, Heisenberg and other overseas physicists than to local Cambridge intellectuals.

Cambridge physicist Nevill Mott once described him as 'quite a total rationalist', Werner Heisenberg referred to the young Dirac as a 'fanatic of rationalism'.

In this chapter, I will first sketch Dirac's life and scientific career. I then discuss his role within the Cambridge environment, paying particular attention to his many years as occupant of the Lucasian chair. I will also discuss his political opinions, but shall insist that these were neither enduring nor important to his career; Dirac was no political activist. He was, however, committed to rationalism, and because of this commitment Dirac's views on religion and philosophy were unconventional and somewhat idiosyncratic. This holds in particular for his ideas about mathematical beauty, a concept that appealed strongly to him and may be taken as another connotation of the purity he sought in science as well as in life. Dirac's conception of beauty, purity and rationality, therefore, are perhaps the principal keys the historian can use to unlock the mystery of his life.

A LIFE IN PHYSICS

Throughout the eighteenth and much of the nineteenth century, most physicists regarded Isaac Newton's dynamical laws as sacred, but during the first decade of the twentieth century many physicists became convinced that phenomena of the microscale, and particularly events associated with radiation, could not be explained in terms of Newtonian dynamics.[3] They recognized that the principles of classical mechanics could not be applied to the behaviour of electrons and nuclei within atoms and molecules. A fundamental change – a change that affected not only the core of physics but the way that individuals viewed their relationship to nature – was afoot. Boldly embracing what to many seemed a heartless, indeterministic view of the world, the physicists developed 'quantum mechanics'.

Central to this new approach to physics was Max Planck's 1900 theory of black-body radiation and Niels Bohr's 1913 theory of atomic structure, which correctly predicted the frequency of hydrogen's spectral lines. But what most historians recognize as bona fide quantum mechanics did not appear until 1926 with the near simultaneous articulation of the matrix theory of Werner Heisenberg, Max Born and Pascual Jordan, the wave mechanics of Louis de Broglie and Erwin Schrödinger and the 1927 transformation theory of Dirac and Jordan.

Physicists did not regard these different formulations as alternative theories, but rather different aspects of a congruous assembly of natural laws. While these quantum theories had no pretence of describing exhaustively the processes within atoms, they helped scientists to understand better many phenomena, from the emission of alpha particles to photodisintegration. In addition, the field theory of quantum mechanics provided some comprehension of electrodynamic processes and the subatomic particles associated with nuclear phenomena.

The new quantum mechanics was sometimes referred to as *Knabenphysik* (boys' physics) because of the young age of the quantum pioneers. Indeed, the expression was coined by the Austrian Wolfgang Pauli, himself one of the quantum boys. Two years younger than Pauli, Paul Dirac was another of the quantum boys and the only founder of quantum mechanics who did not belong to the German-speaking culture of continental Europe. And yet Paul Adrien Maurice Dirac, born in Bristol on 8 August 1902, was not British by birth. His father was Swiss and it was not until 1919 that Paul acquired British nationality.

In 1923, after studying electrical engineering and applied mathematics at Bristol University, Dirac entered Cambridge University as a research student. At that time he knew little about atomic structure and almost nothing about quantum theory, a field that was predominantly cultivated in Germany and Scandinavia. It was therefore opportune that he was assigned Ralph Fowler as a supervisor. Fowler was a versatile theorist and one of the few British physicists who had mastered the quantum theory of atoms. Dirac quickly caught up with the new subject and within two years had himself become an expert in quantum theory. In the short period from 1925, when he developed his own version of quantum mechanics, to 1928, when he published his groundbreaking relativistic quantum theory of the electron, he became internationally recognized as a scientific genius of the first rank. As an indication of his status in the international physics community, he was invited to participate in the prestigious Solvay conference of 1927 which focused on the quantum theory of electrons and photons. Being only 25, he was the youngest of the carefully selected participants.

As a further indication of his growing scientific reputation, in March 1930 he was elected a fellow of the Royal Society, and that on the first occasion after being proposed. He was one of the youngest fellows in the history of Royal Society. The year before his election he had been appointed a university lecturer and also praelector in mathematical physics in St John's College, Cambridge, a post with nominal duties only. In 1928, two years after having completed his Ph.D. thesis, Dirac received his first offer of a chair, namely, to become professor of applied mathematics at Manchester University. ('Applied mathematics' was the nearest British equivalent to the 'theoretical physics' that originated in Germany in the late nineteenth century.) He was asked if he would consider succeeding Edward A. Milne, who had accepted an invitation to be the first Rouse Ball professor at Oxford University. However, Dirac declined the offer, principally because he wanted to be free to cultivate his scientific interests and to avoid the mire of university politics and administration. He was quite content with his position at Cambridge and over the following years he routinely declined offers of chairs of theoretical physics, both in England and abroad. Some of these offers, especially from American universities, included substantial salaries.

Probably one of the reasons why Dirac declined these tempting offers was that he knew that the prestigious Lucasian chair at Cambridge was to be vacant upon the retirement of Joseph Larmor. When Dirac was appointed Lucasian professor of mathematics to commence 30 September 1932, the chair was now occupied by a scientist who was to be regarded as Newton's equal. The chair was also occupied by a scientist who could be as unsociable as his predecessor. According to the *Sunday Dispatch* newspaper, he escaped to the zoo to avoid the many congratulations for his appointment! Dirac was evidently chosen for the position because of his outstanding scientific work, not because of his ability to 'market' either the Lucasian chair or physics at Cambridge University. The committee who elected Dirac must have known that public relations was not among his strengths. While it is debatable who was the better scientist and who was the least sociable, Dirac certainly outlasted Newton's tenure, remaining Lucasian professor for thirty-seven years. Remarkably, although Dirac was pleased to be appointed Lucasian professor, the position itself does not seem to have interested him very much. In his letters to physicists from 1932 to 1933 he did not even mention the new position. Also remarkably, neither did he comment about being Lucasian professor in his later writings and correspondence.

It is worth noting that Dirac's chair was nominally in mathematics and that he was considered an applied mathematician rather than a 'physicist', a term that in Cambridge referred principally to experimentalists. In 1932 there was no department of mathematics at Cambridge University and so students and teachers of applied mathematics or theoretical physics – i.e. 'mathematicians' – had nowhere to meet and work except in university and college libraries or in their own rooms. They did not belong to the well-knit community of 'real' Cambridge physicists – that is, the Cavendish men who worked under Rutherford's direction. Consequently, there was very little cooperation and professional interaction, both between the few theoretical physicists and between them and the Cavendish physicists. Nevill Mott, who became director of the Cavendish in 1954, well expressed the experiences of Cambridge mathematicians and theoretical physicists of Dirac's generation. He found it 'a terribly isolated business' to study theoretical physics at Cambridge and recalled how 'for students there was nowhere to sit, except in the rather small and squalid library'. Similarly, William McCrea, who later became an eminent astrophysicist, recalled that 'We seldom discussed our work with one another.' Dirac seemed to thrive in solitude, but students often felt isolated. McCrea was supervised by Fowler, but lamented that the physicist showed little interest in his studies: 'During the three years Fowler was my supervisor [1926–9], the time I had with him personally probably totalled less than one hour.'[4]

In most people's opinions, the culmination of Dirac's career took place in the year following his appointment to the Lucasian professorship: in 1933, he was awarded the Nobel prize, which he shared with fellow architect of quantum theory, Erwin Schrödinger. The award was given for his 'discovery of new fertile forms of the theory of atoms and for its applications', of which the 1928 theory of the electron was undoubtedly the contribution of greatest weight. When Dirac was notified about the honour, he initially wanted to refuse it owing to the publicity that would inevitably follow it. And Dirac did not like publicity. The London-based newspaper, *Sunday Dispatch,* portrayed the new Nobel laureate as 'shy as a gazelle and modest as a Victorian maid'. In spite of his mixed feelings about the Nobel prize, Dirac did go to Stockholm to receive the prize and to deliver the traditional Nobel lecture.

Dirac's early and most important works dealt with quantum mechanics.[5] Shortly after he became acquainted with Heisenberg's new theory of atomic structure, he developed his own, algebraic version of quantum mechanics which operated with what he called q-numbers and was therefore often known as q-number algebra. In one of his classic papers, the 1926 'On the theory of quantum mechanics', Dirac incorporated Schrödinger's new wave mechanics into his own theory and thereby created a more general and powerful formalism. In the same paper he also introduced the fundamental distinction between particles satisfying Fermi–Dirac quantum statistics (such as electrons) and Bose–Einstein statistics (such as photons). A few months later, while staying at Niels Bohr's institute in Copenhagen, he developed a still more general and abstract version of quantum mechanics which comprised Max Born's probabilistic interpretation within the more general framework of transformation functions. This so-called transformation theory became the recognized mathematical framework for quantum mechanics and the starting point for Dirac's 1927 theory of interactions between matter and electromagnetic fields. The 1927 theory served as the foundation of quantum electrodynamics and initiated a new field of research that would soon occupy centre stage in theoretical physics. Dirac was himself a main contributor to the early phase of quantum electrodynamics, but eventually he came to the conclusion that it was not a satisfactory theory.

The transformation theory was also an important background for what was probably Dirac's most important contribution to physics, his relativistic quantum theory of the electron that was published in *Proceedings of the Royal Society* in early 1928. By 'playing around with mathematics', as he later recalled, but also guided by very general principles of invariance, he found an equation that satisfied the basic requirements of both special relativity and quantum mechanics. The equation had the same formal structure as Schrödinger's wave equation ($H\psi = E\psi$), but operated with wave functions consisting of four components. Most significant, the Dirac equation gave the correct value of the electron's spin and also led to several other empirically correct predictions. Among the startling consequences of the Dirac equation was that it seemed to predict the existence of physically absurd quantities, namely, electrons with positive charge and negative energy. This much-discussed difficulty was brilliantly solved by Dirac in 1930–1 by an imaginative interpretation of the positively charged electrons as 'holes' in an unobservable 'sea' of negative energy. In his remarkable 1931 paper 'Quantised singularities in the electromagnetic field', he not only predicted the existence of positively charged antielectrons (later christened 'positrons') but also of magnetic monopoles – that is, isolated magnetic charges. Although such magnetic charges are not part of classical electrodynamics, Dirac showed that they could well exist within a quantum mechanical framework. Furthermore, he argued that the proton would have its own antiparticle, a negatively charged antiproton. A few years later he speculated that the symmetry between particles and antiparticles would probably imply the existence not only of antiatoms but also of entire antiworlds. In his 1933 Nobel lecture, he suggested the existence of antistars 'being built up mainly of positrons and negative electrons'.

After 1933 Dirac drifted away from mainstream physics and became increasingly heterodox in his views. He never came to peace with the direction quantum theory took after World War II and preferred to stay away from what he called the 'rat race' of modern mainstream physics, which seemed to be characterized by competition, funding and quick access to experimental data from high-energy accelerators and other apparatus. Such 'phenomenology' – in physicists' vocabulary meaning theoretical work depending on data from the laboratory – never interested Dirac. After 1950 nuclear and elementary particle physics witnessed a rapid growth and attracted scores of bright young physicists; but Dirac never showed any interest in the new elementary particles evidenced by these enterprises. While the new generation of competitive particle physicists tended to dismiss him as a cantankerous old man, gripping stubbornly to the past, Dirac dismissed particle hunting as an unappealing enterprise. Compared with the quantum theory of the electron and the relativistic theory of gravitation, it was a messy area. Nevertheless, he was deeply distressed by what he considered the grave mathematical and conceptual problems of standard quantum electrodynamics. In the decades after World War II he suggested several alternatives. Though none of these was accepted by the community of physicists, Dirac was determined to find a solution: 'I have spent many years looking for a good [theory of quantum electrodynamics]... and haven't yet found it', he remarked just two years before his death. None the less, he pledged 'to work on it as long as I can'.

Cosmology was another field to which Dirac contributed in a heterodox way, first in 1937 when he suggested a version of a Big Bang universe that was not governed by Einstein's general theory of relativity. His inspiration came from the no less heterodox theories of Milne and Arthur Eddington. Based on numerological considerations, Dirac suggested what he called the large-number hypothesis which claims that any two of the very large dimensionless numbers occurring in nature are connected by a simple mathematical relation. From this he deduced that the gravitational constant decreased over time and that the number of protons and neutrons in free space increased spontaneously. Because it contradicted both Einstein's theory of gravitation and the time-honoured law of the conservation of energy, Dirac's cosmological model departed radically from standard physics. The theory was not well received, but from about 1970 Dirac and a few others developed it in various ways. For instance, he wrote a large number of papers in which he tried to make the hypothesis of a varying gravitational constant agree with observations. Although experimenters failed to find the slightest support for Dirac's theory, he remained convinced of its essential correctness to his death.[6] It was, he believed, such a beautiful theory that it had to be true.

Although a considerable part of Dirac's postwar work dealt with somewhat unusual topics and approaches, such as cosmology, classical electron theory, magnetic monopoles and an attempt to reintroduce the electromagnetic æther, he remained a highly regarded and productive physicist. Among the lasting contributions of the later Dirac were his work on so-called constrained dynamical systems, his important reformulation of the general theory of relativity and his work on gravitational waves.

CAMBRIDGE CONTEXTS

In spite of Dirac's lifelong association with Cambridge University, he was too much of a loner to be truly a 'Cambridge man'. His Bristol education in engineering was unusual and he was one of the relatively few scientific research students who came to Cambridge without having taken the Mathematical or Natural Sciences Tripos. Considering that Dirac was at Cambridge University for forty-seven years, of which he spent thirty-seven years as a Lucasian professor, his local influence was remarkably small. There were several reasons for this, the most important of which was undoubtedly rooted in Dirac's peculiar personality. Not only was he taciturn, unsociable and reticent, he also avoided public appearances at all costs and tried hard (and successfully) not to be part of committees or take on administrative duties. He was totally uninterested in exerting power or influence of any kind. However, he made use of his right as a Nobel laureate to nominate scientists for the Nobel prize. Many laureates recognized the award as an important instrument of science policy and accordingly took advantage of it. Dirac was no exception and in 1946 and 1950 he nominated, although in vain, the Russian physicist Peter Kapitza for the prestigious honour.[7]

Content with working alone in his chosen fields of fundamental physics, Dirac was just the opposite of the great Cambridge school-builders such as J. J. Thomson and Rutherford. As his impact on Cambridge physics was limited, so his scientific work owed little to the Cambridge environment and his position at the university. Leopold Infeld, a young Polish visiting physicist who in vain sought to work with Dirac, later said that 'He [Dirac] is one of the very few scientists who could work even on a lonely island if he had a library and could perhaps even do without books and journals.' Nevill Mott, Dirac's colleague and himself a successful builder of a Cambridge school of physics (and earlier a Bristol school), shared these perceptions of Dirac's apparent aloofness from pedagogy. He remarked:

> I think I have to say his influence was not very great as a teacher... [H]e never was a man who would advise a student to examine the experimental evidence and see what it means. So his influence would be on the side of the older mathematical development at Cambridge, which results from our educational system. Dirac is a man who would never, between his great discoveries, do any sort of bread and butter problem. He would not be interested at all.

For someone who seemed so indifferent to his environment and professional relationships, it is perhaps surprising that Dirac spent much of his career working at institutions in foreign countries, such as Germany, Denmark, the Soviet Union, India and the USA. Accordingly, his closest colleagues were foreign rather than Cambridge-based physicists. It is noteworthy that his foundational theories of 1925–8 owed very little to Cambridge and much to his stays with Bohr in Copenhagen and with Heisenberg and others in Göttingen and Leipzig. He developed his early theories either abroad or, in splendid seclusion, in his college rooms in St John's. Dirac always thrived best by working alone. His great work of 1928 on the relativistic electron came as a complete surprise even to his Cambridge colleagues. 'All Dirac's discoveries just sort of fell on me and there they were,' Mott recalled. 'I never heard him talk about them, or he hadn't been in the place chatting about them. They just came out of the sky.' When Dirac developed his theory of antiparticles, and was forced to conclude that the antielectron cannot be a proton (as he had first thought), the inspiration and pressure came not from Cambridge colleagues, but from Heisenberg, Pauli, Hermann Weyl and Igor Tamm.

There is no doubt that Dirac benefited from listening to and discussing with foreign experts such as Heisenberg, Pauli, Bohr and Vladimir Fock. Yet even when abroad he largely kept to his Cambridge habits of working alone. He remained reticent but none the less was influenced by the intense intellectual environment in Copenhagen and elsewhere. Typically, Dirac later recalled the protracted conversations he had with Bohr in 1926 as 'long talks on which Bohr did practically all the talking'.

Much of the impact that Dirac had on world physics in general and on Cambridge physics in particular came through his famous textbook, *The Principles of Quantum Mechanics*, which was published by Oxford University Press in 1930 in a series of monographs under the general editorship of Fowler and Peter Kapitza. It was based on a lecture course that Dirac started giving during the Easter term of 1926 in a lecture room in St John's. It was attended by some of the most distinguished physicists of the twentieth century, including Mott, McCrea, Alan Wilson, Douglas Hartree, and J. Robert Oppenheimer. Women were most rare in Cambridge physics at the time, but Dirac's course could boast of one. She was Bertha Swirles, who later became a lecturer at Manchester University and, after her marriage in 1940 to the mathematician and geophysicist Harold Jeffreys, Lady Jeffreys.

Principles of Quantum Mechanics was one of the first textbooks on quantum mechanics and, at the same time, a highly original exposition that very much expressed Dirac's personal taste in physics and unique style. It became a tremendous success, and was used not only by generations of students but also studied by experienced physicists. The fifth edition of 1958 (revised in 1967) was reprinted in 1984 and is still in demand, which is highly unusual for a book of its kind. The Russian translation of 1932, *Printsipy Kvantovoi Mekhaniki*, sold 3000 copies in a few months. It was supplied with a preface from the Soviet Publishing House warning naive readers that Dirac's book 'contains many views and statements completely at variance with dialectical materialism', and also included an additional chapter on the application of quantum mechanics written by Dirac. In the second edition of 1937, the Publishing House similarly stated that 'P. Dirac ... makes some philosophical and methodological generalisations that contradict the only truly scientific method of cognition – dialectical materialism'.[8]

Dirac seems to have ignored this political interference, which was a common feature in Soviet scientific literature of the period. Thanks to the wide distribution of *Principles*, Dirac's interpretation of quantum mechanics was disseminated to a whole generation of physicists, who through it learned about the formal aspects of the Copenhagen school's views of the measurement process and the nature of quantum mechanical uncertainty. However, one significant omission from Dirac's book was the Copenhagen term 'complementarity' which refers to Bohr's notion that there exists complementarity between the wave and particle models of radiation and that a given measurement indicating the wave nature of radiation cannot simultaneously evidence its particle nature. He did not like Bohr's complementarity principle because, as he said in 1963, 'it doesn't provide you with any equations which you didn't have before'.

Regarded as a textbook, *Principles* was and still is remarkably abstract and not very helpful to the reader wanting to obtain physical insight into quantum mechanics. Based on what Dirac called 'the symbolic method', the book eschewed experiments and physical interpretations and emphasized the power of abstract mathematical methods and general concepts. The symbolic method, Dirac argued in the preface, 'deals directly in an abstract way with the quantities of fundamental importance ... [and] seems to go more deeply into the nature of things'. Because 'it enables one to express the physical laws in a neat and concise way ... I have chosen the symbolic method, introducing the representatives later merely as an aid to practical calculation'. He admitted that 'this has necessitated a complete break from the historical line of development', but considered the break an advantage because it made possible a more direct and general approach to the problems of quantum mechanics.

Not only was *Principles* strictly ahistorical, it also contained no illustrations, no index and only very few references. At least from a modern point of view, *Principles* was simply unpedagogical. It reflected Dirac's aristocratic sense of physics – what Born once referred to as his *l'art pour l'art* attitude – and his total neglect of usual textbook pedagogy. Even experts in quantum mechanics could find it a hard read. 'A terrible book – you can't tear it apart!' is how the Dutch physicist Paul Ehrenfest is said to have reacted.

Other physicists found the connectedness and the coherence of the arguments to be among the *Principles'* chief qualities. In his 1931 review of the book, Oppenheimer stated succinctly that the book 'is clear with a clarity dangerous for a beginner, deductive, and in its foundation abstract; ... The book remains a difficult book, and one suited only to those who come to it with some familiarity with the theory'. Yet Dirac's *Principles* is generally accepted as the most influential textbook in quantum mechanics. Although it did not provide the reader with calculational tools for practical problems, it included in a clear and abstract way the foundational features of quantum mechanics.

Arthur Eddington, the Plumian professor of astronomy in Cambridge and one of the best-known physicists in early-twentieth-century Britain (not least because of his popular books and ambitious attempt to establish a philosophical world-view on the basis of the new physics and astronomy), was fascinated by Dirac's approach to physics in general and his wave equation in particular. He praised Dirac's symbolic version of quantum mechanics as 'highly transcendental, almost mystical' and found that it agreed well with his own view of the universe as a dematerialized world of shadows dressed up in mathematical symbols.[9] While Dirac was sceptical of Eddington's grand project of exploiting quantum mechanics in cosmology, he was none the less influenced by Eddington's view of physical science. Indeed, this influence is clearly discernible in the 1930 preface to *Principles* and also in Dirac's later cosmological theory.

Besides Eddington, others – namely, students – at Cambridge might have been deeply influenced by Dirac. Dirac's Cambridge lecture course in quantum mechanics followed the material in *Principles* and changed little over the course of three decades. However, given his lecturing style, this influence is certainly not nearly as great as it could have been. He delivered his lectures tersely and with little engagement with his audiences. If the students did not understand him, they could not expect much help. Asking Dirac to 'reiterate' a point was usually not helpful. In such cases Dirac would repeat exactly what he had said before, using the very same words![10]

Dirac had only a few Ph.D. students and was in general uninterested in acting as a supervisor. He seemed to lack genuine interest in his students, whom he expected to work largely on their own. When young Victor Weisskopf spent a few months in Cambridge in 1933, mainly to work with Dirac, he was unimpressed with the teaching skills of the eminent physicist. 'Dirac is a very great man, but he is absolutely unusable for any student.' 'You can't talk to him, or, if you talk to him, he just listens and says, "yes".' Other students who initially wanted to study under the great Dirac were frightened by the unapproachable character and decided to seek another, more sympathetic supervisor. Those who decided to stick it out with Dirac were often disappointed with what one of his former students described as his 'sink or swim' attitude. Subrahmaniyan Shanmugadhasan recalled that Dirac did not give any guidance on relevant literature and was unwilling to offer his opinion of anyone's work. He did not even always read the papers that he had communicated to journals.[11] None the less, and somewhat strangely, Shanmugadhasan believed that 'Dirac was the best kind of supervisor one could have'. Dennis Sciama, who later became a well-known astrophysicist, had Dirac as his supervisor around 1950 and his experiences testify to the almost comical nature of Dirac's lack of interest in students. On one occasion Sciama went enthusiastically to Dirac's office exclaiming, 'Professor Dirac, I've just thought of a way of relating the formation of stars to cosmological questions, shall I tell you about it?' Dirac's terse answer was 'no' and there the conversation finished.

During the 1930s, Dirac was very reluctant to supervise research students. Among the few were Paul Weiss, Andrew Lees, H. R. Hulme and Fred Hoyle, who later became famous for his work in cosmology. After the end of the war, especially between 1945 and 1952, he accepted several more students, although many of them only for a short period. Apart from Sciama and Shanmugadhasan, they included Christie Eliezer, Richard Eden, Harish-Chandra, R. J. N. Philips and H. J. D. Cole; there was even a woman among the students, the Brazilian Sonya Ashauer. Altogether, Dirac had about fifteen research students for longer or shorter periods of time. It is probably correct to say that, although Dirac's physics was a model of inspiration, Dirac was not himself a role model for any of his students.

Cambridge physics was definitely a 'boys' club', but during Dirac's tenure some dramatic changes in the place of women occurred – for example, in 1948 women were entitled to full degrees. Dirac only lectured to very few women and remained entirely aloof from the debates about the status of women at Cambridge. In his younger days Dirac had, like Newton, a reputation of timidity with regard to the opposite sex. His parents wanted to protect him from girls and the result was that he had, for a long time, a Platonic conception of women. 'I never saw a woman naked, either in childhood or youth', he confided in 1973 to Esther Salaman, an author and good friend of Dirac. 'The first time I saw a woman naked was in 1927, when I went to Russia with Peter Kapitza. I was taken to a girls' swimming-pool, and they bathed without swimming suits.'[12] Yet 'the genius who fears all women', as the *Sunday Dispatch* had labelled him, was (unlike Newton) not the inveterate bachelor that his friends had become used to regard him. In early 1937 he married Margit Wigner Balasz, a divorced Hungarian and the sister of the physicist Eugene Wigner. Margit and Paul had two daughters.

During most of the years between 1925 and 1937 Dirac lived in his rooms in St John's College. Here, the ascetic physicist gained a reputation for his indifference to cold and discomfort. Because of his asceticism and integrity, Mott likened him to Ghandi. 'He [Dirac] is quite incapable of pretending to think anything that he did not really think', Mott reported in a letter of 1931. He added that 'In the age of Galileo he would have been a very contented martyr.' Other visitors often compared him to a monk living in his cell. For example, the author and science journalist James Crowther reminisced that 'When I first called on Dirac he was living in a simply furnished attic in St John's College. He had a wooden desk of the kind which is used in schools. He was seated at this, apparently writing the great work straight off.'

Only marriage and a plea from the college bursar gave Dirac the impetus to leave this stoic environment. He and his wife moved to a house on 7 Cavendish Road and consequently spent less time at the university. However, he had a room in the Arts School, which he used mainly for talking with students and others. Owing to his reticence and infrequent visits to the university, few Cambridge students and physicists knew him well, if at all. Stephen Hawking, the famous astrophysicist who became Lucasian professor in 1980, was a graduate student of Sciama, Dirac's former student. Hawking and Dirac were both members of the Department of Applied Mathematics and Theoretical Physics (DAMTP) and yet Hawking hardly ever saw Dirac in the building. 'Dirac', said Hawking,

> belonged to the old school who didn't believe in these new-fangled departments of pure and applied mathematics, but worked in their college rooms. And I was working on classical general relativity and not quantum theory at that time, so I didn't go to his lectures.[13]

In fact, when DAMTP was formed in 1959, Dirac did not accept the offer of a room in the new building. He preferred to work at home.

APOLITICAL PHYSICIST?

Dirac presented himself as apolitical and simply without interest in or knowledge about political life. If he had such interests he kept them for himself. Political activism, he seems to have thought, was irrational and would only interfere with his scientific work. For example, in May 1926 a period of political and social unrest in England culminated with the declaration of a general strike, and the strike was met by the government and several antistrike volunteers, including many of Dirac's fellow students at Cambridge University. But Dirac, who at the time was completing his Ph.D. thesis, did not want to have anything to do with the strike, either in a supporting or challenging role. 'I concentrated all my energy in trying to get a better understanding of the problems facing physicists', he recalled about his Cambridge student days. 'I was not interested at all in politics, like most students nowadays. I confined myself entirely to the scientific work, and continued at it pretty well day after day.'

The only kind of political interest that Dirac showed was related to his long and close relationship with Russian physicists and his many travels to the Soviet Union. From the late 1920s, when Dirac had firmly established his scientific reputation, he had close associations with Soviet theoretical physicists, including Igor Tamm, Yakov Frenkel, Lev Landau, George Gamow and Vladimir Fock.[14] In 1928 he was invited to the sixth All-Union Conference on Physics and three years later he was elected a corresponding member of the Academy of Science of the USSR, a rare honour for a 27-year-old scientist. Tamm stayed in Cambridge in 1931 and at one occasion Dirac invited him to his college rooms to give a talk about Russia, a subject which evidently interested him. On another occasion Tamm was invited to lecture in London on 'higher education in Soviet Russia'. 'I have a feeling that Dirac is not happy with my talks', Tamm wrote to his wife in Moscow. 'He expressed a hope that the talk would be on education rather than on politics.'[15]

In his famous critique of the two cultures of western society, the Cambridge physicist and novelist C. P. Snow remarked: 'I think that it is only fair to say that most pure scientists have themselves been devastatingly ignorant of productive industry, and many still are.' Although his soul may have been pure, Dirac was not one of these pure-but-ignorant scientists about whom Snow complained. During the 1930s Dirac visited the Soviet Union several times and developed a fascination with the country. He was impressed by the socialist experiment and took great interest in the improvements in industry, living standards and availability of consumer goods. In the summer of 1933 he wrote a letter to Bohr, asking him to join a conference on nuclear physics in Leningrad: 'I think you will find it interesting to see something of the modern Russia', he wrote, adding that 'The economic situation there is completely different from everywhere else.' Two years later, while preparing a new journey to the Soviet Union, he wrote to Tamm that he would like to see a Soviet factory and the Dneproges, the huge hydroelectric power station that served as propaganda for Soviet industrialization in the 1930s.

Dirac's contact with Soviet Russia included cooperation with Russian physicists and, in 1932, co-authorship with Fock and Boris Podolsky of an important paper on quantum electrodynamics which was published in *Physikalische Zeitschrift der Sowietunion*. Of Dirac's almost 200 publications, only four were collaborative papers and, of these, two were with Russian co-authors. In 1937 Dirac wrote a paper in the *Bulletin of the USSR Academy of Science* specially for the commemoration of the October Revolution on its twentieth birthday. Although it could be taken as a kind of political manifestation, its subject – an analysis of the concept of time reversal in quantum mechanics – was limited to pure physics.

From about 1934 the political climate in the Soviet Union hardened; xenophobia, political suppression and arbitrary arrests soon became the order of the day. Foreign scientists were increasingly viewed with suspicion; some were even suspected of espionage. Dirac knew about this new political climate, for in 1935 he got involved in Rutherford's attempt to help Peter Kapitza, the Russian–British physicist who in 1934 was prevented by the Soviet authorities from returning from Moscow to Cambridge. However, despite many warnings, Dirac spent parts of the years 1936 and 1937 in the Soviet Union, hiking in the Caucasus and visiting Kapitza and Tamm in Moscow. When in 1938 he wanted to visit the country once again, his application for a visa was refused. This seems to have had nothing to do with Dirac personally, but was the result of a general decision of the Soviet authorities to stop awarding visas to British scientists – a retaliatory move which took place after the British Embassy in Moscow stopped issuing visas to Russians.

It is difficult to say to what extent Dirac's interest in the Soviet Union went beyond physics, personal friendships and hiking in the mountains. But there is some documentary indication that it was not limited to physics and hiking. He seems to have been associated with, or a member of, the Marxist-dominated Society for Cultural Relations with the USSR, a branch of which was organized in Cambridge in 1936. Thus, at the end of 1941, Vladimir Semenovich Kemenov, a Russian cultural civil servant and chairman of the board of the All-Union Society for Cultural Relations with Foreign Countries, sent a telegram to Dirac which included the line '[The] Society for Cultural Relations sends you hearty new year greetings' and which expressed the Russian confidence that 'all friends [of] world democracy' would soon witness the victory over Hitler's Germany.

Whereas it seems likely that Dirac was for a period sympathetic to the Soviet cause, it is much less likely that his interest implied any direct political or ideological involvement. While some other physicists formulated profound links between quantum mechanics and the moral world, Dirac resisted making connections between physics and politics, and never publicly promoted any political systems. That Dirac should have had 'violently colored political views,' as two American authors have stated, is a claim that lacks documentary support.[16] It is well known that in 1930s Britain there was a group of 'red scientists' who urged the need for improved relations and science planning in connection with communist Russia. The group included John D. Bernal, John B. S. Haldane and Lancelot Hogben but not, to my knowledge, Dirac. Whatever political sympathies he may have had, he always refused to become openly involved in political matters or otherwise make himself a public figure. He was outside the group of Marxist scientists of the 1930s and also showed no interest in the Association of Scientific Workers, a left-wing organization that after 1945 came under increasing influence of the Communist Party.[17]

Socialism, anarchism, liberalism and conservatism were empty labels to Dirac, and his interest in the Soviet system, although genuine, seems not to have been rooted in ideology but rather in what he considered to be socioeconomic logic. From what he considered a rational point of view, Dirac found Soviet methods for organizing society interesting. This view was shared by many other western scientists, whether Marxists or not – after all, Soviet communism was officially labelled 'scientific socialism'. Although Dirac may have had some left-wing sympathies, there is no indication that he was ever a communist or member of any political party. None the less, for a period he seems to have tacitly endorsed, and been impressed by, the communist system.

An example of Dirac's typical unwillingness to be involved in matters outside physics is provided by the debate on 'Modern Aristotelianism' that raged in *Nature* in 1937. It was the result of the astrophysicist and philosopher Herbert Dingle's sharply worded criticism of what he considered the unbalanced a priori methods of the cosmological theories of Eddington, Milne and Dirac. Contrasting Milne and Dirac with Newton, Dingle accused his contemporaries of inventing 'a pseudo-science of invertebrate cosmythology' that was dogmatically based on arbitrary principles rather than on inductions from observations. Typically, whereas Milne and Eddington received the challenge in good form and answered Dingle's attack at great length, Dirac avoided philosophical debate and used most of his brief reply to restate the main points of his controversial theory. Dirac was never much of a debater.

At the heart of the 'Modern Aristotelianism' dispute was not only the proper methods of science, but also, according to Dingle, the question of science policy and cultural standards. Dingle maintained that 'my concern is with the general intellectual miasma that threatens to envelope the world of science' and hinted that 'There is evidence enough on the Continent of the effects of doctrines derived "rationally without recourse to experience".'[18] This was probably an allusion to the non- or antiscientific atmosphere that at the time appeared to be gaining terrain with the dictatorships of Hitler and Stalin. Dingle seems to have implied that the views of Milne and, to a lesser degree, Dirac had greater affinity with dogmatic and authoritarian thought systems than with the critical science characteristic of free and democratic countries. We do not know what Dirac thought of Dingle's accusation. He did not reply. Interestingly, the Marxist biologist Haldane entered the debate on the side of Milne and Dirac. He saw their views as expressions of the fundamental historical dialectics of nature that one would expect according to the world-view of Marx, Engels and Lenin.[19]

Dirac's difficulties in staying outside the world of politics is further illustrated by the years around World War II. Several Cambridge physicists, including Rutherford and Patrick M. S. Blackett, were key figures in the Society for Protection of Science and Learning (and in its predecessor, the Academic Assistance Council) which helped scientist refugees from the Third Reich. Although Dirac occasionally helped people who asked for his assistance, he was not active in the Society and never participated in the political debate concerning science and National Socialism. However, when Britain declared war on Germany in 1939 and British scientists were recruited to fight the conflict, he could not completely withstand the pressure to become engaged in matters outside pure physics. For example, in early 1940 he agreed to accept the presidency of a British committee which was to be part of a British–French organization for scientific cooperation but, owing to the early fall of France, the organization never got started.

Although never a member of the so-called Maud committee, which was established in 1940 to investigate the possibility of constructing an atomic bomb, Dirac did valuable work on a consultancy basis as part of the British bomb project. In 1942–3 he investigated isotope separation by means of centrifuges and also calculated neutron multiplication in various shapes of nonspherical lumps of uranium. This work was relatively important and highly classified. However, although known to the Americans it was not used directly in the Manhattan Project that led to the first atomic bombs in the summer of 1945. The American–British group of Los Alamos physicists wanted Dirac to join them, but he refused. Instead, he spent the rest of the war years in Cambridge and Dublin where he was occupied with attempts to reformulate quantum electrodynamics, a line of pure research that had no connection to military applications.

On the whole, Dirac's involvement in war-related physics was rather sporadic. It did not spring from a desire to contribute to the war efforts, which was the case for many other physicists, including his successor as Lucasian professor, James Lighthill. Dirac enjoyed working with theoretical problems of uranium physics and found these scientifically interesting, irrespective of their possible relation to the war.[20] Neither should his decision not to participate in the Manhattan Project be interpreted as a conscious opposition to using physics for military purposes. This was a question that simply seems not to have occupied Dirac. Nor does he seem to have cared much about the scientists' ethical responsibility and the problems raised by the new nuclear weapons. These topics were widely discussed among British scientists, notably the members of the Atomic Scientists' Association, an organization formed in 1946 with Mott as its first president. Dirac, however, was not even a member.

If Dirac fashioned himself as apolitical, others had their suspicions. After the war, Dirac's contacts with Soviet science caused him trouble when he wanted to go to Princeton to spend a sabbatical year at the Institute for Advanced Study. In 1954, when his application for a visa was denied, it provoked a set of rare statements to the press and caused American physicists to protest vehemently. This was the period of McCarthyism when fear of atomic espionage and political disloyalty made life hard for many scientists living in the USA or those who wanted to visit the country. Many American physicists were brought before the House of Representatives' Committee for Un-American Activities and accused of security violations or having (or having had) affiliations with the Communist Party. Among the better-known victims of the witch-hunt were David Bohm, Frank Oppenheimer, Joseph Weinberg and Bernard Peters. Although these scientists were no longer involved in politics, they were either dismissed from their university positions or their stay in the USA was discontinued – Bohm, for example, had to leave for Brazil, while Peters, a German refugee, went to India and subsequently to Denmark. The 'Dirac case' was neither dramatic nor exceptional. Dirac was only one among numerous foreign scientists who were denied visas under the McCarran Immigration and Nationality Act of 1952. British scientists who were either denied a visa or granted one only after long delay included not only Dirac, but also Michael Polanyi, Rudolf Peierls, Mark Oliphant, Paul Erdös and E. A. Guggenheim, none of whom had any sympathy for the communist cause.[21]

Dirac was an élitist rationalist and an ivory-tower scientist for whom the so-called larger aspects of life – religion, art, literature, philosophy and politics – held no appeal and with which he did not want to engage. To a very limited extent, he occasionally expressed some interest in these matters, but it was in his own peculiar way, tending to see such matters from a scientific and logical point of view. Any other point of view was outside his perspective and the result was a basic lack of appreciation of social, aesthetic and religious values. Dirac was not a religious man and had little understanding of the cause of religion. Although his wife, Margit Dirac, denied that he was an atheist, in his younger days he did seem to have favoured an atheistic or perhaps agnostic view.[22] During the 1927 Solvay Congress, he became involved in an informal discussion of religion with some of the other young physicists. According to Heisenberg's recollections, he rejected any religious idea and held religion to be based on irrationalism and foolish postulates. It should have no appeal to the man of science. Religion, Dirac maintained – again, according to Heisenberg – was nothing but a collection of myths, an opium for the people. However, Dirac never publicly expressed such sentiments and neither did he, until much later, comment on religion and its relationship with science.

Cosmology is one area of science that almost naturally invites religious comment. During the 1930s, cosmology and creation were discussed by both theologians and scientists, including James Jeans, Robert Millikan, Arthur Eddington, E. A. Milne, Ernest W. Barnes and Georges Lemaître. The expanding universe and Lemaître's Big Bang theory were sometimes interpreted as support of the Christian view of divine creation. As early as 1929, Jeans described the cosmic creation as 'the finger of God agitating the æther', which implied that any explanation of the origin of the world would ultimately lie outside the realm of science. In none of his cosmological works from either before or after World War II did Dirac comment on religious or other extrascientific questions. He also remained silent in the debate in the 1950s concerning the religious aspects of the controversy between the Big Bang and the steady-state models of the universe.

Dirac was well acquainted with Lemaître, the Belgian priest and pioneer of the Big Bang theory, with whom he occasionally discussed the grander aspects of cosmology. 'I told him I thought cosmology was the branch of science that lies closest to religion', Dirac wrote in a 1968 memorial article on Lemaître. At that time he was a member of the Pontifical Academy of Science in Rome, an institution of which Lemaître was president from 1960. However, membership in the Pontifical Academy does not indicate any particular faith or interest in Roman Catholicism, but is solely an indication of scientific excellence. Yet late in life Dirac began to express some interest in religion and conceived the subject in his own rationalistic and somewhat naive way. In an address of 1971 he discussed the old question of the existence of God, not from a human or religious perspective but from what he considered a strictly scientific perspective. He explicitly ruled out views based on faith or philosophical principles, which were said to be 'really just sort of guessing or expressing one's feelings'. According to Dirac, the existence of God could be scientifically justified only if the emergence of life required a highly improbable event to have taken place in the past:

> If physical laws are such that to start off life involves an excessively small chance, so that it will not be reasonable to suppose life would have started just by blind chance, then there must be a God, and such a God would probably be showing his influence in the quantum jumps which are taking place later on. On the other hand, if life can start very easily and does not need any divine influence, then I will say that there is no God.

This was the closest Dirac came to religion in a public address – and it may be significant that the address was never published. It must be concluded that Dirac tried to banish political, cultural and religious views from his scientific work. Whereas other Cambridge professors used their positions to further political or religious causes, Dirac used his status as a Nobel laureate and Lucasian professor as a means of self-protection, to avoid the involvements that might disturb his chosen lifestyle that was so completely organized around problems of fundamental physics.

THE METAPHYSICS OF SCIENCE AND MATHEMATICS

Much of Dirac's success in physics rested on his style of mathematics and the peculiar way in which he used mathematical reasoning. As a student, he relied on a few standard texts, in particular Henry Baker's *Principles of Geometry* and Whittaker's *Analytical Dynamics of Particles and Rigid Bodies*. When he encountered Heisenberg's quantum mechanics, he initially found its mathematical methods difficult and complicated, but he quickly learned to translate them into the more algebraic formulation that would become a hallmark for Dirac's version of quantum mechanics. In 1926, when he studied Schrödinger's new wave mechanics, he was again faced with mathematical methods (such as the theory of eigenfunctions and eigenvalues) which were not part of his education and which he therefore had to learn. Later in his career he needed to educate himself in mathematical fields with which he was not well acquainted from Bristol and Cambridge. He preferred mathematical ideas that were simple and directly useful for the solution of physical problems. In spite of his mastery of mathematics, he was only interested in pure mathematics if it could be applied to physics. He did not believe that exact equations, rigorous proofs and axiomatic structures should be of prime concern in theoretical physics. What mattered was good mathematical intuition and convenient tools.

Dirac's relaxed attitude toward mathematical rigour was probably a result of his early training in engineering, when he had learnt to appreciate the value of theories based on approximations and perhaps tenuous mathematical theories. Engineering problems in, for example, electrical technology cannot be solved by means of exact solutions to the equations of physics. Although exact equations may be used, solutions can only be obtained by approximation methods adapted to the problem under consideration. In contrast to many other theoretical physicists, Dirac not only tolerated such approximations, but also considered them to be endowed with their own aesthetic qualities. In 1933, having attended the physicist's course on quantum mechanics in Cambridge, the American mathematician Garrett Birkhoff testified to this feature of Dirac's work. He was surprised to see the 'mathematical liberties' that Dirac permitted and added that 'He impresses me as being at least comparatively deficient in appreciation of quantitative principles, logical consistency and completeness, and possibilities of systematic exposition and extension of a central theory'. Dirac's derivation of the relativistic wave equation was brilliant and impressed the quantum physicists but from a mathematical point of view his arguments were far from convincing. With a more conventional grounding in differential equations, he probably would not have followed the approach he did, but then perhaps he would not have discovered the equation. Pauli criticized Dirac's derivation of the relativistic equation from the somewhat different point of view of quantum field theory. 'The success seems to have been on the side of Dirac rather than on logic', he said in 1935. 'His theory consisted in a number of logical jumps.' But it was precisely such unconventional reasoning, based on his highly developed intuition rather than on logic, that was the secret of some of Dirac's greatest discoveries.

Characteristically, when Dirac introduced the so-called δ-function in 1927, he did so in a manner that must have shocked many a pure mathematician. The δ-function was useful, but not mathematically well defined. Likewise, Dirac regarded the 'Dirac matrices' of his 1928 theory of the electron to be convenient quantities that worked in practice and left the task of establishing the mathematical properties of the matrices to the pure mathematicians. The result of this work was a new mathematical subdiscipline – spinor analysis and the δ-function being given a respectable mathematical foundation in the distribution theory developed in the 1940s. Dirac's arguably considerable impact on pure mathematics is further illustrated by the fact that the differential operator which he introduced in a loose way in 1928 has continued to be of importance for mathematicians.

Although in the preface to *Principles of Quantum Mechanics* Dirac stressed that 'mathematics is only a tool', there was another side of his attitude to mathematics. From about 1935 he began to develop a strongly aesthetic view of mathematics which insisted that

Figure 48 Pencil drawing of Dirac in 1963, by R. Tollast. Despite a lifelong agnosticism, he held that, since truth and beauty were the same in mathematical theories of the cosmos, then God had to be 'a mathematician of a very high order'.

the fundamental laws of nature were characterized by a high degree of mathematical beauty. The doctrine of mathematical beauty was not Dirac's invention but he arrived at it independently, primarily as a rationalization of his own work in quantum physics. He believed there was a deep harmony between beautiful mathematical equations and the way the physical world operated. 'As time goes on', he wrote in 1939, 'it becomes increasingly evident that the rules which the mathematician finds interesting are the same as those which Nature has chosen.' Again, in 1963 he described God as 'a mathematician of a very high order [who] used very advanced mathematics in constructing the universe'. That there is a preestablished harmony between physics and mathematics is an old idea, but Dirac emphasized that the search for beautiful mathematics was the way to obtain knowledge of the fundamental laws of nature. This was the main theme of the James Scott lecture he delivered in Edinburgh in 1939, and a philosophy that increasingly occupied him in later years. Unfortunately, he did not define his concept of mathematical beauty in a satisfactory way, nor did he think this was necessary. 'Mathematical beauty', he wrote, 'is a quality which cannot be defined, any more than beauty in art can be defined, but which people who study mathematics usually have no difficulty in appreciating.'

The identification of beauty with truth led Dirac to a one-sided emphasis on the mathematical–aesthetic method at the expense of the empirical–inductive method. As far as fundamental physics was concerned, he wanted to subordinate experimental tests to the admittedly vague idea of mathematical beauty.[23] Extraordinarily, he insisted that if a mathematically beautiful theory was contradicted by experiments, it meant that the experiments were wrong. That is, mathematical beauty should at certain occasions take priority over agreement with experiment and in this way act as a criterion of truth. In contrast, a mathematically ugly theory could not possibly be correct. It could work well and agree excellently with experiments, but Dirac denied that such a theory could be fundamentally true. His preoccupation with the doctrine of mathematical beauty was primarily a result of both his lifelong search for a correct theory of quantum electrodynamics and his dissatisfaction with the predominant theory that had first appeared in 1947. He found the theory of quantum electrodynamics proposed by Julian Schwinger, Richard Feynman and others to be 'complicated and ugly' because it was based on questionable mathematical procedures and was really only a set of working rules. That these rules resulted in predictions in excellent agreement with high-precision experiments did not impress him the slightest. As quantum electrodynamics was his paradigm of a mathematically unsound theory, so he often emphasized Einstein's general theory of relativity as the quintessence of beauty. He was deeply impressed that Einstein did not try to account for observational results, but developed the theory in accordance with internal criteria of mathematical beauty and elegance.

There is little doubt that Dirac's preoccupation with beautiful mathematics further distanced him from mainstream research areas of theoretical physics. Thus, he was convinced that the foundation of his cosmological theory with a gravitational constant varying in time was 'beautiful' and therefore must describe nature correctly. Very few astronomers and physicists agreed or took this view seriously.

There was also an ontological aspect of Dirac's belief in the power of beautiful mathematics. Without ever explicitly discussing his view, at several occasions he inferred the existence of physical quantities from mathematical theories. He tended to believe that many mathematical quantities occurring in fundamental theories must have a counterpart in nature, that is, be endowed with physical existence. Or, to put it differently, he assumed that entities must exist in nature as far as they are subject to mathematically consistent description and are not ruled out by laws of fundamental physics. This assumption – a modernized version of what historians of ideas call the principle of plenitude – was particularly clear in his 1931 prediction of the positron and the magnetic monopole. He deduced that monopoles were physically justified in the sense that they were not forbidden by either quantum mechanics or electrodynamics. Since there was no theoretical reason barring the existence of monopoles, he was inclined to believe that they would probably exist somewhere in nature. For, as he argued about the existence of the monopole, 'Under these circumstances one would be surprised if Nature had made no use of it.' In the

1930s Dirac's suggestion was ignored, but when magnetic monopoles later became interesting quantities similar plenitude reasoning guided the monopole hunters. However, in spite of many attempts to detect the hypothetical particle, it was not found. Physicists had to conclude that nature had not after all 'made use of' the monopole.

Although Cambridge was crammed with world-famous philosophers from Bertrand Russell to Ludwig Wittgenstein, Dirac was not interested in philosophy, a discipline of which he had only a very superficial knowledge. Though he tended to ignore his philosophical colleagues at Cambridge, he was unable completely to avoid philosophical questions. Especially in his younger years (roughly before 1945), his attitude to the epistemological problems of quantum mechanics was essentially instrumentalist. He believed that quantum mechanics was devoid of ontological content and that its value lay solely in supplying a consistent mathematical scheme that would allow physicists to calculate measurable quantities. This, he felt, was the principal goal of physics and all that can be said about the empirical world. Whether the world exists in an objective sense or not was a question he considered 'philosophical' or meaningless from the point of view of physics. As a physicist he was more interested in the mathematical structure of quantum theory. To his mind, this was a problem on a deeper level than the interpretation issue. His indifference to the great philosophical questions of quantum physics stemmed from an early age. During the 1927 Solvay Congress he listened to the famous debate between Einstein and Bohr over the interpretation of quantum mechanics. He later recalled: 'I listened to their arguments, but I did not join in them, essentially because I was not very much interested. I was more interested in getting the correct equations.'

Dirac's view of quantum mechanics, which in a general way can be labelled positivistic, implied that he was often considered an antirealist or even an 'idealist'. As mentioned, this was the reason for the ideological preface inserted in the Russian editions of *Principles of Quantum Mechanics*. At that time, in the 1930s, the Copenhagen philosophy was not yet branded 'reactionary' in the Soviet Union and was defended by several leading Soviet physicists, among whom the most important was Dirac's friend and co-author, Vladimir Fock. In contrast to Dirac, Fock was intensely interested in the philosophical consequences of quantum mechanics and continued to defend the Copenhagen interpretation (in a modified version) even after 1947, when Bohr's philosophy came under heavy ideological fire.[24] The debate among communist scientists and philosophers at the time was essentially concerned with Bohr's complementarity principle and the lack of determinism in quantum physics as given by Heisenberg's uncertainty relations. Many left-wing physicists, both in the Soviet Union and in the west, were attracted to deterministic 'hidden variable' theories of the kind proposed by Bohm in 1952. Bohm's theory, they felt, resonated much better with the principles of dialectical materialism than the orthodox interpretation of quantum mechanics.[25] As far as Dirac is concerned, his dislike of the complementarity principle was unrelated to philosophy and political ideology; and he never expressed any interest in causal interpretations à la Bohm.

Dirac's relaxed attitude to the philosophical–ideological divide in quantum physics may be illustrated by his collaboration with Soviet physicists, which included both pro- and anti-Copenhageners. Thus, in his important 1932 paper with Fock and Podolsky he collaborated with a Marxist protagonist of the Copenhagen school (Fock) and a non-Marxist antagonist of this school. Podolsky is best known for his contribution to the famous Einstein–Podolsky–Rosen (EPR) paper of 1935, written after he had left the Soviet Union for the USA. The paper, in which the authors argued a realist interpretation of quantum mechanics against the views of Bohr and his allies, became the focus of intense discussions concerning the philosophical implications of quantum mechanics. Characteristically, Dirac remained silent and never commented in public on the so-called EPR paradox. However, according to a 1962 interview with Bohr, Dirac initially considered the EPR argument devastating to the Copenhagen interpretation. 'Now we have to start all over again', he is reported to have said. Only in the late 1970s did he return to the question, now to express some sympathy with Einstein's view of determinism in quantum theory.

CONCLUSION

About 1930, a general feeling of optimism pervaded much of physics. Relativity and quantum mechanics had been united in Dirac's wave equation of the electron and quantum mechanics had been successfully applied to the structure of simple chemical compounds. The problems that would later beset quantum electrodynamics had not yet become manifest. Dirac seems to have believed not only that physics was in principle complete, but also that the other sciences were on their way to be (in principle) reduced to fundamental physics. In a paper of 1929 he expressed this reductionistic view: 'The underlying physical laws necessary for the mathematical theory of a large part of physics and the whole of chemistry are thus completely known, and the difficulty is only that the exact application of these laws leads to equations much too complicated to be soluble.' In a partly autobiographical novel, the author C. P. Snow referred to Dirac's reductionism. Snow, who at the time worked in crystallography at Cambridge University, remembered hearing Dirac at one of Cambridge's scientific clubs:

> Suddenly, I heard one of the greatest mathematical physicists say, with complete simplicity: 'Of course, the fundamental laws of physics and chemistry are laid down for ever. The details have got to be filled up: we don't know anything of the nucleus; but the fundamental laws are there. In a sense, physics and chemistry are finished sciences.' ... This man who spoke of 'finished sciences' was Newton's successor.

Dirac's youthful confidence in the completeness of physics was fleeting. The troubles with formulating a satisfactory quantum electrodynamics led him to a pessimistic view of scientific progress. By the 1950s he was convinced that fundamental physics needed drastic revision, possibly as drastic as the replacement of classical physics by quantum mechanics. Dirac concluded that the ambitious attempts to create a unified theory of all physics based on existing theories would inevitably fail. The fundamentals of physics would mutate indefinitely.

Figure 49 The Dirac memorial stone, before being laid in Westminster Abbey not far from the tomb of fellow Lucasian, Newton.

Other physicists think differently. In 1980, four years before Dirac's death, Stephen Hawking gave his inaugural lecture as the new Lucasian professor. Whereas his predecessor, Dirac, had no confidence in 'theories of everything', Hawking optimistically suggested that a complete, consistent and unified theory of all physical interactions might well become a reality within a few decades. The completion of this immense unifying project, Hawking prophesied, would signal the end of physics. But will Hawking change his mind too?

Notes

1 A. Zichichi, 'Dirac, Einstein and physics', *Physics World,* 13 (2000), 17–18.

2 Helge Kragh, *Dirac: A Scientific Biography* (Cambridge, 1990), p. 10. Other works on Dirac include Behram N. Kursunoglu and Eugene Wigner

(eds), *Paul Adrien Maurice Dirac. Reminiscences about a Great Physicist* (Cambridge, 1987), J. G. Taylor (ed.), *Tributes to Paul Dirac* (Bristol, Adam Hilger, 1987), Abraham Pais *et al.* (eds), *Paul Dirac. The Man and his Work* (Cambridge, 1998), and Olivier Darrigol, 'Dirac, Paul Adrien Maurice', in *Dictionary of Scientific Biography* (New York, 1990), vol. 17, suppl. II pp. 224–33. For biographical details, see Richard H. Dalitz and Rudolf Peierls, 'Paul Adrien Maurice Dirac', *Biographical Memoirs of Fellows of the Royal Society*, 32 (1986), 139–85.

3 H. Kragh, *Quantum Generations: A History of Physics in the Twentieth Century* (Princeton, 1999).

4 W. McCrea, 'How quantum physics came to Cambridge', *New Scientist*, 96 (1985), 58–60. See also McRea, 'Cambridge physics 1925–1929: diamond jubilee of golden years', *Interdisciplinary Science Reviews*, 11 (1986), 269–84.

5 For details of these works, see Jagdish Mehra and Helmut Rechenberg, *The Historical Development of Quantum Theory* (New York, Springer-Verlag, 1982), vol. 5, part 2. Dirac's scientific works are reprinted in R. H. Dalitz (ed.), *The Collected Works of P. A. M. Dirac 1924–1948* (Cambridge, 1995).

6 H. Kragh, 'Cosmonumerology and empiricism: the Dirac–Gamow dialogue', *The Astronomy Quarterly*, 8 (1991), 109–26.

7 Elisabeth Crawford, *The Nobel Population 1901–1950* (Tokyo: Universal Academy Press, 2002), pp. 174, 190.

8 Dalitz, *Collected Works*, pp. 472–5.

9 A. S. Eddington, *The Nature of the Physical World* (Cambridge, 1928), p. 204.

10 For the following recollections, see Kragh, *Dirac*, pp. 253–4.

11 Fellows of the Royal Society could submit their own papers to *Proceedings of the Royal Society* and also 'communicate' papers from other authors.

12 E. Salaman and M. Salaman, 'Remembering Paul Dirac', *Endeavour*, 10 (1986), pp. 66–70, 69.

13 Pais *et al.*, *Paul Dirac*, p. xiv.

14 V. Y. Frenkel, 'Professor Dirac and Soviet physicists', *Soviet Physics, Uspekhi*, 30 (1987), 816–22.

15 Alexei B. Kojevnikov, *Paul Dirac and Igor Tamm: Correspondence, 1928–1933* (Munich, 1993), p. 54.

16 Robert Crease and Charles C. Mann, *The Second Creation* (New York, 1986), p. 81.
17 Gary Werskey, *The Visible College. A Collective Biography of British Scientists and Socialists of the* 1930s (London, 1978). See also Greta Jones, 'The mushroom-shaped cloud: British scientists' opposition to nuclear weapons policy, 1945–57', *Annals of Science*, 43 (1986), 1–26, and David E. H. Edgerton, 'British scientific intellectuals and the relations of science, technology, and war', in Paul Forman and José M. Sánchez-Ron (eds), *National Military Establishments and the Advancement of Science and Technology* (Boston, 1996), pp. 1–36.
18 H. Dingle, 'Deductive and inductive methods in science. A reply', *Nature*, 139 (1937), 1011–12. The debate is discussed in H. Kragh, 'Cosmo-physics in the thirties: towards a history of Dirac cosmology', *Historical Studies in the Physical Sciences*, 13 (1982), 69–108.
19 J. B. S. Haldane, *The Marxist Philosophy and the Sciences* (New York, Macmillan, 1939), pp. 68, 76.
20 Dirac's war work is analysed in Richard H. Dalitz, 'Another side to Paul Dirac', in Kursunoglu and Wigner, *Paul Adrien Maurice Dirac*, pp. 69–92 and five of the reports reproduced in Dalitz, *The Collected Works*.
21 See Victor F. Weisskopf, 'Visas for foreign scientists', *Bulletin of Atomic Scientists*, 10 (1954), 68–9 and the special issue of the *Bulletin* (no. 7, vol. 8, 1952) on 'American visa policy and foreign scientists'. For background, see Jessica Wang, *American Science in an Age of Anxiety: Scientists, Anticommunism, and the Cold War* (London, 1999) and Silvan S. Schweber, *In the Shadow of the Bomb: Oppenheimer, Bethe, and the Moral Responsibility of the Scientist* (Princeton, 2000).
22 Pais *et al.*, *Paul Dirac*, p. 29.
23 See Kragh, *Dirac*, pp. 275–92. Subrahmanyan Chandrasekhar, *Truth and Beauty: Aesthetics and Motivations in Science* (Chicago, 1987), and James W. McAllister, 'Dirac and the aesthetic evaluation of theories', *Methodology and Science*, 23 (1990), 87–102.
24 Loren R. Graham, *Science and Philosophy in the Soviet Union* (New York, 1972), pp. 93–101.
25 Andrew Cross, 'The crisis in physics: Dialectical materialism and quantum theory', *Social Studies of Science*, 21 (1991), 735–59. Olival Freire, 'Quantum Controversy and Marxism', *Historia Scientarium*, 7(1997), 137–52.

10 Is the end in sight for the Lucasian chair? Stephen Hawking as Millennium Professor

Hélène Mialet
Cornell University, New York, USA

HAWKING'S THREE BODIES

A holder of a professorial chair, such as the Lucasian chair of mathematics, is, to paraphrase the historian Ernst Kantorowicz, 'a mystical person by perpetual devolution whose mortal and temporary incumbent [is] of relatively minor importance as compared to the immortal body corporate by succession which he represent[s]'. Though Kantorowicz was discussing medieval kingship, what he says about the divine nature of sovereignty in the Middle Ages also holds true for the Lucasian professorship. Indeed, to hold a chair means that you are not just an individual, but also the representative of a long line of predecessors and potential successors. These predecessors and successors are 'present and incorporated in the actual incumbent of the [chair's] Dignity'. And this 'corporation by succession' is 'composed of all those vested successively with that particular Dignity'. Thus, on one hand, the professor has 'a Body natural . . . as every other Man has, and in this he is subject to Passions and Death as other Men are'. On the other, he represents a corporation which 'is not subject to Passions as the other is, nor to Death, for as to this Body the [Professor] never dies'. At the heart of this double nature is a problem of continuity. If the professor dies or retires, the chair will not disappear with

From Newton to Hawking: A History of Cambridge University's Lucasian Professors of Mathematics, ed. Kevin C. Knox and Richard Noakes. Published by Cambridge University Press.

him, but will be 'transferred and conveyed over from the Body natural now dead, or now removed from the Dignity . . . to another Body natural'. The identity of the chair *qua* chair will remain intact regardless of changes in tenure. In so doing it has attained a certain form of immortality.[1]

In October 1979 the Lucasian chair of mathematics was incarnated in the form of a new individual: Stephen Hawking. Unlike his predecessor, Sir M. James Lighthill, who stated publicly that he desired the chair when Paul Dirac retired, Hawking not only refrained from applying but openly asserted that he did not wish to fill the prestigious post. Already professor of gravitational physics at Cambridge and professorial fellow at Caius College, he felt that he was unlikely to gain financially from the chair. He hoped – more or less secretly – that the vacancy would be used to lure a brilliant scientist from outside Cambridge. Hawking recalled that he was 'rather disappointed' when the head of the Department of Theoretical Physics and Applied Mathematics, George Batchelor, announced that he had been elected to the chair.[2] He reminisced:

> *I think I was appointed as a stop-gap to fill the chair* as someone whose work would not disgrace the standards expected of the Lucasian chair, but I think they thought I wouldn't live very long and then they [could] choose again by which time there might be a more suitable candidate. Well, I'm sorry to disappoint the electors. I have been Lucasian Professor for 19 years and I have every intention of surviving another 11 to the retiring age. Even so, I won't match Dirac who was Lucasian Professor for 37 years and Stokes who was for 54.[3]

Thus, he who was chosen 'as a stop-gap', partly 'because he was more mortal than other men', is still, today, the incumbent professor. Paradoxically, and partly for the same reason, he is undoubtedly the most famous Lucasian professor of mathematics since Newton.

The premature death from which Hawking has escaped is now part of his illustrious history. In 1961, during his third year of a physics degree at Oxford, Stephen Hawking realized that he was becoming increasingly clumsy and was often falling for no apparent reason. In the following year, when he was furthering his studies at Cambridge, his father, a biologist specializing in research on tropical diseases for the British Medical Research Institute, decided to take him to the family doctor. The physician subsequently referred them to a specialist. After performing a number of tests on Stephen, it turned out that he had amyotrophic lateral sclerosis, more commonly known as Lou Gehrig's disease. The disease is characterized by the slow destruction of all the body's muscles, while the mind remains unaffected. Doctors gave Hawking no more than two years to live. That was in 1963. Forty years later, in defiance of the awful fate assigned to him and despite total paralysis, Hawking is still alive.

Thus, the body of Stephen Hawking – with all the imperfections that make it human and mortal – has transcended its own limits. On the one hand, by embodying the Lucasian chair, Hawking has gained immortality by virtue of the timeless professorship. As Hawking himself likes to recall, the last visible trace of his active body was a signature that introduced him into the pantheon of the ever-lasting great when he was elected to 'Newton's Chair':

> In 1979 I was elected Lucasian Professor of Mathematics. This is the same chair once held by Isaac Newton. They have a big book which every university teaching officer is supposed to sign. After I had been Lucasian Professor for more than a year, they realised I had never signed. So they brought the book to my office and I signed with some difficulty. That was the last time I signed my name.[4]

On the other hand, Hawking has also acquired a certain form of immortality as an individual. His history (as he recounts it) has become a standardized account (available on his website): the subject of countless readings and interpretations, it is constantly repeated, reproduced and commented upon, whether in articles written by those who have known him or by others who have never even seen him. He is glorified

Figure 50 Portrait of Stephen Hawking by Yolanda Sonnabend [1985].

and esteemed, at least in part, because he has transcended the conditions imposed upon him by his own body.

At the same time, this story of personal triumph over physical adversity makes Professor Hawking an emblem for the ideology that dominates our understanding of science – namely, that science is practised by disinterested scientists who are able to transcend the political, social and cultural spaces that their bodies inhabit in order to live in the unadulterated world of the pure mind. More than the mere representative of the Lucasian chair of mathematics, Hawking incarnates 'the perfect Scientist' as we have come to imagine him or her in the rationalist Cartesian tradition. The man who can do nothing but 'sit there and think about the mysteries of the Universe', this intellect liberated from his body, seemingly emancipated from everything that clutters the mundane mind (for example, emotions, values and prejudices), is thus in a position to contemplate and understand the ultimate laws of the universe. Fearing neither age nor death, Hawking has become a kind of angel who is at once immortal (he is still alive despite being condemned to an early death), immaterial (he doesn't need a body to think and can be regarded as pure spirit) and ubiquitous (he is everywhere and nowhere so it is difficult to know where he is). These qualities, normally attributed to the corporate body that he incarnates, also derive from Hawking himself. In other words, he has acquired a double immortality. The mortal and temporary incumbent that was to be 'of relatively minor importance' has become equally as, if not more important than 'the immortal body corporate by succession which he represents'.

One might say, then, that Hawking is simultaneously endowed with three bodies. The first is his natural body subject to imperfections and to death, like that of all men. The second is a natural body that has access to immortality in so far as it is a natural body that, miraculously, did not die when it should have. The third is a natural body that has access to immortality as a corporate body. These bodies are what makes him so exceptional – it is *his* singularity, so to speak – and it is around these bodies that this chapter revolves. Thus, though Hawking has acquired new visibility by holding the Lucasian chair of mathematics, he has also given a new visibility to the chair he holds, leaving open the question of his succession: what will happen to the Lucasian professor of mathematics after Hawking?

This chapter explores the creation, sustenance, manipulation and interdependence of Hawking's 'different' bodies. I begin by describing Hawking at work in Cambridge, as well as the labour that must be done, and the machines that must be deployed, in order to keep him working. I then consider the content of his work, including the achievements which made him a candidate for such a distinguished academic chair. This work is contrasted with the 'practical' labours of his predecessor, the late Sir James Lighthill, who doubted that Hawking's scientific investigations could in any way be construed as 'applied mathematics'. One reason for Lighthill's sentiments is Hawking's interest in cosmology and I therefore examine some of the diverse ways in which Hawking's cosmological physics intersects with various religious issues. Since Hawking's work is so entangled with the 'big questions' regarding our existence, it has attracted enormous attention. This chapter therefore shows not only how Hawking has become a kind of prophet, but how he has also come to represent (literally, to act for) diverse social, political, cultural and business interests, from the disabled to Intel. I thus describe how these groups, in attaching their own interests to Hawking's identity, are able to extend their presence in Cambridge and transform its intellectual and physical environment, while simultaneously allowing Hawking to translate his own presence in and beyond the city. The chapter ends with the examination of how Hawking has become a kind of living saint.

THE COLLECTIVE BODY BEHIND THE MORTAL BODY AND THE SACRED BODY

Imagine Hawking, this immortal, immaterial angel endowed with the gift of ubiquity, sitting in his office in Cambridge University's Department of Applied Mathematics and Theoretical Physics (hereafter DAMTP), a location that he inhabits by virtue of his professorship. His attention is fixed on the computer that has been an indispensable part of his life since he lost his voice following a tracheotomy in 1986. It is an ordinary day. As usual, he has started it by making his way from home, crossing the streets of Cambridge in his synthesizer-equipped wheelchair. He arrives at his office at 11 a.m. and will stay until 7 p.m. Thanks to his voice synthesizer and its 'Living Centre' communication software, he is able to work. He consults new articles sent directly to his electronic archives, responds to e-mails and answers a specially equipped telephone. His nurses take care of him and attend to his physical needs. His secretary, whom he shares with the DAMTP's General Relativity Group, sorts through his abundant mail, putting to one side his fan mail, and answering numerous invitations from the press and the scientific world.

If he agrees to give a public lecture, she checks the contract with him and negotiates his fees. His secretary also helps to recruit the graduate assistant, who is financed by Cambridge University and hired to 'help the Professor in all areas in which he has difficulty due to his handicap', not least the organization of his journeys. When the graduate student is not doing this, he helps Hawking prepare diagrams for lectures and replies to Hawking's numerous e-mails. He directs authors of e-mails concerning Hawking's research to the Los Alamos website where all Hawking's papers are listed, while e-mails concerning specific aspects of Hawking's research are redirected to his students. Hawking replies to e-mails only very rarely, even though they are sent to an address bearing his name. One notable exception to this was when one day his graduate assistant received a desperate e-mail from someone who wanted to commit suicide because she had Lou Gehrig's disease. The student redirected the e-mail to Hawking's personal address, which is known only to his close colleagues, family and friends. Hawking's reply was immediate. Another important duty of the graduate assistant is to polish Hawking's web page, adding, for instance, the title of the latest conference to which Hawking has been invited, a photograph of Stephen Hawking and Bill Clinton taken during the professor's visit to the White House, and the cartoon of Hawking that was used in the television series *The Simpsons*. Throughout the day he also ensures that Hawking's computer and medical equipment are functioning properly.

A considerable part of Hawking's working day is spent mobilizing this collective of humans and machines on which he depends entirely and which in turn depend on him. The perfect coordination of this collective serves not only to relieve Hawking's suffering, but also enables him to work, to communicate with students and colleagues and to travel across the globe. This collective body allows Hawking to act; it also plays a crucial role in the maintenance of his identity as 'Stephen Hawking, Lucasian Professor of Mathematics at Cambridge University'. To paraphrase Kantorowicz again, the defining feature of any corporation is 'a conjunct or collection in one body of a plurality of persons,'[5] and it is only through the smooth operation of this collective body that it has become possible to maintain the immortality of Hawking's mortal body and the immortality of the corporative body (the Lucasian chair) which he represents. In this instance, however, we are dealing with more than people; it is also a question of machines.

Before his tracheotomy, Stephen Hawking – already paralysed – spoke, albeit with great difficulty. Thanks to his secretary or to students capable of understanding his speech, he was able to dictate scientific articles and give lectures. His succession of student assistants lived with him for a long time, under the same roof, helping him with his work and taking care of his physical needs. Following his tracheotomy he lost the use of his voice and could only communicate by facial gestures. To spell words he had to twitch his eyebrows every time an assistant pointed to the right letter on an alphabet. As he once lamented, 'it is pretty difficult to carry on a conversation like that, let alone write a scientific paper.' Shortly after his operation, he heard about a computer system that enabled seriously handicapped people to write and even to talk by moving a single button. The designer of this system, Walter Woltosz, after learning of Hawking's condition, presented him with one. Hawking thereby regained the ability to write scientific articles and give lectures. He is able to operate this entire system on his own, using a simple switch (the commutator) which somebody places in his hand, and around which his fingers are curled.

The device consists of a highly sensitive on-off switch that operates a cursor which he can see on a computer screen mounted on the arm of his wheelchair. Since the cursor automatically moves up and down, Hawking can select words by pressing a switch. When he has constructed what he wants to say, he can send it to a speech synthesizer. Hawking can either speak what he has written or save it to disk. He can then print it out, call it back and speak it sentence by sentence. As he has declared, 'Using this system, I have written a book [*A Brief History of Time,* 1988], and dozens of scientific papers. I have also given many scientific and popular talks.'[6]

As we have seen, the plurality of individuals following one another in time is an essential factor structuring the continuity – and immortality – of the corporation. Like his predecessors and successors, Stephen Hawking incarnates this plurality of individuals. He has also become a kind of collective body himself by virtue of attendants and machines that enable him to function as professor; this collective body has similarly undergone radical transformations in time. From 1980 several nurses were hired, a change following Cambridge University's decision to pay for certain medical expenses of the newly elected Lucasian professor. In 1986, the computer allowed Hawking to regain the capacity to 'talk' and 'write' which meant that his students no longer had to act as interpreters or nurses. Since being appointed reader in 1975, he has also enjoyed the help of a personal secretary. Nevertheless, the immense workload generated by Hawking's fame, especially following the publication of *A Brief History of Time* in 1988, required the recruitment of a new assistant to take charge of equipment and public relations.

A Brief History of Time was to modify both the climate for publication of popularized scientific books and Hawking's environment. Following its appearance, strangers found that he was very difficult to approach. To gain any information about him, it was now necessary to contact his personal secretary 'who deals with the outside world'.[7] As one Hawking graduate assistant pointed out: 'From where I'm sitting, Stephen's not famous, because he's the guy who sits in the office next door, but, to somebody who walks past the window and goes 'that's where Stephen Hawk— [whispers],' he's actually quite well-known really.'[8]

Hawking's fame is attributable to the severe nature of his disability, to his virtuosity as a scientist and to his capacity to present the content of his work in a form accessible to the public. Interestingly, his good friend, Kip Thorne, the Feynman professor of theoretical physics at the California Institute of Technology, links Hawking's physical condition to his way of solving mathematical problems:

> Stephen does his mathematical side of research in his head, using techniques of geometry, where he can manipulate pictures rather than manipulating equations. He finds ways to redefine the questions in the geometrical manner. He can solve and understand much more deeply than anybody else those kinds of problems which can be formulated geometrically.[9]

For Thorne, Hawking's singularity as a thinker resides in his development of special mental techniques. As we have seen, however, his singularity also derives from the fact that he has had to surround himself with a 'collective body'. The perfect coordination of these individuals and machines has paradoxically given him a large degree of freedom to work in a creative way. Really to collaborate with someone meaningfully, he argued, it is necessary to be in the same office.[10] That is one of the reasons why he usually works with the students or colleagues at Cambridge with whom he has developed very close ties. When Hawking agreed to be Lucasian professor, he did so only on condition that the university create a new readership in mathematics and that this position be given to one of his students, Gary Gibbons. It is normal, he recalled, that individuals elected to the Lucasian chair negotiate with the university to ensure that their research subject is properly represented. His students are thus spread across different research fields. This enables him to integrate different information from diverse sources, allowing him to work on many different kinds of problems at the same time. He thus knows that if he is stuck on one problem he can make progress on another. Furthermore, Hawking believes that he performs better when he can talk to someone else about the problems on which he is working. Even if the person facing him makes no suggestion, the imperative of describing his ideas means that he has to organize his thoughts logically. This helps him to see things he might have overlooked. Indeed, Hawking's students are integral to his work. One word from him might entail students spending days trying to interpret what he means, or can totally redirect a conversation. His students are also responsible for manipulating equations on paper, blackboards or on computer screens. His research techniques resemble those of his Ph.D. supervisor, Dennis Sciama, albeit magnified by his physical disability. Sciama allegedly sent his students to scientific conferences and meetings to attend all the presentations and draw up a report about all that were relevant. Thus, Sciama is said to have acquired basic knowledge on many subjects, and formed 'a generation of exceptional cosmologists, relativists, astrophysicists, applied mathematicians and theoretical physicists, . . . [whose] identities seemed to change depending on the title of the conferences' they attended.[11]

THE SCIENTIFIC ACHIEVEMENTS OF THE 'STOP-GAP' PROFESSOR

Stephen Hawking has said that 'It is nice to feel that one holds the same position as Newton and Dirac, but the real challenge is to do work that is even a small fraction as significant.'[12] Hawking's status as a Lucasian professor on a par with Newton and Dirac rests on the startling cosmological theories that he began to develop in the 1960s when he was a graduate student at Cambridge. In collaboration with the Oxford mathematician Roger Penrose (whom he met through Sciama), he showed that the possible existence of singularities – regions of space–time which were infinitely small and infinitely dense – did not indicate a flaw in Einstein's relativity equations, but, rather, indicated the specific conditions where all the laws of science, including Einstein's general theory of relativity, would break down.

Penrose also showed that singularities existed at the centre of black holes – stars in an extreme state of gravitational collapse – and, with Hawking, went on to make the more startling prediction that the universe must have begun with a singularity which then expanded dramatically in the Big Bang. Hawking's second important discovery was made in the 1970s when he showed that black holes can effectively radiate energy. This work is considered extremely important since it unifies three fundamental theories of physics: quantum mechanics, relativity and thermodynamics. Today, Hawking is continuing to investigate the origin of our universe. His most recent work focuses on the no-boundary proposal that he elaborated with Jim Hartle and indicates that it might be possible, and indeed essential, to predict how the universe began by taking into consideration quantum mechanics.[13]

In order to understand the significance of Hawking's achievements we need to examine those of Einstein and his followers. In the early decades of the last century, Einstein showed that space and time were not separate entities but a single amalgamated whole: space–time. Einstein's General Theory of Relativity was used to challenge Newtonian cosmology, Einstein showing that space–time itself was curved, and interpreters of relativity arguing that the universe was not static but expanding. By the early 1930s, thanks to new and powerful telescopes, astronomers (notably Edwin Hubble) had firmly established by observations that the universe was indeed expanding and that the many other observable galaxies beyond our Milky Way were moving away from one another. The conclusion was inevitable: Einstein's equations were telling a story about the beginnings of the universe, for if space–time was now expanding, there must have been a point when galaxies were vastly closer together. At this 'point of origin', space–time was a highly compressed superdense singularity composed of matter, energy, space and time.

In the early 1960s, astronomers already knew that any star which contains more than about three times as much matter as our sun, when it runs out of fuel, would most probably end its life by collapsing inward to form a black hole. The gravitational weight of its constituent matter falling in from all sides would reduce the dying star to a point of zero volume and infinite density, called a singularity. More than twenty years previously, researchers had used Einstein's theory of general relativity to calculate the characteristics of space–time in a black hole. They showed that such an object would bend space–time completely round upon itself, cutting itself off from the rest of the universe. Even light rays passing near such an object would be deflected and absorbed – hence their blackness. A black hole, then, is in essence a region of extremely intense gravity from which nothing, not even light, can escape.

Although it was well known that the simplest solution of Einstein's general theory predicted that the universe expanded from a superdense state, few, at the time, took the notion of black holes seriously. Indeed, relativity and cosmology were considered exotic branches of science, investigated by only a few. What captured most researchers' attention were the interactions on infinitesimally small scales governed by quantum theory. By focusing on the interactions and behaviour of the smallest subatomic particles, physicists were in the process of transforming our understanding of the universe. Yet there seemed to be a contradiction between the microphysics of quantum theory and the laws describing macro, gravitational phenomena; what worked for one, failed for the other. The Holy Grail, then, would be to discover a grand theory capable of unifying the cosmological predictions made by general relativity and quantum theory. Once these two theories were brought together in a grand unified theory (or GUT, as it is known), presumably one could explain everything in the cosmos.

It was not until 1971 that satellite-borne instruments first identified a plausible black hole candidate in one of the component stars of the binary X-ray star system Cygnus X-1. In the late 1960s, Hawking was amongst the few people who believed that black holes might exist; Roger Penrose was another. Working at Birkbeck College in London, Penrose argued that at the heart of every black hole was a singularity which generated such intense gravitational pull that it even embraced space–time itself in its crushing grip. In the singularity of a black hole, there was nothing left to support the edifice of modern physics – its laws were no longer applicable. Hawking and other members of Sciama's research group (which included Brandon Carter, George Ellis and Martin Rees) were tremendously excited by this. Indeed, Hawking was so captivated that in 1965 he joined forces with Penrose.

During the middle and late 1960s, Penrose's ideas inspired Hawking, Robert Geroch, George Ellis and other physicists to develop a powerful set of topological and geometrical tools – now called global methods – for general relativity calculations. In 1970 Hawking and Penrose used these methods to prove that if the general theory of relativity is the correct description of the universe, then there must have been a singularity at the beginning of time. The Big Bang appears to be the exact opposite of the gravitational processes inaugurated by the collapse of a dying star into a black hole. If this is true, the cosmic censor (a term coined by Penrose to illustrate the bashful nature of singularities hidden within black holes) had at least overlooked the exposure of one naked singularity – that which formed the point of origin of our expanding universe some 15 billion years ago. What thus began as esoteric mathematical research conducted on the fringes of physics had, by the end of the 1960s, become a momentous scientific achievement. The mathematical essay that grew out of Hawking's and Penrose's work, 'Singularities and the geometry of space time', won Cambridge University's prestigious Adams prize, prompting Dennis Sciama to remark that Hawking had a career worthy of Newton ahead of him.

In the 1970s, Hawking shifted his focus from the singularity hidden at the heart of a black hole to the horizon surrounding it. Using the same global methods he had helped develop in the 1960s, Hawking formulated the concept of the black hole's event horizon and proved that the surface areas of such horizons always increase. Any object swallowed by a black hole has its energy transferred inwards because it is impossible for this energy to escape. Since all forms of energy produce gravity, this means that the hole's gravitational field is continually being strengthened, and, correspondingly, its surface area is continually growing. As Hawking recalls, this discovery meant 'that a black hole had a quantity called entropy, which measured the amount of disorder it contained; and if it had an entropy, it must have a temperature. However, if you heat up a poker in the fire, it glows red-hot and emits radiation. In other words, black holes were thermodynamic systems. But a black hole cannot emit radiation, because nothing can escape from a black hole. Yet at the time it was held that since a hot black hole could not emit radiation, the laws of thermodynamics did not apply to it[14]. It was for this reason that Hawking believed that these analogies to thermodynamics were merely curious.

Some time between 1973 and 1974, Hawking was looking at what would happen when quantum mechanical effects were introduced into the region outside the event horizon of a black hole. What materialized from this calculation was the insight that even a black hole, from which no light could escape, still gave off residual radiation (now known as 'Hawking radiation') with a thermal spectrum. Black holes radiated like any other hot body. In using the general theory of relativity, Hawking showed that, while the surface area of a black hole might expand, it could not contract; yet, by then applying quantum principles, he argued that black holes could lose energy as radiation, reducing their mass, and thus shrink. Indeed, the smallest holes would shrink most rapidly and evaporate in a burst of gamma-radiation.

Hawking's investigation of black holes in the early 1970s is often regarded as one of the most important pieces of scientific research ever carried out. According to John Gribbin, not only did it 'succeed in partially uniting the General Theory of Relativity and quantum theory, but also in bringing into the fold the great developments of nineteenth century science – thermodynamics'. It changed our understanding of physics in fundamental ways.

Hawking subsequently applied his insights to his work with Penrose. We have seen that by using the general theory of relativity Hawking and Penrose conjectured that the universe was born some 15 billion years ago out of a point of infinite density and zero volume.

Now Hawking asked, what would occur if – as he had done with his work on the surface area of black holes – he now applied quantum rules to his equation describing the baby universe.

If the universe has its origins in a Big Bang, where space–time expands outwards from an initial singularity, then there might plausibly be a countervailing process where space–time implodes back in upon itself, a kind of 'Big Crunch'. In this view, there are beginning and end points of time which Hawking called 'edges'. With regard to space, however, there appear to be no edges, because space is folded round itself like a ball. Hawking found this asymmetry disturbing and sought to remove the edges in time and space, so as to produce a model of the universe which had no edges – or boundaries – at all. The 'no boundary proposal' that Hawking developed in 1982–3 with Jim Hartle of the University of California, Santa Barbara, claims that in the quantum-mechanical framework, there is no spatial or temporal boundary beyond which it is impossible to make calculations regarding the state of the universe. If proved, this proposal would mean that there were no singularities or origins in the universe, and that the laws of science could be applied to all times and all places, including the beginning of the universe.

HAWKING'S SINGULARITY: APPLIED MATHEMATICS OR RELIGION?

The content of the physics on which Stephen Hawking works is fundamental to an understanding of what constitutes the specific characteristics of this Lucasian professor of mathematics, especially from the point of view of the general public. Hawking likes to point out that he was elected Lucasian professor because he represented the applied branch of mathematics, unlike Michael Atiyah who, in the opinion of the DAMTP head George Batchelor, represented a 'pure' mathematician disguised as a physicist. 'Giving Atiyah the Chair,' Hawking added, 'would have meant that the Lucasian professor was lost from applied mathematics to pure.' For this reason 'Batchelor, the head of the department made it clear that the Lucasian chair would

Figure 51 Portrait of James Lighthill c. 1990.

go to Atiyah over his dead body.'[15] Nevertheless, there remains some disagreement over the status of Hawking as an applied mathematician. For James Lighthill, Hawking's predecessor, the preoccupations of the current Lucasian professor of mathematics have nothing to do with applied mathematics. He expressed negatively what many of Hawking's fans see as one of his best qualities: 'I like practical applications in mathematics, rather than speculating about the first ten to the minus something seconds of the universe. Cosmology seems to be almost too close to theology to be interesting. To me, it is not quite science, but more like creation myth.'[16]

While theology and myth did not interest Lighthill, most everything else – from Portuguese poetry to the lift mechanisms of insects – did. Born in Paris in 1924, Sir James spent a lifetime describing mathematically thousands of mundane phenomena. His fascination with practical applications for mathematics was apparent at both Winchester and Trinity College, Cambridge, where he read the natural sciences as an undergraduate, but really started to bear fruit during the final years of the Second World War when he began work on supersonic flight for the Ministry of Aircraft Production. This interest in aeronautics never waned; but as a lecturer, and then professor, at the University of Manchester he also delved into fields such as gas dynamics, ionization processes and blast waves.

Among other research and teaching duties, Lighthill launched two important new fields during the 1950s: aeroacoustics and nonlinear acoustics. A key feature of the first of these fields was 'Lighthill's eighth power law for jet noise', while his work on nonlinear acoustics was applied to such diverse things as machines for the disintegration of kidney stones, the analysis of flood waves in rivers and the prediction of highway traffic flow. From 1959, as director of the Royal Aircraft Establishment, Lighthill investigated countless other research quandaries, including the delta wing for the Concorde, communications satellites and biofluid dynamics (the study of the flow of blood and gases in animals).

By 1968 Paul Dirac was on the verge of retirement, and by October 1969 Lighthill was back in Cambridge – as Lucasian professor. As he speculated, he was offered the chair 'because there had not been anyone in fluid dynamics since Stokes had been Lucasian professor'. Here, in Cambridge – and like that 'grand genius', Stokes, for whom he had 'boundless enthusiasm' – Lighthill 'concentrated entirely on fluid dynamics', making great advances in the fields of sound, antisound, waves, oceanography and atmospheric dynamics.

By the time of his tragic death in 1998 (he retired from the Lucasian professorship in 1979), Sir James had published six books and approximately 150 papers; he had sat on innumerable committees in academia, in industry and in government; he had garnered some twenty-four honorary doctorates; he had been given memberships in the world's most prestigious scientific academies; and, like his hero Stokes, he had obtained a knighthood. Yet during his entire career he had always worked hard to make his research accessible to men and women who were not gifted theoreticians and mathematicians, something for which countless designers and engineers were exceptionally grateful. Partially for this reason, D. G. Crighton was led to observe that he is undoubtedly 'acknowledged throughout the world as one of the great mathematical scientists of this century'.[17]

Except for their desire to make their work comprehensible to those outside their immediate research specialty, it therefore seems few things are as dissimilar as the careers of Hawking and his predecessor. But Hawking would agree with Lighthill that it is difficult to discuss the beginning of the universe without mentioning the concept of a Creator. Yet, although Hawking is said to have smuggled Bibles into Russia to help Baptists, and is reported to be helping the Jewish orthodox L'Chaim Society establish itself in Cambridge, he is no less reserved as to his own beliefs:

> It is quite possible that God acts in ways that cannot be described by scientific laws. But in that case one would just have to go by personal belief.[18]

> We are such insignificant creatures on a minor planet of a very average star in the outer suburbs of one of a hundred thousand million galaxies. So it is difficult to believe in a God that would care about us or even notice our existence.[19]

Hawking recognizes that his work on the origins of the universe is on the boundary between science and religion, but he tries to remain on the scientific side of that boundary: 'In the past, cosmology was an area where wild theoretical speculation was unconstrained by observations', Hawking insists. 'But now accurate observations place strong limits on theoretical models. I'm glad to say the observations so far seem to be consistent with the no boundary proposal.'[20]

As we have seen, the 'scientific' proof of the no boundary proposal would mean that there were no singularities or origins in the universe, and that the laws of science could be applied to all times and all places, including the beginning of the universe. More significantly, it would cast into doubt the possibility of a Creator. Oddly enough, Hawking started contemplating the origin and destiny of the universe during a 1981 conference organized by the Jesuits of the Pontifical Academy, a small group of eminent scientists who advise the Pope on scientific subjects. On this occasion, distinguished cosmologists were invited to discuss the evolution of the universe since the Big Bang, excluding the Big Bang itself on the grounds that it was 'God's dominion'. During the conference, Hawking put forward his 'no boundary proposal', with its stipulation that time and space are finite but without limits. 'So long as the Universe had a beginning', Hawking suggested, 'we could suppose it had a creator.' However, Hawking then laid out a scarier prospect for Christians: 'if the Universe is really completely self-contained, having no boundary or edge, it would have neither beginning nor end: it would simply be. What place, then, for a creator?'[21] Hawking would thus have succeeded in fulfilling his ambition to discover how the universe began (albeit independently of why it began) and, in doing so, he would have violated the Pope's 1981 ban on research concerning the Big Bang 'because it was the moment of creation and thus the work of God'.

By embracing such a 'heretical' view, Hawking in many ways seems to be recapitulating the career of Galileo. A number of times he has pointed out that he 'was born on January 8, 1942, exactly three hundred years after the death of Galileo'. Despite the coincidence he is usually careful to quell any speculation about the astrological significance of that particular day: 'However, I estimate that about two hundred thousand other babies were also born that day. I don't know whether any of them were later interested in astronomy.'[22] Similarly, his graduate student, Bernard Carr, stresses the Galilean significance of Hawking's birthday. In 1975, he remembers that when Hawking went to the Vatican to receive the Pius XII Medal from Pope Paul VI for remarkable work by a young scientist, he 'was very keen to go into the archives and see the document which was supposed to be Galileo's retraction, when he was put under pressure by the Church to recant on his theory that the earth went around the sun. . . . I think it gave us some pleasure that the Church fully announced . . . that in fact Galileo was right. But whether or not the Pope would have approved of what Stephen had discovered, if he actually had understood it, I'm not quite sure.'[23]

Hawking's award of the Pius XII medal was but one of many conciliatory gestures made towards him by the Catholic Church. In 1981, following the Pontifical Academy conference, the Pontiff himself knelt down at Hawking's feet, much to the surprise and indignation of other guests. Perhaps equally surprising, five years later Hawking himself became a member of the Pontifical Academy. This unexpectedly warm reception of Hawking by the Church might also be attributed to the enigmatic nature of some of Hawking's own writings. For instance, Hawking's ambiguous religious position is illustrated by remarks in his *A Brief History of Time* which concluded with meditations on the possibility of one day knowing the thoughts of the Creator:

> if we do discover a complete theory, it should in time be understandable in broad principle by everyone, not just a few scientists. Then we shall all, philosophers, scientists, and just ordinary people, be able to take part in the discussion of the question of why it is that we and the Universe exist. If we find the answer to that, it would be the ultimate triumph of human reason – for then we would know the mind of God.[24]

BETWEEN HEAVEN AND EARTH, PAST AND FUTURE, EXPERT AND PUBLIC

Hawking's remarkable statement about knowing the mind of God has been reproduced throughout the world. *A Brief History of Time*, initially written because Hawking needed more income to support his ailing body and his family, sold ten million copies across the globe, and has been reprinted eleven times in Britain alone and translated into thirty languages. It is a best-seller in which the author – realizing that pages filled with complex equations would hamper sales – allowed himself only one equation, Einstein's famous $E = mc^2$. In addition to the film by Errol Morris (where an exact replication of Hawking's office was reconstructed in a disused church) and the book that was later to accompany it (the book of the movie of the book), there was also a stage adaptation of *A Brief History of Time*, entitled *God and Stephen Hawking*, which was performed in London. According to its promotional flyer,

> Stephen Hawking is one of the icons of our age . . . With a cast of characters that includes Hawking himself, his wives and associates, Newton, Einstein, Pope John Paul, the Queen, and God, *God and Stephen Hawking* is a major new theatrical production, an extraordinary revelation of the life and vision of one of the most famous and revered men on the planet.

Since the publication of *A Brief History of Time*, Hawking has indeed become 'one of the icons of our age', receiving hundreds of letters, drawings and poems each month. His huge volume of mail includes meticulous, well-written letters; wild, fanatic writings; postcards from the Himalayas; letters with highly complicated diagrams in which God and black holes mingle; packets containing CVs and IDs; letters from Europe, Asia, India, Africa and America; letters from teenagers searching for the meaning of life and others penned by lonely, retired lecturers; letters from science students and lecturers of Urdu, others from chemists or even windsurfers. Nearly all these correspondents want solutions to the problem of the origin of the universe, the end of time and the existence of God.

Ironically, the man who says openly that he believes neither in God nor in the possibility of a Creator is today a sort of spiritual leader in the public eye. For this reason, the following letter to Hawking is not atypical:

> There are two or three people in the world who I would most like to meet and you are at the top of the list with the Dalai Lama ... If I never have the privilege of meeting you then the next best thing is to put pen to paper to ask you the question I would like to ask you face to face. Along with a million others I have read your book *A Brief History of Time* and mathematics never being my strong point, some parts I comprehended dimly. I read or heard that you have changed your opinion as to time/space travel? What I want to know is, what do you think happens to past times? Scientific thinking begins to match religious thought and vice versa i.e. is it true that in 'empty' space-matter (or life?) is inherent? When all the worlds have ceased to be – will all things then return to thee? Or as it is put in the Uparishads; 'the end of the worlds all things sleep then from his being new worlds arise and awake' – a universe which is a vastness of thought. Do you think that from nothing (the Buddhist nothingness?) all things come in turn? So is it a return to pure consciousness? I'm sorry to bother someone as busy as you must surely be with my questions but I would so much like to know what you think about these things. To possess an intellect and understanding such as yours must be amazing ... I find it fascinating to wonder what direction you think your life would have taken if circumstances in your life had been different? The world would have been the poorer I think.

The image of Hawking as an oracle and having access to extraordinary information has been evoked by his ex-wife when she reflected on the inaugural lecture consecrating his position as Lucasian professor of mathematics entitled 'Is the end in sight for theoretical physics?':

> Hypnotised, as though receiving the words of an oracle, the audience of scientists, many of them young hopefuls, strained to catch his utterances. The words were not designed to offer the comfortable prospect of a secure future for, in his lecture, Stephen gleefully predicted that the end of physicists, if not of physics was in sight.[25]

As she recalls, Hawking was accompanied that day by a student who stood next to him on the stage of Cambridge University's Babbage lecture theatre to interpret his words that had become so weak that only his closest students, colleagues and family members could still understand them. Hawking calmly predicted that the increasing sophistication of computers would help solve all the major problems confronting physicists within the following twenty years – that is, by the end of the twentieth century. In short, it would not be long before 'there would be nothing left for physicists to do. He himself would be all right, he declared jovially, as he would be retiring in the year 2009.'[26] Hawking repeated his prophesy in 1998 when he was invited by Bill Clinton to the White House for the 'Next Millennium' conference: 'I am confident we will discover the [Theory of Everything] by *the end of the 21st century* and probably much sooner', he declared. 'I would take a bet at 50–50 odds that it will be within twenty years starting now'. Nevertheless, he pointed out that 'the ultimate theory will place no limit on the complexity of systems that we can produce and it is in this complexity that I think the most important developments of the next millennium will be'. Emphasizing the inevitable emergence of human genetic engineering, he concluded that

> I think the human race, and its DNA, will increase its complexity quite rapidly. In a way the human race needs to improve its mental and physical qualities if it is to deal with the increasingly complex world around it and meet new challenges like space travel. And it also needs to increase its complexity if biological systems are to keep ahead of electronic ones.[27]

The programme was broadcast by satellite for the public at large. This time the computer served as his interpreter.

We have seen that Hawking may be regarded as an incarnation of three bodies: the mortal man; the mortal man who has overcome death; and the representative of the Lucasian chair. While pursuing his interests in the singularities and origin of the universe, Hawking has combined his unique knowledge and all three of his bodies to become a kind of mediator between (1) heaven and earth (since he is questioned about the possible existence of God); (2) the past, present and future (since he is questioned on the origin and the fate of the universe); and (3) the experts and the public (since he allows the common mortal access to the complex world of science). Moreover, he has become a sort of role model, a voice of authority and an inspiration for young people 'to make the world a better place'. He who says he was chosen because he embodied the applied dimension of mathematics has thus, for many, come to embody religious and moral values. The oracular powers bestowed on him – he is even referred to as the 'Prophet of the Big Bang – is simply, perhaps, the product of a society craving religion, eager to believe.[28] For Rabbi Botteach, who recently invited Hawking to give a talk at Oxford for the L'Chaim Society, the present Lucasian professor is capable of reviving religious values through his presence alone:

> Hawking's life is an embodiment, a living personification of great religious ideals. He is a living inspiration to hundred[s] of millions of people the world over, of what we are capable of achieving under a physical disability. That human will – or what Stephen Hawking might call human will and what I call soul, is capable of triumph over the body. You have in him, the phenomenal spectacle of the world['s] greatest scientist coupled with an absolutely tortured and broken body . . . Not only has it not stopped him from pursuing science, it has not stopped him from being the absolute best. To me that's very spiritual. And I think, one of the great messages of religion is to inspire hope in people. And I think he's given

> hundreds of millions of people hope. . . . We're trying to inspire young people to hope – to look forward to the future – to believe that mankind can overcome every obstacle, every impediment, to make the world a better place.[29]

It was also as a voice of authority, an inspiration to young people 'to make the world a better place', that he was invited to take part in an episode of the famous television cartoon series *The Simpsons*.

Hawking agreed to contribute to *The Simpsons* because he considered it to have a good moral and to be intelligent. The episode in which he agreed to partake focused on intelligence. In the story, the character of Lisa Simpson joins the Mensa Society. The club for geniuses overthrows a municipal authority in order to run their town in a more 'intelligent' manner. However, the self-proclaimed geniuses go too far and their clever but impractical reforms lead to disastrous consequences. According to the show's writer, Matt Selman, 'we needed someone smarter than them, someone they could respect, who could shame them into seeing the error of their ways'.[30] Hawking, 'representing the most intelligent man in the world', was announced in the show as having an IQ of 280, and fitted the part perfectly. His cartoon incarnation scolds the Mensa maniacs with the comment 'I don't know which is a bigger disappointment: my failure to formulate a workable unified theory or you.' And when the school principal tries to interpret the Lucasian professor, Hawking interrupts him and exclaims: 'Silence, I don't need anyone to talk for me – except this voice box.' Proving that actions speak louder than words, Hawking punches Principal Skinner with a spring-loaded boxing glove that is concealed in his wheelchair.

CO-CONSTRUCTION OF HAWKING AND OF CAMBRIDGE

Whereas the boxing glove remains a total fiction, when Hawking's synthetic voice 'speaks' people listen, whether this voice resonates in the fictional world of TV or in real life. This ability to make people listen is strikingly apparent from his struggles to improve facilities for physically handicapped people in Cambridge. Thanks to his efforts, DAMTP gained a wheelchair ramp and the kerbs of the city were lowered, especially along the roads that Hawking negotiates every day to and from work. After these successes, he challenged Cambridge City Council on the question of the accessibility to public buildings for the physically handicapped. Furthermore, in 1989 a project was launched to create a special residence for the university's handicapped students. In pleading the case of handicapped students, he is reported to have said that 'the attitude of the University towards the handicapped was appalling', and that the university was 'flouting the law by ignoring an Act of Parliament dating back to 1970, which made it illegal not to provide appropriate access to disabled persons'. In the year when Hawking was elected Lucasian professor, buildings in Cambridge were being constructed without any access for handicapped persons; but it was largely due to Hawking's struggle against university and civic authorities that such plans were amended. As his ex-wife observes, 'By a curious coincidence, the attitude of the City Council towards access for the disabled and the lowering of kerbs mellowed rapidly as Stephen's fame grew, but that was long after those strenuous years during which I pushed the wheelchair with two small children in tow.'[31]

Disabled people identify with Stephen Hawking: he has become their spokesperson. But just as Hawking has used his fame – especially as Lucasian professor – to improve facilities for the disabled in Cambridge, so the university has used Hawking's fame. Indeed, his reputation has enabled the university to raise substantial funds for a number of different projects, from buildings to scientific equipment. At the same time, as representative of the university, Hawking has expanded his presence within Cambridge and beyond (for example, the Cambridge University Development Office organized Professor Hawking's trip to the White House).

As we have seen, Hawking's disability and the world of machines to which he is connected are both part of his identity. Accordingly, he likes to emphasize the connections between machines, computing and his own identity in his various bids to woo potential donors:

> I'm also proud to have Babbage as one of my predecessors, even if he wasn't Lucasian Professor for very long. I do a certain amount of fund raising for Cambridge University in places like Silicon Valley, and it is a good line to throw in that the father of computers was one of my forebears.[32]

Indeed (and though he may be considered an oracle), Hawking is said to have been a key player in the deal orchestrated between Cambridge and Microsoft from which the university gained about £50 million to give Microsoft the right to build a research centre in Cambridge. Jane Hawking points out, 'The fact that it had granted one of the world's richest men [Bill Gates], at current estimates worth around $40 billion, the rights to all its invaluable computer research in the future appeared to have been underestimated in the equation.'[33] Hawking also promoted the decision of Dr Gordon E. Moore, Chairman Emeritus of Intel and also a prophet of computing, who donated £7.5 million for the construction of a new physical sciences and technology library. The Betty and Gordon Moore Library will house all the collections of scientific data presently scattered in Cambridge, including material Hawking wrote before 1973 and the typescript of *A Brief History of Time*. Eventually, the library will contain digitized archives of information stored in Hawking's computer. The long tradition of interaction between the professorship and the university library, originally inaugurated by Henry Lucas's donation of his 4000 volumes, will thus be perpetuated. The new library will also provide access to Newton's papers, including those concerning the invention of the general binomial theorem. As Hawking said:

> As we approach the millennium, we are making progress in science and technology, at an ever-increasing rate. Central to these endeavours, is information on what we have already achieved. That is why the new library will be so important. It is funded by a

> generous gift from Gordon and Betty Moore, and will include a small archive of my work. Gordon Moore and Intel have greatly speeded up science by improving computers. *I know because I'm Intel inside myself.*[34]

HAWKING'S BEATIFICATION

This new voice, created by Intel, which announces that it is part of the identity of the man who uses it, validates Intel's donation. While Intel has enabled Hawking to work and communicate with the world, the chip-maker – owing to the construction of Cambridge's new library – has also made it possible for the professor to leave traces of his production from before and after the neuromotor functions of his body wasted away. Thanks to the development of a new and more sophisticated voice, Intel has also allowed Hawking to leave his former voice and telephone system to the new Wellcome Wing of London's Science Museum. Stephen Hawking's equipment can now be seen in one of the museum's Digitopolis galleries of digital technology, where it is exhibited in the 'Sound' section on synthesizers. Apart from 'digital sound', the visitor can also learn about 'digital beings' (computers), 'individual networks' (communication) and 'future machines' (artificial intelligence/robots).

Meanwhile, Hawking's new voice, far from being artificial, is now part of him. As he says: 'I'm Intel inside myself'. It has become his real voice. It is also a voice that is an authority and which acts as mediator between heaven and earth, the past, present and future, the experts and the public. As suggested by the caption to the Hawking voice exhibit at the Science Museum, 'Speech is difficult for computers but Stephen Hawking's voice is recognised throughout the world; digital technology has enabled him to communicate with the whole world.'

The old equipment, formerly his 'real' voice, may have been superseded, but in doing so it has also become a pseudoreligious relic of sorts. It also enables us to forge another link between Hawking and Galileo, for it is common for historians to talk of 'the cult of Galileo who came to be venerated rather than an author to be studied

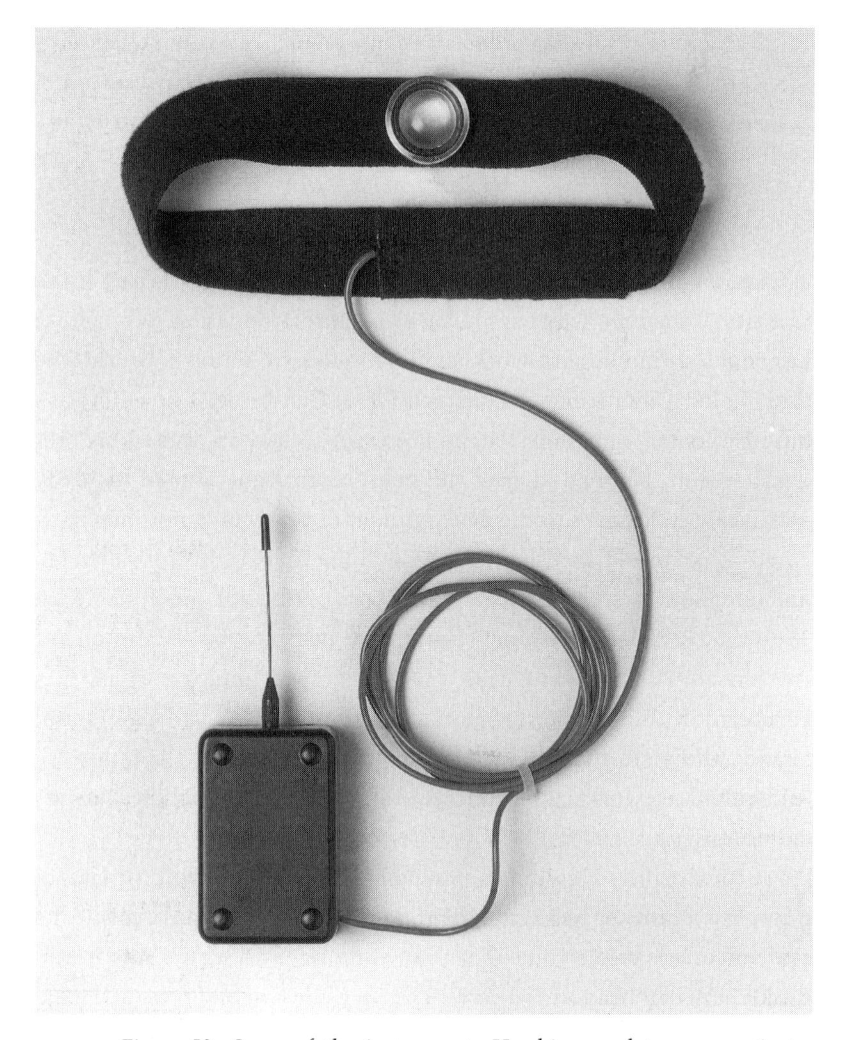

Figure 52 Some of the instruments Hawking used to communicate, from the Stephen Hawking display at the Science Museum, London. This shows the sound sensor (in the throat strap) that enabled Hawking to connect his voice synthesiser to a mobile phone (bottom of picture).

and understood'. This veneration is evidenced by the detachment of Galileo's middle finger from his body by Anton Francesco Gori in 1737. Eventually, the finger became the property of the Museo di Storia della Scienzia in Florence; at the bottom of the marble stand on which it is placed are inscribed the words of Tommaso Perelli, an astronomer of the University of Pisa: 'This is the finger with which the illustrious hand covered the heavens and indicated their immense space. It pointed to new stars with the marvelous instrument, made of glass, and revealed them to the senses. And thus it was able to reach what Titania could never attain.'[35]

Like the cult of a saint, Hawking's fame has been translated by the distribution of other 'relics'. While his archives in the Betty and Gordon Moore Library will undoubtedly obtain the same kind of sacred status as Newton's manuscripts, his book has become something of a hallowed relic, granting access to the great mysteries of the universe. As his biographer John Gribbin suggests:

> Owning his book is like having a piece of the 'true cross'. You know that just by owning the book, you imagine that you're put in touch with the great mysteries of the universe, even if you don't actually read it, but that somehow, magically, it's going to be good for you to own it.[36]

Furthermore, like a saint, Hawking's seeming ubiquity appears to transcend the chair he represents. The features of Hawking's identity have mediated the transformation of Cambridge's physical and intellectual environment, thus making it possible for him further to extend his presence and visibility. All those parts of his extended body (notably, the archived material he wrote before and after he could write, his 'real/artificial' voice, and the draft and published versions of his book) are distributed and made public (for example, in the Science Museum, in the library and in the press). As such, they make the world of science accessible. Moreover, they simultaneously make visible and augment both the aura of Hawking and the 'World of Science'. In the case of the archives, the 'World of Science' means laws and theories about nature or how the universe functions. In the case of the voice, the 'World of Science' means the 'scientific statement' and messages delivered through the voice, but also the object itself (the voice synthesizer) which represents what new technology can do. Hawking's presence is also extended through the countless biographies written about him, including John Boslough's *Beyond The Black Hole: Stephen Hawking's Universe* (1985), Kitty Ferguson's *Stephen Hawking: Quest for the Theory of Everything* (1992), Michael White and John Gribbin's *Stephen Hawking: A Life in Science* (1992), and J. P. McEvoy and Oscar Zarate's *Stephen Hawking for Beginners* (1995). His portrait hangs in the National Portrait Gallery in London, and there are, according to one popular internet search engine (in February, 2003), more than 108 000 sites which refer to him. But, as a consequence, does the chair that he both represents and transcends become something altogether different?

CONCLUSION: WILL HAWKING'S PREDICTIONS COME TRUE?

In 1975 the Lucasian professor of mathematics, Sir James Lighthill, was commissioned by the Science Research Council to judge whether Britain ought to invest in the development of artificial intelligence. He was neither confident in the possibility of creating intelligent robots nor of reconciling the differences between humans and machines. The work achieved hitherto, he wrote, 'casts doubts upon whether the whole concept of artificial intelligence as an integrated field of research is a valid one'.[37] After considering Lighthill's words, it is strange to think that in the year 2003 Lighthill's successor in the Lucasian chair cannot talk, write or move his body without the (partial) mediation of a computer.[38] It is also strange to think that today his synthetic voice, now an emblem of the power of current technologies, is also held up as a relic. It is even stranger to see how much power is granted to computers, whose increasing sophistication should, according to Hawking, in the near future solve all the enigmas facing today's physicists. Hawking has prophesied the obsolescence of the chair he holds, although he noted in passing that he would not really be affected by that since he would already have retired. For if the *raison d'être* of physics and physicists disappears, what about the Lucasian chair of mathematics? Will people talk about the chair held by Newton and by Hawking or only by Hawking? Will there still be a chair? What is the next Lucasian professor of mathematics after Hawking likely to be like? If Hawking transcends the chair he represents, he also challenges the immortality of the chair as a corporation. Hawking hopes that somebody worthy will succeed him, but thinks that in any case this particular chair will not have the same value given that the number of established chairs and professorships has increased substantially. To paraphrase Kantorowicz again, this chair, if it is a 'corporation by succession', is a 'fiction which makes us think of the witches in Shakespeare's *Macbeth*... who conjure up that uncanny ghostly procession of Macbeth's predecessor kings whose last one bears the "glass" showing the long file of successors'.[39] If Hawking is right about the future role of computers, his successor will perhaps pay homage to the predictions of Hawking, and of Babbage, who tried to establish how and to what extent computers would be capable of mimicking human actions. Babbage thought that his analytical engine would have characteristics of intelligence: specifically, memory and foresight. In other words, he envisaged that a computing engine could extrapolate future developments from past information.

In the year 2395, Data, hero of *Star Trek: The Next Generation* – a science fiction series in which Hawking himself once appeared playing cards with Newton and Einstein – accepts the Lucasian chair of mathematics, 'the chair previously held by Sir Isaac Newton and Professor Hawking'. Data is a cyborg, an ultrasophisticated machine made by humans and that resembles a human. The spiralling progress and sophistication of computers will not make the chair disappear, but it will, perhaps, give it a new incumbent who is like the chair itself, theoretically, immortal. Will a future Lucasian professor of mathematics, then, be a person, a brain-machine or a robot capable of mimicking the action of a person, an oracle, or perhaps even a demigod?

Notes

1 Ernst Kantorowicz, *The King's Two Bodies. A Study in Medieval Political Theology* (Princeton, Princeton University Press, 1957), pp. 313, 387, 13.

2 Interview conducted by the author with Professor Hawking in July 1998.

3 *Ibid.* (my emphasis).

4 Stephen Hawking and Gene Stone (eds) *A Reader's Companion to A Brief History of Time* (New York, Bantam Books, 1992), pp. 151–2.

5 Kantorowicz, *The King's Two Bodies*, p. 304.

6 'The computer communication system' at http://www.hawking.org.uk/text/disable/disable.html.

7 It is interesting to note that Hawking has his own press relations office and does not use the Cambridge press office.

8 Interview conducted with Stephen Hawking's assistant by the author in 1998.

9 Interview conducted with Professor Kip Thorne by the author in August 1999.

10 Interview conducted with Stephen Hawking by the author in 1999.

11 Jane Hawking, *Music to Move the Stars. A Life with Stephen* (London, Macmillan, 1999), p. 91.

12 Interview conducted with the author in July 1998.

13 For good summaries of Stephen Hawking's contributions to science see, for example, Hawking and Stone, *Reader's Companion*; Kip Thorne, *Black Holes & Time Warps. Einstein's Outrageous Legacy* (London, W. W. Norton, 1994), and Michael White and John Gribbin, *Stephen Hawking. A Life in Science* (Middlesex, Plume Books, 1993).

14 Hawking and Stone, *Reader's Companion*, p. 92.

15 Interview conducted with Professor Hawking by the author in June 2000.

16 Robert Bruen and Jean Flanagan, 'The Lighthill path', available at: http://www.tiac.net/users/bruen/lighthill-interview.html.

17 See D. G. Crighton, 'Obituary of M. James Lighthill', *The Independent*, Wednesday Review, 22 July 1998 p. 6; for Lighthill's comments concerning the professorship and Hawking see Robert Bruen, 'A brief history of the Lucasian professorship of mathematics at Cambridge University' (unpublished Ph.D. thesis, Boston College, 1995).

18 White and Gribbin, *Stephen Hawking*, p. 167.

19 *Ibid.* p. 166.

20 Interview conducted with Professor Hawking in July 1999.

21 White and Gribbin, *Stephen Hawking*, p. 183.

22 Stephen Hawking, *Black Holes and Baby Universes and Other Essays* (New York, Bantam Books, 1994), p. 1.

23 Stephen Hawking and Gene Stone, p. 119.
24 White and Gribbin, *Stephen Hawking*, p. 167.
25 See White and Gribbin, *Stephen Hawking*, p. 173. See also Jane Hawking, *Music to Move the Stars*, p. 372.
26 *Ibid.*
27 http://clinton4nara.gov/textonly/initiatives/millennium/shawking.html.
28 Interview conducted with John Gribbin by the author in 1996.
29 Interview conducted with Rabbi Botteach by the author in 1997.
30 Laurie Mifflin, 'Homer meets Dr. Hawking' in the *New York Times*, TV Notes, Section the Arts/Cultural Desk, May 12 1999 p. E9.
31 Jane Hawking, *Music to Move the Stars*, p. 178.
32 Interview conducted with Professor Hawking by the author in 1998.
33 Jane Hawking, *Music to Move the Stars*, p. 589.
34 Available at: http://www.admin.cam.ac.uk/news/pr/1998100101.html.
35 Available at: http://galileo.imss.firenze.it/museo/4/eiv10.html#eiv10a.
36 Interview conducted with John Gribbin by the author in 1996.
37 Available at: http://coldrain.net/lucas/lighthill-interview.html. In marked contrast to Lighthill, Hawking has argued that '[w]e must develop as quickly as possible technologies that make possible a direct connection between brain and computer, so that artificial brains contribute to human intelligence rather than opposing it'. Cited in Nick Paton Walsh, 'Alter our DNA or robots will take over, warns Hawking: special report: the ethics of genetics', *The Observer*, Sunday, 2 September, 2001, p. 1.
38 That Professor Hawking bridges the gap between men and machines is not to say that his intelligence is in any sense 'artificial', but rather that it is – to one extent or another, like that of all scientists – distributed across (and through) diverse groups of humans and machines.
39 Kantorowicz, *The King's Two Bodies*, p. 387. For Babbage, see Alison Winter, 'A calculus of suffering. Ada Lovelace and the bodily constraints on women's knowledge in early Victorian England', in C. Lawrence and S. Shapin (eds), *Science Incarnate. Historical Embodiments of Natural Knowledge* (Chicago, University of Chicago Press, 1998), pp. 202–39, p. 226.

Appendix
The statutes of the Lucasian professorship: a translation

Ian Stewart
University of King's College, Halifax, NS, Canada

TRANSLATOR'S INTRODUCTION

The documents translated below are the founding statutes of the Lucasian professorship (drawn up in December 1663 by Henry Lucas's executors) and the letters patent of Charles II (written the following month), which hereafter are referred to collectively as the 'Lucasian statutes'. Their inclusion in this volume may need a word of explanation, since, as the chapters of this book make clear, what determined the story of the professorship was not the formal strictures of the statutes, but rather the individual genius, interests and agendas of the chair's incumbents. Moreover, as time went on, these statutes became increasingly irrelevant to that story, eventually being formally abrogated by the Senate in 1857. However, for the early decades of the chair's existence at least, they served as the legal basis for the foundation and subsequent interpretation of the terms and conditions defining the person, remuneration and conduct of the Lucasian professor.

As such, they defined the incumbent as one who held a *university* office, as distinct from a college office, a distinction which the period itself made by referring to university offices and events as 'public' versus the more 'private' (*privatim*) aspects of life within one's college or study. As Mordechai Feingold explains in Chapter 1, figures at Cambridge who understood themselves primarily as university officers were still in a distinct minority in the seventeenth and

From Newton to Hawking: A History of Cambridge University's Lucasian Professors of Mathematics, ed. Kevin C. Knox and Richard Noakes. Published by Cambridge University Press.

eighteenth centuries, given the collegiate character of Cambridge (and Oxford). Thus, in a quite general sense, the Lucasian statutes are a step in the journey the university took from its quasimonastic, medieval roots to its modern form, where faculty usually regard themselves first as members of academic departments and second as members of colleges.

But understand that 'public' identity was not merely a matter of defining it by way of statute: it was the result of an interpretative process on the part of both the university and of the court. This is made clear in what was perhaps the most important – indeed perilous – moment in that process, namely the case of Isaac Newton. The story of that moment has been told elsewhere, and those not familiar with it should turn to Rob Iliffe's chapter in this volume. Here I wish to stress an aspect of that story that may have eluded historians perhaps not so familiar with the Lucasian statutes. In their original formulation, the statutes, in and of themselves, did not give Newton *any* grounds for assuming he could avoid taking up holy orders in the Church of England, a requirement of his fellowship at Trinity College. The university did not, in this respect, 'trump' college.

The statutes contain two specific clauses that are relevant to this issue. The first clause is found in the context of limitations on the professor's absence from the university and, in general, of prescriptions against distractions from executing his office. 'And for the same reason' the clause states, the holder was to be prohibited from 'enjoying any kind of ecclesiastical preferment that includes the cure of souls, or that demands residency contrary to these statutes, under penalty of forfeiture.' This was hardly a dispensation from taking up holy orders, as the context makes clear. It was clearly intended to prevent the Lucasian professor from using his influence, whether as a professor or as college fellow, to acquire further ecclesiastical livings that would encourage absences contrary to the statutory teaching duties of his office. As such, this is a clarification and intensification of the equivalent clause in the statutes for the Savilian professorship of mathematics at Oxford, which in general formed the model for the Lucasian statutes.[1] But on the question of whether or not the professor should or should not be an ordained clergyman, the clause is silent.

The second clause concerns the different but related question of the professor's right to hold or take up concurrently a college fellowship. Here the statutes are also silent on the question of ordination, but their sense bears careful reading: 'Further, we will and enact that, the said Professor may be elected fellow in any college, his professorship notwithstanding, *and that he not, alone for that reason or title,* be deprived of his fellowship (nor of any its emoluments or privileges), whether he obtained it [the fellowship] before taking up his office, or whether he shall obtain it thereafter, whatsoever college statute notwithstanding' (emphasis mine). Here the sense is clearly that the Lucasian professorship, *in itself*, may not be used as a justification either for ejecting a newly appointed professor from his college fellowship, or for preventing an existing professor from seeking one. The reason a college should wish to do either was that fellowships were, almost as a rule, relatively short-term, usually serving as stepping-stones to future ecclesiastical – and sometimes to other – preferments.[2] To be Lucasian professor *and* a college fellow meant one had little apparent financial motivation to be part of the cycle of what was called 'succession', by which individuals moved in and out of college fellowships. For the swelling ranks of would-be fellows, particularly at Trinity, who sought fellowships as steps in their ecclesiastical advancements, every fellow who was not part of that cycle represented one less career opportunity – and perhaps therefore one less reason for attending Trinity. This is the first reason that Isaac Barrow, master of Trinity during this period, cited when resisting the dispensation requested by the court on behalf of another Trinity fellow seeking freedom from the demand to become ordained: it 'hindered succession'.[3]

In short, the statutes protected the office of the Lucasian professor from the clutches of college divines seeking either to use the chair in order to accrue ecclesiastical livings to the detriment of their professorial duties, or to prevent the Lucasian professor from enjoying the benefits of a college fellowship. Not only were the statutes silent on the precise question of whether ordination was required, they were, strictly speaking, also silent on whether a college could demand of its incumbent that he be ordained in order to become or remain a college fellow. It is likely for this reason that in January 1675 – the year that his college fellowship would have demanded that he take up holy orders – Isaac Newton *assumed* that he would have to give up his fellowship. That he did not do so was due to the interpretation of a second set of letters patent issued later that year by the court of Charles II, and likely occasioned by Newton's own entreaties. These stipulated that the Lucasian professor was exempted from taking holy orders 'unless he himself desires to' and that he could not be ejected from his college fellowship on the grounds of non-ordination.[4] The effect of *that* interpretation, which one might be tempted to call a *modification*, was a clearer distinction of the person holding the office of Lucasian professor as such from the person holding the office of a college fellowship.

Newton was successful in occasioning the interpretation of the statutes to imply these distinctions, and his case illustrates the complex relations between official documents, interpretation and actual practice that constituted the official identities of university officers such as the Lucasian professor. An instructive epilogue to Newton's success in this regard is William Whiston's notorious failure. The story of his expulsion from Cambridge, recounted by Stephen Snobelen and Larry Stewart in this volume, was one that Whiston regarded to his death as a texture of injustices. Not the least of these was his being deprived of the Lucasian professorship because of his heretical views and publications. In fact, the church's efforts to convict Whiston of heresy failed, and this would have given the university ample cause to deprive him *ipso facto* of the professorship, according to the Lucasian statutes.[5] In the end the university's case against Whiston rested on university statute forty-five, which governed, amongst other things, the orthodoxy of lectures given 'publicly' (*publice*), namely a lecture delivered to the university or in the context of a university office. Whiston's published defence of himself rested in part on the insistence that as a 'public' person, that is, as Lucasian professor, his pronouncements were entirely mathematical in character, having nothing to do with theological questions, whatever else he may have published in London, preached outside the university as a priest or said in conversation in his college common room or in a coffeehouse.[6] Whiston's understanding of the 'public' character of his person as Lucasian professor was thus, to a certain extent, an inheritance from Newton (as was almost certainly his unorthodoxy).[7] And he had a point. That he failed spectacularly in getting the university to agree to his reading of the 'public' character of the chair belongs to the story of the diminishing role that the statutes played in the affairs of the Lucasian professor, but in an ironic sense. They turned out to be more prophetic than the university, at that time, could handle.

Lastly, a note about the translation itself. The challenge of translating these statutes lies in choosing between on the one hand an exact literal translation and, on the other, rendering them into modern English style and legal form. In general, I have erred rather on the side of literal translation, with the drawback that the sentences are usually uncomfortably long, and with formulations that seem unnecessary to the modern reader versed in law. However, it is hoped that I have thereby preserved some of the seventeenth-century legal 'feel' of the text. I have included notes to the text where I deemed they might provide further detail of use to historians, or explain the choice of translation.

THE LUCASIAN STATUTES[8]

To all such faithful in Christ to whom this present document has come, Robert Raworth of Gray's Inn, Esquire in the County of Middlesex and Thomas Buck of Cambridge, Esquire in the County of Cambridge, executors of the last will and testament of the most worthy Henry Lucas of London, Esquire, recently deceased, send greeting in the Lord everlasting.

Understand that, since the aforementioned, venerable and wise Henry Lucas, out of his favourable disposition toward the University of Cambridge, and out of his good will toward learning, ordered in his last testament that his executors, the aforementioned Robert Raworth and Thomas Buck, should purchase lands of the yearly value of one hundred pounds to be left as a yearly stipend or perpetual salary for the professor of mathematics in the said university, under those constitutions and rules that – in consultation with the Vice-chancellor and heads of colleges of the said university – they shall have judged to be most suitable to the honour of that great body and to the increase and promotion of these studies, which hitherto have not been provided for, we the aforementioned executors, anxious of carrying out the desire of this famous benefactor on account of the trust invested in us, and having first sought and obtained the advice of the said Vice-chancellor and heads of colleges, and especially out of respect to that advice, will and declare the following rules, which, we warrant, shall tend to the promotion of these mathematical studies, and which are determined by the authority granted us through the last will and testament of the said Henry Lucas.

Therefore we first enact and ordain that, in perpetuity, whatsoever annual income from the abovementioned lands that have been or are to be acquired – the necessary expenses having been deducted – by whatever just means that income shall have been gained, the entire amount should go, in the way specified below, to the subsidy and reward of the professor of mathematics, elected and established under conditions hereafter to be set forth.

Concerning the office of the said mathematics professor, in order that he have reason to cultivate these studies, both publicly and privately, we will and enact that the said professor be obliged to lecture and expound some part of geometry, astronomy, geography, optics, statics, or of any other part of the mathematics (according to his choice, unless the Vice-chancellor deem it fit to declare otherwise), at least once a week for one hour, in a place and time to be assigned by the Vice-chancellor, under penalty of forty shillings for each missed lecture, to be subtracted from his stipend by the Vice-chancellor, and put to the university library for the purpose of buying books or mathematical instruments. An exception is to be made when, on account of grievous bodily ailment, he is not able to fulfil his office, an excuse, however, that we do not wish to remain available beyond three weeks, at which time he shall substitute another competent reader, to be approved by the Vice-chancellor. Let him further understand that for each missed lecture twenty shillings shall be subtracted from his stipend by the Vice-chancellor, and put to the abovementioned uses. However, so that the said professor be obliged to perform assiduously his office of reading, and that there be more certain evidence of the performance of his office, and also that in some measure the fruit of present studies be derived to posterity, we enact that each year before the next feast of Michelmas the said professor produce for the Vice-chancellor at least ten lectures, neatly written out, taken from what he held publicly in the preceding year, which are to be deposited in the university archives. If he fail to do so before the prescribed time, let his stipend be withheld until he does. And let him be obliged to pass on to the Vice-chancellor that part of his profits or annual salary, proportionate to the elapsed time until he shall have done so, to be put to the university library for the abovementioned uses.

Moreover, we decree that the said professor during term shall maintain free access to all who might come to his rooms on two days each week (appointed by the Vice-chancellor) for the length of two hours, and on one day per week outside of term if the said professor be at the university. He should respond willingly to their questions and difficulties, and have ready to hand for this purpose globes and other suitable mathematical instruments. In all things pertaining to what has been asked he should encourage to the best of his ability the endeavours of the studious. If he intentionally neglect any of these things, he is to be reproved by the Vice-chancellor, and if he shall have been often warned of his neglect, and still does in nowise amend his ways, he shall incur the penalty of intolerable neglect, as specified below.

Further, that the observation of these things might be more strongly secured, and lest by the absence of the said professor a neglect should arise,[9] we enact that the said professor not depart from the university during term, nor abide anywhere outside the university for six continuous days, except for the gravest of reasons to be approved by the Vice-chancellor, and only after having sought and gained leave from him. Otherwise, he shall be deprived of that part of his salary proportionate to the duration of his absence. But if the necessity of his absence should arise, approved by the Vice-chancellor and two senior doctors who are masters of colleges, and if the absence exceed the duration of half a term, then he shall substitute another competent person who is to read in his place, and to discharge his other duties in the abovementioned manner and subject to the same penalties. And for the same reason, lest the said professor be distracted from due execution of his office, we wholly will and prohibit his enjoying any kind of ecclesiastical preferment that includes the cure of souls, or that demands residency contrary to these statutes, under penalty of forfeiture, *ipso facto*, of every right that might extend to this his professorship.

Concerning the person and quality of the professor of mathematics, we will and impose this injunction, that whoever is led to this task be a man of good character and reputable life,[10] at least a Master of Arts, soundly learned and especially skilled[11] in the mathematical sciences. The right and power of electing him belongs to us, the aforementioned executors to the venerable Henry Lucas, as long as we shall live; or at the death of one of us, it shall pass to the other.[12] When both have deceased, the full power of election shall in perpetuity rest with the Vice-chancellor and with all the masters of colleges of the said university, or that number of masters that shall have been present at the election.

Let the election proceed as follows. After the place of the professor of mathematics for whatever reason becomes vacant, it shall fall to the Vice-chancellor, as quickly as possible, to give notice of the said vacancy and of the time of future election, by posting a bill on the door of the public schools for eight continuous days. The time of election we do not wish to be extended past the thirtieth day after first giving such notice. At the time of the election the said electors, gathered together in the public schools, and each having laid aside any private consideration and any perfidious hope of himself being nominated, shall engage themselves under oath to approve by vote that person from among those seeking the office (or from among those nominated by the electors) whom, according to the witness of their conscience, they shall judge best suited for taking up this office, according to the aforementioned qualities. He on whom the majority has agreed shall be regarded as the elected. But if it should happen that two or more should equally have won the most votes, it shall be the right of the Vice-chancellor to elect according to his judgement one from among that number who equally won the most votes. The elected shall at the next opportunity be brought to the Vice-chancellor, and before admission to the professorship he shall take an oath to execute faithfully, to the best of his ability, the office of professor of mathematics established in this university by the most honourable Henry Lucas, according to the rules and statutes concerning his office.

Finally,[13] so that the said professor remain within the due limits of good reputation and behaviour, and that he not presume any impunity with respect to crimes willingly committed, we enact and decree that if the said professor be convicted of any serious crime, either on his own confession, or by competent witnesses, or through evidence of some more serious crime (such as treason, heresy, schism, voluntary manslaughter, notable theft, adultery, fornication or perjury), or if he should be notoriously incompetent, or past amendment by means of the penalties abovementioned, he shall be removed from his professorship by the Vice-chancellor and masters of colleges (or a majority of them), without hope of returning or of perceiving any further benefit. But if he becomes inadequate to the discharge of the duties of the professorship (in the aforesaid manner and form) due to old age, chronic disease or incurable impotence or weakness that breaks up the body or mind, that professorship shall be declared void by the Vice-chancellor and the said masters (or a majority of them); yet with this mitigation: since he was thus dismissed from his professorship through no fault of his own, and assuming his conduct to have been praiseworthy during the period of his professorship, and if he have no private means amounting to one hundred pounds per year, then a third part of his stipend shall continue to be paid to him until his death. His successor must be content until that time with the remainder, and shall receive the full yearly stipend on his predecessor's death.

In witness of which, we the aforenamed Robert Raworth and Thomas Buck place our seal before these present, on the date of the nineteenth of December, in the fifteenth year of the reign of our King Charles the Second, by the grace of God King of England, Scotland, France and Ireland, Defender of the Faith, in the year of our Lord, 1663.

signed by
Robert Raworth
Thomas Buck

Sealed and resolved upon in the presence of
James Windet, M.D.
Isaac Barrow
Richard Spoure
Michael Glyd
William Player

LETTERS PATENT OF CHARLES II CONCERNING THE LUCASIAN PROFESSORSHIP

Charles II, by the grace of God King of England, Scotland and Ireland, Defender of the Faith, &c: to each and everyone who has seen these letters, greetings.

Since our beloved Robert Raworth and Thomas Buck, Esquires, executors of the last will and testament of Henry Lucas, Esquire, lately deceased, made known to us that the said Henry Lucas has to the benefit of learning established a professorship of mathematics in our University of Cambridge, and since his said executors, with the advice of the Vice-chancellor and masters of colleges of the said university, have taken upon themselves the charge of ordaining those things that by custom are seen to tend to the good establishment and due execution of the said office; and likewise since the said executors are the more eager concerning the aforementioned office that it be well earned, and since in good faith to their commission, having first sought and obtained the advice of the Vice-chancellor and masters, they humbly beseeched us that these things be established and ratified by our royal authority, and since, in order that the aforementioned office should benefit the studious, the said executors humbly entreated us according to the advice of the Vice-chancellor and masters of colleges: that a certain audience for the lectures be appointed according to our royal sanction; that there be a power of exacting the oath from the electors of the said professor, and from the professor elect; that, if he shall have held a fellowship of any college, he be granted the privilege of keeping that fellowship along with the place of said professor; and that if before taking up this office he shall not have been a fellow, he be allowed the possibility thereafter of being elected fellow; and finally, that the possibility of taking up certain public offices be taken away from the same professor.

So, desiring the good of the university, as well as a statute to the advantage of the studious, and graciously granting the wishes of the said executors, first we ratify and confirm by our royal authority these rules framed by the said executors, and we will and order full compliance to be owed and performed in each and every thing.

Likewise, we will that all undergraduates past the second year and all Bachelors of Arts up to the third year attend those lectures, which the said professor of mathematics is bound to hold according to the aforementioned rules, and that they be subject to the same penalties under the university statutes to which those are subject who are absent from the other university lectures.

Moreover, we bestow on the Vice-chancellor the power of exacting and administering the oath both to those who through the said rules obtain the right of election, as to the professor elect hereafter to be elected, according to the senses expressed and prescribed in the said rules.

Further, we will and enact that the said professor may be elected fellow in any college, his professorship notwithstanding, and that he not, alone for that reason or title, be deprived of his fellowship (nor of any its emoluments or privileges), whether he obtained it before taking up his office, or whether he shall obtain it thereafter, whatsoever college statute notwithstanding.

We refuse and prohibit that the said professor take up the office of dean, bursar, steward or of any lecturer in his college, or that he act as tutor (except perhaps to fellow-commoners of high or noble birth), or hold the office of proctor, steward, taxator or any other public office in the university, under penalty, *ipso facto*, of forfeiture of every right that should proceed to this his professorship. For which reason we will that the said professor be free and exempt from each and every of the aforementioned offices.

In witness of which, we have caused to be drawn up these letters patent, myself as witness at Westminster, the eighteenth day of January, in the fifteenth year of our reign

by writ of privy seal
Hastings[14]

Notes

1 'But I expressly forbid my professors to accept any ecclesiastical benefice after their admission, with cure or without cure, or any prebend, canonry or archdeaconry . . . ; and if any person previously to his admission holds a benefice or any other of the aforesaid places or offices, I will that he shall resign it . . . within the space of six months from his first admission. . . .' Oxford University Statutes, G. R. M. Ward, tr., (2 vols, London, 1845–51), vol. I, p. 281.

2 John Twigg, *The University of Cambridge and the English Revolution, 1625–1688* (Woodbridge, 1990), p. 99.
3 The other reason Barrow cited was that the college's main purpose was to 'breed divines', which raises the question of how the college fellows reacted to Newton's dispensation. The incident and Barrow's role in it is discussed by Mordechai Feingold, 'Isaac Barrow: divine, scholar, mathematician' in Mordechai Feingold (ed.), *Before Newton. The Life and Times of Isaac Barrow* (Cambridge, 1990), p. 88; see also Richard Westfall, *Never At Rest. A Biography of Isaac Newton* (Cambridge, 1980), pp. 330–3.
4 For Newton's assumption, see A. R. Hall and Laura Tilliry (eds) *Correspondence of Isaac Newton* (Cambridge, 1977) vol. VII, p. 387. The letter continues, 'And particularly we will and declare that he shall not take up holy orders unless he himself desire to, and that he not, by failing to take up holy orders, be considered to give up his fellowship, nor that he be forced to do so. Rather, he should enjoy this immunity as long as he hold his office, just as the fellow professing medicine and either civil or canon law enjoys this immunity, whatever college statute, custom or interpretation notwithstanding.' The 1675 letters patent in John Willis Clark (ed.), *Endowments of the University of Cambridge* (Cambridge, 1904), pp. 171–3. The reference to medicine and law concerns the dispensation from holy orders for college lecturers in medicine and law in the Trinity statutes, chapter XIX: *De tempore assumendi gradus Scholasticos et sacros ordines*; cf. *Documents relating to the University and Colleges of Cambridge* (3 vols, London, 1852), vol. III, pp. 441–2.
5 The story of this attempt and its failure is amply recounted by Eamon Duffy, 'Whiston Affair': the trials of a primitive Christian, 1709–1714, *Journal of Ecclesiastical History*, 27 (1976), 129–51.
6 *An Account of Mr Whiston's Prosecution at, and Banishment from, the University of Cambridge . . . now Reprinted. With an Appendix: Containing Mr Whiston's farther Account . . . Never before Printed* (London, 1718), pp. 7–18.
7 On Whiston's theological debts to Newton see Stephen Snobelen, 'William Whiston, Isaac Newton and the crisis of publicity', forthcoming in *Studies in History and Philosophy of Science* (2003).
8 For the following translation I have used the text of the statutes and letters patent of Charles II as found in John Willis Clark (ed.), *Endowments of the University of Cambridge* (Cambridge, 1904),

pp. 171–2. The statutes can also be found reprinted, with only a few slight differences in Tom Whiteside (ed.), *The Mathematical Papers of Sir Isaac Newton* (8 vols, Cambridge, 1967–80), vol. III, pp. xx–xxvii. Whiteside's version is taken from Packet E of the Lucasian papers, ULC. Res. a. 1893, which Whiteside quite reasonably judged to be the original version. I have used the Clark version because the slight variations in Clark from Whiteside are clearly corrections.

9 Reading with Clark '*oriatur*' instead of '*aboriatur*' in Whiteside.

10 '*bonae famae et conversationis honestae*'; for this rather general expression I have followed Ward's translation of the identical phrase in the Savilian statutes: Ward, *Oxford University Statutes*, vol. I, p. 277.

11 '*peritia*', which also has the sense of a practical knowledge.

12 This means that both Isaac Barrow and Isaac Newton were in fact elected by the executors.

13 This paragraph is almost entirely taken verbatim from the Oxford statutes for the Savilian professorship of mathematics: J. Griffiths, Statutes of the University of Oxford, Codified in the Year 1636 (Oxford, 1888), pp. 251–2. In ULC Res. a. 1893, this addition appears to be in Isaac Barrow's hand, suggesting that it was he who used the Savilian statutes as model at this point, either at the direction of the executors or on his own initiative. The only significant deviation is that the Savilian statutes give the power of judging and removing the professor to the university, whereas here it rests with the Vice-chancellor and a majority of the heads of colleges.

14 Theophilus Hastings, seventh Earl of Huntingdon, then Custodian of the Rolls for Warwickshire.

Index